Capitalism: An Unsustainable Future?

The four decades of neoliberalism, globalisation and financialisation have produced crises – financial and pandemic – and rising inequality. The climate emergency threatens the future of the planet. This book explores many dimensions of the background to these crises. There is the development of policy agendas to address the climate emergency. The rise in inequality is studied in terms of impacts of financialisation and the relationships between growth and inequality. The record of the neoliberal experiment in the USA is critically examined. The roles of financial institutions including public banks and micro-finance are explored, as is the need for improved financial oversight in the Economic and Monetary Union. The growth of global value chains has been a major aspect of globalisation, and the question is examined of whether such chains provide a ladder for development. Globalisation has also featured trade imbalances and large capital flows, and their causes and effects are examined with respect to China and South Africa respectively.

This volume will be of great value to students, scholars and professionals interested in political economy, economic thought, climate change, sustainability and business studies.

The chapters in this book were originally published as a special issue of the journal, *International Review of Applied Economics*.

Malcolm Sawyer is Emeritus Professor of Economics, University of Leeds, UK and retired Managing Editor of *International Review of Applied Economics*. He is author of 12 books, co-editor of over 30 books and over 250 academic journal articles and book chapters.

Jonathan Michie is Professor of Innovation and Knowledge Exchange at the University of Oxford, UK where he is also President of Kellogg College. He is Managing Editor of *International Review of Applied Economics*, and Chair of the *Universities Association for Lifelong Learning*.

Capitalism: An Unsustainable Future?

Edited by
Malcolm Sawyer and Jonathan Michie

Routledge
Taylor & Francis Group

LONDON AND NEW YORK

First published 2022
by Routledge
4 Park Square, Milton Park, Abingdon, Oxon OX14 4RN

and by Routledge
605 Third Avenue, New York, NY 10158

Routledge is an imprint of the Taylor & Francis Group, an informa business

Introduction, Chapters 1, 2 and 4–14 © 2022 Taylor & Francis
Chapter 3 © 2021 Gary A. Dymski and Annina Kaltenbrunner. Originally published as Open Access.
Chapter 15 © 2021 Jonathan Michie. Originally published as Open Access.

British Library Cataloguing in Publication Data

A catalogue record for this book is available from the British Library

ISBN: 978-1-032-21143-5 (hbk)
ISBN: 978-1-032-21146-6 (pbk)
ISBN: 978-1-003-26696-9 (ebk)

DOI: 10.4324/9781003266969

Typeset in Minion Pro
by Newgen Publishing UK

Publisher's Note
The publisher accepts responsibility for any inconsistencies that may have arisen during the conversion of this book from journal articles to book chapters, namely the inclusion of journal terminology.

Disclaimer
Every effort has been made to contact copyright holders for their permission to reprint material in this book. The publishers would be grateful to hear from any copyright holder who is not here acknowledged and will undertake to rectify any errors or omissions in future editions of this book.

Contents

Citation Information

The chapters in this book were originally published in the journal, *International Review of Applied Economics*, volume 35, issue 3–4 (2021). When citing this material, please use the original page numbering for each article, as follows:

Introduction

The first 30 years of the International Review of Applied Economics, *and the future of capitalism*
Jonathan Michie
International Review of Applied Economics, volume 35, issue 3–4 (2021), pp. 331–337

Chapter 1

Financialisation, industrial strategy and the challenges of climate change and environmental degradation
Malcolm Sawyer
International Review of Applied Economics, volume 35, issue 3–4 (2021), pp. 338–354

Chapter 2

UK and other advanced economies productivity and income inequality
Philip Arestis
International Review of Applied Economics, volume 35, issue 3–4 (2021), pp. 355–370

Chapter 3

Financial oversight, the third flawed pillar of the European Union: the missing piece in the Arestis-Sawyer critique of EMU macropolicy design
Gary A. Dymski and Annina Kaltenbrunner
International Review of Applied Economics, volume 35, issue 3–4 (2021), pp. 371–388

Chapter 4

The industrial policy requirements for a global climate stabilization project
Robert Pollin
International Review of Applied Economics, volume 35, issue 3–4 (2021), pp. 389–406

Chapter 13

Sovereign currency and long-term interest rates
Hongkil Kim
International Review of Applied Economics, volume 35, issue 3–4 (2021), pp. 577–596

Chapter 14

Government expenditure and economic growth: a post-Keynesian analysis
Pintu Parui
International Review of Applied Economics, volume 35, issue 3–4 (2021), pp. 597–625

Chapter 15

Interpreting the world, in various ways – and changing it
Jonathan Michie
International Review of Applied Economics, volume 35, issue 3–4 (2021), pp. 626–631

For any permission-related enquiries please visit:
www.tandfonline.com/page/help/permissions

Notes on Contributors

Philip Arestis, Department of Land Economy, University of Cambridge, Cambridge, United Kingdom.

Rob Davies, Research Fellow, School of Finance and Management, SOAS, University of London, UK.

Marwil J. Dávila Fernández, Department of Economics and Statistics, University of Siena, Siena, Italy; Department of Economics, Bucknell University, Lewisburg, PA, USA.

Petra Dünhaupt, Berlin School of Economics and Law (HWR Berlin), Institute for International Political Economy, Germany.

Gary A. Dymski, Leeds University Business School, University of Leeds, Leeds, UK.

Laurence Harris, Emeritus Professor, School of Finance and Management, SOAS, University of London, UK.

Hansjörg Herr, Berlin School of Economics and Law (HWR Berlin), Institute for International Political Economy, Germany.

George Joseph, Department of Accounting, University of Massachusetts Lowell, One University Avenue, Lowell, MA, USA.

Annina Kaltenbrunner, Leeds University Business School, University of Leeds, Leeds, UK.

Hongkil Kim, Department of Economics, University of North Carolina at Asheville, Asheville, NC, USA.

Konstantin Makrelov, Lead Economist, South African Reserve Bank, South Africa.

Jonathan Michie, Kellogg College, University of Oxford, Oxford, UK.

André de Melo Modenesi, Department of Economics, Federal University of Rio de Janeiro, Rio de Janeiro, Brazil.

Pintu Parui, Centre for Economic Studies and Planning, Jawaharlal Nehru University, New Delhi, India.

Nikolas Passos, Faculty of Political and Social Sciences, Scuola Normale Superiore, Florence, Italy.

Robert Pollin, Department of Economics and Political Economy Research Institute (PERI), University of Massachusetts Amherst, Amherst, USA.

Lionello F. Punzo, Department of Economics and Statistics, University of Siena, Siena, Italy; INCT/PPED, Federal University of Rio de Janeiro, Rio de Janeiro, Brazil.

Shakil Quayes, Department of Economics, University of Massachusetts Lowell, Lowell, MA, USA.

Samuel Rosenberg, Department of Economics, Roosevelt University, Chicago, IL, USA.

Malcolm Sawyer, Emeritus Professor of Economics, University of Leeds, Leeds, UK.

Anwar Shaikh, Department of Economics, The New School for Social Research.

Remzi Baris Tercioglu, The New School for Social Research, USA.

Isabella Weber, Department of Economics, University of Massachusetts Amherst, Amherst, MA, USA.

Introduction

The Future of Capitalism

Jonathan Michie

This Special Issue is a Festschrift for Professor Malcolm Sawyer, the Managing Editor of the *International Review of Applied Economics* for thirty years, from its launch in 1987 until his retirement in 2017, when he passed the baton on to Professor Jonathan Michie. Professor Sawyer continues as an active referee for and advisor to the *Review*.

1. History and purpose of the *International Review of Applied Economics*

The *Socialist Economic Review* had been published annually by Lawrence and Wishart (under the Merlin Press imprint) in the early 1980s. The papers arose from an annual conference. The *Socialist Economic Review* was organised initially around the Alternative Economic Strategy, attracting economists from the left of the political spectrum (particularly from the left of the Labour Party and from the Communist Party). When Lawrence and Wishart decided to stop publishing the *Socialist Economic Review*, other publishers were approached. Edward Arnold (a publisher of books and journals for students, academics and professionals) responded with a suggestion for a journal, which then became the *International Review of Applied Economics*. Malcolm Sawyer and Saziye Gazioglu were instrumental in arranging this, and were asked to be the lead editor and associate editor respectively of the *International Review of Applied Economics*, having both served previously as co-editors of the *Socialist Economic Review*. The editorial board of the *International Review* reflected its origins in the *Socialist Economic Review*, and welcomed the opportunity to internationalise and broaden the contributors and audience beyond the UK.

The *International Review of Applied Economics* was launched in 1987 with two issues. Shortly after, it was joined at Edward Arnold by the *Review of Political Economy* (whose first issue was in 1989) as something of a companion journal, in that both journals promoted the development of heterodox economic analysis.

The *International Review of Applied Economics* invited papers which involved the 'application of economic ideas to the real world. Applied economics is interpreted to include both the publication of empirical work, and the application of economics to the evaluation and development of economic policies. The interaction between empirical work and economic policy is an important feature of the *International Review*.' Papers and book review articles would be published 'in the area of applied policy-oriented economics. It [the *IRAE*] adopts a broadly left-of-centre perspective on economic policy, but within that perspective is not identified with any specific theoretical or political position. It encourages articles dealing with policy issues and adopting theoretical positions on current issues which are neglected in mainstream journals.'[1]

This was later revised to be expressed as 'Although the *International Review of Applied Economics* associates itself broadly with the non-neoclassical tradition, it does not identify itself with any specific theoretical or political position.'[2]

The journal went through several change of ownership of the publishers: Edward Arnold itself was acquired by Hodder & Stoughton in 1987. It eventually came to Carfax – a specialist journal publisher – which in turn was acquired by Taylor and Francis. The number of issues of the journal per annum gradually increased from the initial two, up to three in 1991 (Volume 5), then four in 2000 (Volume 14), five in 2006 (Volume 20) and six – bi-monthly – from 2008 (Volume 22) onwards.

2. Capitalism: an unsustainable future?

The 'Call for Papers' for this Festschrift was entitled *Capitalism: an unsustainable future?*, reflecting Professor Malcolm Sawyer's work on the nature of the capitalist economy, through to economic policy measures to make the economy more equitable and sustainable, whether this might develop as a reformed or post-capitalist economy:

> The last four decades have been variously viewed and conceptualised in terms of neo-liberalism, globalisation, and financialisation. During those decades, Soviet style communism collapsed and the 'end of history' was proclaimed. The financial crisis of 2007/09, though located in the North Atlantic area, had global consequences of recession and in general slower growth than hitherto. Alongside perceptions of secular stagnation, there is the mounting urgency to tackle climate change. There have been indications of 'peak globalisation' (at least in terms of international trade) and 'peak financialisation' (at least in terms of the scale of the financial sector in the industrialised world). Papers are invited within the broad themes of the sustainability of global capitalism, and the current varieties of capitalism; recent and future trends and policies in: neo-liberalism, globalisation, financialisation, inequality, and issues of breaks within the past four decades; and alternative futures for ownership forms, environmental sustainability, and economic organisation.

This has echoes of the work around the 'Alternative Economic Strategy' of the 1970s and 1980s, from which the *International Review of Applied Economics* emerged, but in the current context of financialisation and the climate crisis, and hence necessarily with a global perspective, both because the 2007–2009 financial crisis and recession were global, and because the required Green New Deal needs to be global, as would any post-capitalist economic system.

In 'Financialisation, industrial strategy and the challenges of climate change and environmental degradation', Malcolm Sawyer discusses the nature of the present era of financialisation, outlining the changes in the financial sector and its relations with the real sector which are particularly relevant for the climate emergency. The relationship between growth of the financial sector ('financial development') and economic growth is reviewed, and the relevance of recent empirical findings for the role of the financial sector in addressing the climate emergency is drawn out. It is argued that the policy approach to the climate emergency and environmental degradation should be embedded within an industrial strategy. Further, it is argued that the structures of the financial sector need to be changed to encourage financial institutions which are more favourably disposed towards the allocation of funds to 'green investment'. It is also argued that the central banks should act in ways that are supportive of environmental policies, but that their role would necessarily be a rather limited one.

In 'UK and other advanced economies productivity and income inequality', Philip Arestis focuses on weak productivity growth – in the UK and other advanced economies – which has been slowing down since around 2000, and on increasing income inequality. There are a number of factors, including 'secular stagnation' causing these two phenomena of weak productivity growth and increased income inequality. According to Arestis, recent evidence clearly suggests that labour productivity and income inequality have been closely and significantly related; this is so since there is a strong relationship between productivity, inequality, economic growth and real wages. Productivity growth is the key determinant of how demand can grow without inflation, thereby reducing inequality of income. The slowdown in productivity growth and increase in inequality have been in evidence in the UK and other advanced economies. Indeed, both have become more pronounced following the Global Financial Crisis. While weak productivity growth and increased income inequality predate the Global Financial Crisis, and the Great Recession, they have clearly worsened following them.

In 'Financial oversight, the third flawed pillar of the European Union: the missing piece in the Arestis-Sawyer critique of EMU macropolicy design', Gary A. Dymski and Annina Kaltenbrunner present a chronological survey of the twenty academic papers that Malcolm Sawyer authored or co-authored between 1997 and 2017 on the flawed design – and hence flawed implementation – of the European Monetary Union (EMU)'s macroeconomic policy pillars. Dymski and Kaltenbrunner report how Sawyer had consistently argued that the EMU's two defining macroeconomic-policy pillars were fundamentally flawed, in featuring an independent central bank focused solely on inflation targeting to the exclusion of employment or growth targets, and in placing tight constraints on member-state fiscal policy with no adequate plan for Euro-area fiscal policy.[3] Dymski and Kaltenbrunner augment Sawyer's analyses by pointing out a third – complementary – design flaw, namely the EMU's two-tiered structure of financial regulation and oversight. While this financial pillar aimed at reconciling Europe's historically bank-based financial systems with large European banks' entry into global financial competition, it created a combustible mix when combined with the EMU's macroeconomic policy pillars. The Global Financial Crisis lit the fire: member states, forced to rescue their domestically-chartered too-big-to-fail megabanks, had to adopt austerity policies that both slowed the pace of post-crisis economic growth and eroded support for pro-Union political leaders. Only marginal changes have been made to these policy pillars post-crisis. Consequently, Europe faces a financial bifurcation point: either to continue 'whatever it takes' support for its megabanks, or to rethink both its financial architecture and its macroeconomic and financial policy pillars.

In 'The industrial policy requirements for a global climate stabilization project', Robert Pollin presents an industrial policy approach for advancing a global climate stabilization project. The centrepieces of the project are firstly, to dramatically improve energy efficiency standards in the stock of buildings, automobiles and public transportation systems, and industrial production processes; and secondly to equally dramatically expand the supply of clean renewable energy sources – primarily solar and wind power –

available at competitive prices. Global investment spending in these areas will need to average about 2.5% of global GDP over 2024–2050 to achieve a net zero CO_2 emissions global economy by 2050. The paper works within a policy approach similar to that advanced by Malcolm Sawyer that integrates industrial and macroeconomic policies targeted at achieving full employment.

With the significant changes in the economy and society under neo-liberalism as his backdrop, Samuel Rosenberg in his paper on 'Challenges to neo-liberalism in the United States' analyzes the extent to which government policies – including federal, state, and local – along with labour activism since the Great Recession constitute challenges to neo-liberalism in the United States. Rosenberg investigates the legacy of neo-liberalism – including the ineffective federal governmental response to the COVID-19 economic and health crisis – and the emerging discourse within the Democratic Party calling for a major reorientation of government policy away from neo-liberalism. The paper concludes by discussing the effects of the neo-liberal agenda on economic well-being, and evaluating whether the neo-liberal agenda has been successful in its own terms.

The US–China trade imbalance is commonly attributed to a Chinese policy of currency manipulation. However, empirical studies have failed to reach a consensus on the RMB misalignment. In 'The U.S.–China trade imbalance and the theory of free trade: debunking the currency manipulation argument', Isabella Weber and Anwar Shaikh argue that this is not a consequence of poor measurement but of theory. At the most abstract level the conventional principle of comparative cost advantage suggests real exchange rates will adjust so as to balance trade. Therefore, the persistence of trade imbalances tends to be interpreted as arising from currency manipulation facilitated by foreign exchange interventions. By way of contrast, the absolute cost theory explains trade imbalances as the outcome of free trade among nations that have unequal real costs. Weber and Shaikh argue that a disparity in real costs is the root cause of the US–China trade imbalance.

For countries of the Global South, global value chains offer an opportunity to integrate into international trade and to industrialise relatively easily. However, in 'Global value chains – a ladder for development?', Petra Dünhaupt and Hansjörg Herr argue that this is not sufficient for a catching-up development – that is, to reach the GDP per capita levels of the countries of the Global North. On the contrary, there is a risk that countries will remain trapped in low value-added activities. The theoretical argument is supported by case studies of four industries in six countries. For catching-up, countries need comprehensive horizontal and vertical industrial policy.

South Africa has a very well-developed financial sector and high reliance on capital flows. The country saw large capital outflows as the Covid-19 crisis developed, accompanied by a large depreciation of the rand and spikes in bond yields. In 'The impact of capital flow reversal shocks in South Africa: a stock-and-flow-consistent analysis', Konstantin Makrelov, Rob Davies and Laurence Harris employ a 'stock- and flow-consistent' model to study the impact of capital flow reversal shocks on the South African economy. The model includes a richer representation of institutional balance sheets than other models. The financial sector's behaviour in the model draws on the theoretical frameworks, which highlight the relationship between bank capital, the risk-taking behaviour of the financial sector, lending spreads and economic activity. The paper specifies a dynamic adjustment model of household expectations with properties that differ from the way in which expectations are formed in either 'stock- and flow-consistent' or DSGE models. Household expectations resemble bounded rationality. The financial accelerator mechanism operates through the balance sheets of all institutions in the economy. Makrelov, Davies and Harris find that a reversal in capital flows can affect the domestic economy through its impact on domestic liquidity, on the risk-taking behaviour of the financial sector, and on the demand for assets.

In 'Do public banks reduce monetary policy power? Evidence from Brazil based on state dependent local projections (2000–2018),' Nikolas Passos and André de Melo Modenesi test the hypothesis that public banks reduce monetary policy power. Previous studies have shown that companies with access to government-driven credit present smaller falls in investment and production after a contractionary monetary policy shock. Nevertheless, these studies are based on microeconomic data and ignore macroeconomic effects, especially the cost-push effects of monetary policy. The paper employs state-dependent local projections to compare monetary policy power – the sensibility of inflation to changes in policy interest rates – between periods of high credit of public banks and periods of high credit of private banks. Passos and de Melo Modenesi do not find evidence that monetary policy is less powerful in periods of high credit of public banks. Even though periods of high credit of public banks present a lower effect over output, those periods present less persistent price puzzles than periods of high private credit. Passos and de Melo Modenesi attribute their results to a lower flexibility in interest rates in relation to credit from public banks, which leads to a lower transmission of financial costs, lower reduction in capital stock and lower variation in the exchange rate.

In 'Some new insights on financialization and income inequality: evidence for the US economy, 1947–2013', Marwil J. Dávila Fernández and Lionello F. Punzo study the relationship between income distribution and financialization in the United States between 1947 and 2013. Financialization is introduced as a two-fold process. On the one hand, it implies an increase in the contribution of the financial sector in the composition of production. On the other hand, it is related to an increase in the importance of financial assets in terms of the composition of wealth. The paper takes the share of financial employment as a *proxy* for the contribution of the financial sector in the composition of production, and financial assets as a share of corporations' total assets as a *proxy* for the importance of financial assets in terms of the composition of wealth. Applying cointegration techniques, the paper identifies a positive long-run relationship between financialization and income inequality. Causality goes from employment to income inequality, and from the latter to wealth. Nonlinear estimators suggest the existence of certain asymmetric effects such that changes in income distribution cannot be reversed by simply reversing financialization.

In 'Determinants of social outreach of microfinance institutions', Shakil Quayes and George Joseph analyse the determinants of the social outreach of micro-finance institutions (MFIs), using three measures of outreach – depth of outreach, breadth of outreach, and outreach to women – and its possible complementarity with financial performance. The paper uses an unbalanced panel of 1,219 MFIs over a period of twenty years to investigate the effect of firm-specific characteristics and the impact of the prevailing legal system on social outreach of MFIs. Rejecting the notion of a trade-off between financial performance and social out-reach, their empirical results show that better financial performance has a positive association with social outreach. Furthermore, they observe that a common law legal system is more conducive in facilitating social outreach, and MFIs operating under common law legal systems achieve better depth of outreach, breadth of outreach, and outreach to women, than MFIs under code law legal systems and mixed law legal systems. While Quayes and Joseph had expected non-profit MFIs to exhibit better social outreach than for-profit MFIs, they found empirical evidence of this only in the case of outreach to women. Finally, they found empirical evidence that unregulated MFIs achieve better social outreach than regulated MFIs.

Five sectors have increased their contribution to US GDP growth since 1973: professional-business services (PBS), finance, information, healthcare, and arts-entertainment. In 'Rethinking growth and inequality in the US: what is the role of measurement of GDP?', Remzi Baris Tercioglu argues that of these, finance, healthcare, and PBS should not be regarded as final consumption of households. Contra published National Income and Product Accounts, treating expenditures on finance, healthcare, and PBS as intermediate consumption reveals a significantly different picture of US economic growth, including firstly a deeper slowdown of real output growth since 1973; secondly, a more moderate rise in consumption share since 1980; and thirdly a sharper decline in labour share – defined as the compensation of employees over GDP – since 1985. The paper thus contributes to the literature on secular stagnation and rising inequality in the United States.

In 'Sovereign currency and long-term interest rates', Hongkil Kim investigates the effects of government debt and deficits on long-term interest rates in seventeen advanced economies over the period 1973–2016 from the perspective of currency sovereignty. Kim's empirical findings suggest that the market penalizes non-sovereign nations for the same amount of fiscal deficit with higher interest rates than sovereigns. In addition, non-sovereign countries face higher interest rates for an increase in the debt-to-GDP ratio beyond a certain threshold (49% of GDP) while such a pattern is not obvious among sovereign nations. Overall, the results support the argument of Modern Monetary Theory (MMT) that a monetarily sovereign government, as a monopoly issuer of currency, can influence the prices of their liabilities to a significant extent, somewhat independent of existing public debt and market sentiment.

In 'Government expenditure and economic growth: a post-Keynesian analysis', Pintu Parui uses a post-Keynesian growth model – with positive saving propensity out of wages – to analyse the implication of different kinds of government expenditures on aggregate demand and economic growth. Parui distinguishes between government expenditure on consumption and investment. The basic idea is that certain kinds of government investment expenditure influences labour productivity. In a formal model, Parui incorporates this idea by assuming labour productivity is an increasing function of government investment expenditure. When the economy is in a profit-led demand regime, under the balanced budget assumption, Parui shows that a shift in government expenditure from consumption to investment leads to an unambiguous rise in both aggregate demand and economic growth. However, the result is ambiguous in the wage-led demand regime. Once the balanced budget assumption is dropped, while in a wage-led demand regime a rise in government investment expenditure may decrease aggregate demand and growth, it unambiguously raises both aggregate demand and growth in a profit-led demand regime. On the other hand, in the absence of a balanced budget assumption, a rise in government consumption expenditure has a positive effect in both regimes. Parui also shows that allowing the government to run a deficit and incur debt does not necessarily lead to the public debt rising without bounds.

As noted above, the founding purpose of the *International Review of Applied Economics* was the 'application of economic ideas to the real world' and 'to the evaluation and development of economic policies', with papers and book review articles to be published 'in the area of applied policy-oriented economics. . . . dealing with policy issues and adopting theoretical positions on current issues which are neglected in mainstream journals'. This Special Issue concludes with just such a review article, with 'Interpreting the world, in various ways – and changing it' by Jonathan Michie reviewing Milanovic (2019) *Capitalism Alone*, Blakeley (2019) *Stolen: How to Save the World from Financialisation*, and Pettifor (2019) *The Case for the Green New Deal*. The review article notes that Malcolm Sawyer's work has played and continues to play an important role in both developing a better understanding of the economy and how it functions, and in developing policies to change it for the better – that is, to change the dominant economic system for the better, and hence thereby improve the state of the economy and society. These grand challenges – of analysing the current economic system, and considering how it might be changed for the better – are ones that in various ways are taken up by the three books here reviewed, with a common conclusion that the current form of capitalism needs to be either reformed or replaced, because of the corruption it engenders (stressed by Milanovic) and the financial crises it creates (analysed by Blakeley), and because of the need for an economic system in which incentives and behaviours create environmental as well as socially sustainable outcomes – which is the focus for Pettifor, as well as for several of the papers in this Special Issue, most notably those by Malcolm Sawyer and Robert Pollin.

Notes

1. Editorial Policy included from the first issue of the *International Review of Applied Economics*.
2. 'Aims and Scope' of the *International Review of Applied Economics*.
3. Citing Sawyer (2013, 2015, 2018).

Disclosure statement

No potential conflict of interest was reported by the author.

References

Blakeley, Grace. 2019. *Stolen: How to Save the World from Financialisation*. London: Repeater Books.
Milanovic, Branko. 2019. *Capitalism Alone: The Future of the System that Rules the World*. Cambridge, Massachusetts: Harvard University Press.
Pettifore, Ann. 2019. *The Case for the Green New Deal*. London & New York: Verso.
Sawyer, Malcolm. 2013. "Alternative Economic Policies for the Economic and Monetary Union." *Contributions to Political Economy* 32 (1): 11–27. doi:10.1093/cpe/bzt005.
Sawyer, Malcolm. 2015. "Can Prosperity Return to the Economic and Monetary Union?" *Review of Keynesian Economics* 3 (4): 457–470. doi:10.4337/roke.2015.04.02.
Sawyer, Malcolm. 2018. *Can the Euro Be Saved?* Cambridge: Polity Press.

Financialisation, industrial strategy and the challenges of climate change and environmental degradation

Malcolm Sawyer

ABSTRACT

The paper discusses the nature of the present era of financialisation, outlining the changes in the financial sector and its relations with the real sector which are particularly relevant for the climate emergency. The relationship between growth of the financial sector ('financial development') and economic growth is reviewed, and the relevant of recent empirical findings for the role of the financial sector in addressing the climate emergency drawn out. It is argued that the policy approach to the climate emergency and environmental degradation should be embedded within an industrial strategy. Further, it is argued that the structures of the financial sector need to be changed to encourage financial institutions which are more favourably disposed towards to the allocation of funds to 'green investment'. It is also argued that the central bank should act in ways that are supportive of environmental policies but that their role is a rather limited one.

1. Introduction

The challenges posed by climate change and environmental degradation are well-known, as are the needs for urgent actions to de-carbonise economies and address damage to the natural world and to bio-diversity.[1] The major structural changes in shifts towards low (or no) carbon production and consumption will have to feature strongly. It would likely require not only de-coupling of environmental degradation from gross domestic product (GDP), but a much slower rate of growth of GDP, and shifts in the location of economic activity away from the market. The questions are then whether the financial system can adjust to a lower growth rate and what that implies, and whether the financial system can ensure the required industrial re-structuring with its present structures. Further questions involve the financial institutional arrangements which are required, and roles for government in terms of State Green/Development Bank, and regulations (e.g. classifying what is deemed to be 'green investment').

Financialisation, alongside globalisation and neo-liberalism, have been the major economic and social forces in the past four decades. Here, I seek to indicate the ways in which the processes of financialisation have impacted on climate change, and to understand the ways in which financialisation and the financial sector have to change in order to address the climate emergency. The paper is structured as follows. In the next section, the nature of the present era of financialisation is discussed, outlining the particular changes in the financial sector and its relations with the real sector which are particularly relevant for the climate emergency. Section 3 reviews the relationship between growth of the financial sector ('financial development') and economic growth, and draws out the relevant of recent empirical findings for the role of the financial sector in addressing the climate emergency. Section 4 argues that the policy approach to the climate emergency and environmental degradation should be embedded within an industrial strategy. Section 5 argues that the structures of the financial sector need to be changed to encourage financial institutions which are more favourably disposed towards to the allocation of funds to 'green investment'.

2. Features of financialisation

There is a reflection on the significance and consequences of financialisation for the design and implementation of policies to address climate change and environmental degradation. A widely quoted general perspective on financialization is provided by Epstein (2005, 3) that 'financialization means the increasing role of financial motives, financial markets, financial actors and financial institutions in the operation of the domestic and international economies'. Financialisation in that broad sense has been a central feature of industrialised economies since the late 19th century, with a long pause in the 1930s and 1940s. In the present era of financialisation (broadly since the late 1970s), the processes of financialisation have involved the further expansion of the banking sector and of equity markets and the growth of what is often termed 'shadow banking', growth of a range of complex financial instruments with securitization and derivatives, the engagement of non-financial corporations in financial dealings, and the growth of consumer borrowing and household debts.

Van der Zwan (2014) identifies three themes within financialisation studies: the emergence of a new regime of accumulation, the 'pursuit of shareholder value' and the 'financialization of the everyday'. Financialisation has involved greater involvement of the general population with the financial sector, and has re-inforced financial motives – that is emphasis on undertaking decisions and actions based on monetary returns rather than on broader calculations such as the well-being of those affected as well as one's self.

The expansion of the financial system requires commodification of activities on the basis of financial calculations and the ability to trade in corresponding assets. The processes of financialisation were well exposed in the global financial crises 2007/09 where the financialisation of housing, sub-prime mortgages and the securitization of those mortgages was central to the financial crisis. Further, oil and food prices had displayed volatility and dramatic rises based on securitization, notably in the late 2000s.

There are many changes in the financial and banking system associated with the processes of financialisation, and two notable ones have been securitization and the growth of 'shadow banking'. Securitisation involves the pooling various types of contractual debt such as residential mortgages, commercial mortgages, loans, credit card debt obligations and other assets which generate receivables and selling their related cash flows to third party investors as securities. Shadow banking system is viewed as a collection of non-bank financial intermediaries that provide a range of services which are similar to those of commercial banks but subject to different less demanding banking regulations and without the relationships with the central bank which commercial banks have, including the central bank being lender of last resort. The shadow banking system raises many concerns in association with financial instability and regulation of the financial system.

Industrialisation and capitalism extended the logic of the market 'from agricultural land to built environment and the biosphere in general, transforming nature from an end in itself to a mere instrument. In particular, the process of financialisation has captured since long the vital activity of energy production, distribution and consumption' (Vercelli 2017), who notes the ways in energy resources and primary commodities have become subject to financial speculation. Clark and Hermele (2014) argue that '[o]nce commodified, environments are increasingly securitized, treated as pure financial assets, and, turned liquid, enter the orbit of rent-seeking financial capital: as potential sites for investment, or disinvestment, depending on their valuation in the calculations of financial capital (potential yield to shareholders). The penetration of financialisation into the fabric of socio-ecological systems works through – i.e. generates and subsequently builds upon – processes of commodification, privatization and securitisation of environments.'

In a similar vein, Sullivan (2013) shows 'how business and finance sectors, in collaboration with conservation organisation, conservation biologists and environmental economists, are engaging in an intensified financialisation of discourses and endeavours associated with financialisation more generally. This tendency permits capital accumulation to be generated through the movement of interest-bearing capital into new areas of social and economic (re)production, even as other areas of production are stagnating.' (p.199). The financialisation of the everyday 'extends into discourses of environmental conservation and sustainability and atmospheric change'. As Gabbi and Ticci (2013) point out, there is a push from the proponents of financialisation to place monetary values on nature and environmental services. It is then argued that 'pricing environmental resources (natural capital assets, environmental risks and environmental free-access or public goods and services) can mobilise financial resources and business practices towards investment in environmental conservation and towards incorporation of sustainability principles in economic and financial activities and decisions.' Thus, financialisation is conducive to an agenda attempting to save nature through a commodification of its resources, services, perceived values, but the complexity of ecosystems cannot be narrowed down, compressed and summarised in a single metric or in a single service." (Gabbi and Ticci 2013).

The processes of financialisation raise many issues with respect to environmental degradation, and two are highlighted here. First, there are the need for changing the structures of the financial sector in ways which are conducive to the channelling of funds into 'green investment' and away from 'brown investment'. The evidence discussed in the next section indicates that the size of the financial sector is no longer positively correlated with economic growth, and the challenge is to focus the financial sector's funding activities on the transition to a sustainable economy. This is a theme which is picked up in section 5 below. Second, financialisation has involved financial calculations and motives being applied to a widening range of human activities, and notably to the environment. Addressing the climate emergency will require diminishing the role of financial calculations and focus on the reductions of environmental damage.

3. Financialisation, economic growth and the allocation of funds

There is a long-standing set of literature on the relationship between the size of the financial sector and the pace of economic growth (in terms of gross domestic product). The growth of the financial sector has often been evaluated under terms such as financial development, financial deepening, and the perceived role of financial development as a promoter of savings and investment.

Financial deepening, often measured by variables such as bank deposits to GDP, focuses on the growth of the formal financial sectors and is also to be viewed as a dimension of financialisation. The first phase of the literature had generally found a positive relationship between financial development and economic growth, though the causal relationships involved are matters of debate. A more recent literature has tended to find a much weaker relationship, and often finding an inverted U-shaped relationship such that industrialised countries are often operating on the negative part of the curve. For example, 'up to a point, banks and markets both foster economic growth. Beyond that limit, expanded bank lending or market-based financing no longer adds to real growth' (Gambacorta, Yang, and Tsatsaronis 2014, 21).[2] Others, for example, Crotty and Epstein (2013), have argued that the financial sector has become too large and dysfunctional; and that the ways in which the financial sector has expanded in recent times (through securitisation for example) have aided instability and have engaged in activities which will resource-using do not contribute to the financing and funding of investment.

The growth of the financial sector and financialisation being no longer (at least for industrialised countries) positively related with growth of GDP comes from a combination of two factors. First. the propensity to save (relative to GDP) has reached an upper limit and further increases are not facilitates by a larger financial sector. There have been changes in savings behaviour, notably the role played by retained profits and by demographic change. There may indeed be a sense in which saving propensities have become 'too high', a modern-day counterpart to an 'underconsumptionist approach'. The widespread existence of budget deficits (being the counterpart of private net savings) is suggestive of private savings exceeding private investment. A high portion of savings takes the form of retained earnings by corporations and not directly passing through the financial markets.

The 'pursuit of shareholder value', one of the key dimensions of financialisation (as identified by Van der Zwan 2014 for example), has generally been viewed as placing emphasis on short-term profits and dividends at the expense of the re-investment of profits. Hein (2012) argues that 'Financialization has been associated with increasing shareholder power vis-à-vis management and labourers, an increasing rate of return on equity and bonds held by rentiers, and decreasing managements' animal spirits with respect to real investment, which each have partially negative effects on firms' real investment' (p. 116). Hein (2012) summarises a range of arguments on the effects of 'shareholder value' under financialization on investment. It is argued that shareholders (most of whom are financial institutions) impose on corporations a larger distribution of profits and hence a higher dividend payment ratio. The lower retention of profits ratio, and on occasions share buybacks mean reduced internal finance for real investment. Hein labels this the 'internal means of finance channel' A further channel, labelled 'preference channel', arises from the weakening of the preference of managers for growth (which translates into firms pursuing growth) as managerial remuneration schemes are based on short-term profitability and share price. A lower rate of investment (which I would broaden to include research and development by firms) would tend to be associated with a lower rate of economic growth (in so far as the capital-output ratio does not change significantly). In that respect, financialisation may well be associated with lower growth, rather than the positive relationship between financial development and economic growth which had once been postulated.

Second, the growth of financial sector has been particularly pronounced in terms of the developments in securitization and derivatives. Securitization involves the creation of financial assets whose value depends on other financial assets. Such creation of financial assets and the trading in those financial assets are not (directly) related with savings nor with the funding of investment.

There is the general requirement for a banking and financial system which serves the rest of the economy rather than the rest of the economy serving the interests of the financial sector. Apart from the provision of a well-functioning payments technology, the key requirements which for a socially beneficial financial system are that it develops in ways which are consistent with the environmentally sustainable rate of growth, and that it channels funds into the socially desirable types of investment. This requires social and State intervention, as "there is no clear evidence from experience that the investment policy which is socially advantageous coincides with that which is most profitable" (Keynes 1936, 157)

The first part of such requirements includes seeking to construct a stable banking and financial system. A more significant element is that the financing of investment, initial and final in the terminology of the circuitist approach (Graziani 2003), is channelled towards socially desirable and environmentally friendly investment, and not towards financial asset accumulation and speculation. The thrust of the operations of the banking and financial system should be on the financing and funding of real investment, and not financial investments.

The transition to a low carbon sustainable economy would involve a shift of resources including capital equipment from carbon-intensive industries into low carbon industries. Lewney (2020) argues that 'because the net zero technologies are more capital-intensive, the overall scale of investment in the energy system would be higher, especially in the period to 2050 when the entire new system needs to be put in place'. However, he points out that the estimates of the level of investment which would be required vary quite widely and depend on the ways by which the emissions reductions are achieved. For example, investment in energy efficiency may have lower investment requirements than investment in alternative energy sources. 'For a 1.5°C pathway, IPCC (2018) draws on existing studies to present an average estimate that annual global investment in the energy system amounting to 2.4 USD trillion (at 2010 prices) would be needed between 2016 and 2035, equivalent to 2.5% of world GDP. This is a gross figure, meaning that the energy system would in any case require substantial investment in a current-policies baseline case (about 1.8% of global GDP (IPCC, op. cit., p. 373), but the additional investment is still substantial.'

Although significant investment is likely to be required to enable a transition to low carbon economy, it is a matter of conjecture what the overall level of investment would need to be and how that compares with the potential level of savings. A tendency for slower growth of GDP in industrialised countries has already been occurring would suggest that the requirements of investment (relative to GDP) may well be lower than in the past. Further, The lower growth and indeed decline of carbon emissions regarded as vital to avoid global warming above 1.5°C, which requires a combination of slower growth of GDP and the de-coupling of carbon emissions from GDP (Jackson and Victor, 2019).

The simple formula that the ratio of increase of capital stock (=net investment) to GDP is equal to capital-output ratio times growth rate of GDP is an indicator of the degree to which the net investment (to GDP) would be lower. The average capital-output ratio may, of course, change. Further, this formula relates to net investment whereas the use of resources for investment purposes relates to gross investment. The depreciation allowances can potentially provide funds for investment in the transition. In terms of the overall availability of funds, it is necessary to think in terms of the balance between the potential level of savings and the requirements for investment. The major requirement is not likely to be the stimulation of the levels of savings and investment expenditure in total, but rather to ensure that investment is environmentally friendly and is not promoting economic activities which are, for example, carbon-intensive.

4. An industrial strategy approach to the climate emergency

It can be readily agreed that a structural transformation of economic activities is required to address the climate emergency and environmental degradation. A structural transformation requires appropriate funding as investment and economic activity shift between sectors. In order to enable such a transformation, what may be termed an industrial strategic approach is required. The major component of an industrial strategy approach to map out the broad contours of the development path of the economy, which in the context of the climate emergency, is the transformation to a low carbon economy. However, a green industrial strategy would need to be much more encompassing confronting the use of nature, the destruction of bio-diversity.

There was a short-lived burst of interest in industrial strategy in the UK in 2017/2018, which was rather overwhelmed by negotiating Brexit (and now the COVID 19 pandemic). The UK government stated the approach to industrial strategy was, in the words of the then Prime Minister, Theresa May, in her introduction to HM Government (2017), 'a new approach to how government and business can work together to shape a stronger, fairer economy. At its heart it epitomises my belief in a strong strategic state that intervenes decisively wherever it can make a difference. It is rooted in the conviction that a successful free-market economy must be built on firm foundations: the skills of its workers, the quality of the infrastructure, and a fair and predictable business environment' (p.4). It is rather questionable that an economy can be labelled as free market when there is a strong strategic state, and also highly doubtful whether there has ever been or could be a free market economy. The essential vision here is of an industrial strategy approach which ensures the construction and delivery of an environmental sustainable economy (and society) which addresses issues of the climate emergency, development of a low carbon economy, confronting environmental degradation and preserving bio-diversity.

In Sawyer (2000), I wrote that what I termed the industrial strategy approach 'does not lead to the advocacy of central planning: in part because of the informational and incentive requirements for successful planning are impossible to achieve. Instead, the government accepts a strategic role under which a broad view on future developments is evolved, and in which public support is forthcoming for productive activities (rather than exchange or financial ones). Much of the co-ordination of economic activity is undertaken through the market, though it is recognized that substantial parts of such co-ordination take place within firms and within households.' The industrial strategy approach does not see a sharp dichotomy to be drawn between allocation through markets and allocation through planning. In the present context, the particularly significant element of what may be termed a climate emergency industrial strategy is that the re-structuring of the economy in an environmentally sustainable direction becomes the centre piece. This requires, *inter alia*, mapping out the shifts in the structures of economic activity which would be required, the identification of the types of investment which are consistent with addressing the climate emergency and environmental degradation (and which may be labelled 'green investments') etc. An important feature of a climate emergency industrial strategy would be setting out what is regarded as 'green', and what is not, and the degrees of 'greenness'. A government formulated and transparent assessment of what would constitute 'green investment' is required to inform the government's own funding policies, the central bank's asset purchase policies and also as guidance for those seeking to support 'green investments' through their own savings. Pollin (2020) sets out a green industrial strategy focused on dramatic improvements in energy efficiency and the development of renewable energy sources.

An industrial strategy approach stands in some contrast with the (mainstream) industrial policy approach and an Austrian economics industrial policy approach. The mainstream industrial approach can be simply perceived in terms of the correction of market failure, based on the presumed allocative and technical efficiency of perfect competition. From that perspective, industrial policy focuses on monopoly and mergers policy and the correction of externalities – the classic policy approach to environmental issues being the use of taxes and subsidies to correct for externalities (e.g. pollution). The Austrian approach promotes the market mechanisms and property ownership – and specifically in the context of environmental issues on the 'Coase theorem'. The 'Coase theorem' describes the economic efficiency of an economic outcome in the presence of externalities. The theorem states that if trade in an externality is possible and there are sufficiently low transaction costs, bargaining will lead to a Pareto efficient outcome regardless of the initial allocation of property. In practice, obstacles to bargaining or poorly defined property rights can prevent Coasean bargaining.

The mainstream perspective has addressed environmental issues in terms of externalities imposed by production and consumption and the correction of those externalities through taxes and subsidies. This may have some usefulness in the context of, for example, localised pollution, where the negative externality of pollution can potentially be addressed by tax on the generators of pollution. But this type of approach is not capable of addressing the gravity of the climate emergency.

Lewney (2020) argues that the neo-liberal agenda in principle 'does not preclude the adoption of policies intended to meet ambitious targets to prevent and reverse environmental degradation, but it does circumscribe tightly the permissible policy tools used to pursue those targets (limited essentially to market-based instruments).' He further argues that there are two reasons why neoliberal economics ends up with a very limited policy intervention for environmental goals. These are first 'its philosophical standpoint gives primacy to the individual as the arbiter of value' and second 'it places a very high priority on individual freedom as against state action that limits such freedom, even if the action itself uses a market-based instrument rather than, say, regulation as the tool.' Lewney (2020) argues that mainstream approaches seek to value climate and environmental damages on the same metric as consumption and then they can be traded off against each other. In contrast, an alternative approach treats them as incommensurable, and sets targets for climate change mitigation and then explores the required 'time profile of net emissions and [ranks] alternative pathways to achieve that objective according to their economic and social impacts'

Chang and Andreoni (2020) provide a discussion of industrial policy in the 21st century, in a way which fits with what I would have termed industrial strategy. The particularly significant feature of industrial strategy is that it goes far beyond the market failure, control monopoly and promote competition approaches of neo-classical industrial policy.[3] Chang and Andreoni (2020) first point to industrial policy tools which 'reduce uncertainty by guaranteeing demand'. These include infant industry protection which enables infant firms to survive and learn, and restrict competition from foreign producers, limiting competition among domestic firms and government procurement policies. Further, 'the government can provide a clear platform for technological evolution of an industry by taking a lead in the development of the basic technologies …. Second, the government can push firms to form research consortia to develop basic technologies, which they will share and use in developing more applied technologies … . Third, during the early stage in the development of an emerging industry, where different technological standards compete with each other, the government can reduce uncertainty about the path of future technological evolution by imposing a technological standard … . Fourth, the government can subsidize or directly provide technology-related "public goods" (such as data, metrology, prototyping and testing facilities'.

These arguments can be applied to a 'green industrial strategy' with regard to, for example, the emergence of alternative energy supplies. Lewney (2020) notes the importance of fundamental uncertainty, which is particularly acute in the context of climate change. 'If there is an uncertainty penalty for new, clean technologies, we can no longer interpret low responses to carbon price signals as indicating the preferences of fully informed, rational individuals. In other words, what looks like a high mitigation cost in a world of perfect information (agents will not act unless the price signal is very high) becomes a case of herding behaviour (agents will not act until they see other agents doing so)'.

The development of new products and processes which can aid the transformation to a low carbon, nature friendly economy will often have features which have been long known to require State involvement. The overall social benefits from the new processes will often be in excess of private benefits. The pay-offs from the new products and processes will often be long-term (measured in decades) and subject to fundamental uncertainty. There will be path dependency 'implied by technological lock-in [which] is strengthened by the endogenous nature of technological change and the role played by radical uncertainty in decisions to invest in innovation and to adopt new technologies' (Lewney 2020).

Chang and Andreoni (2020) observe that 'the industrial policy debate has historically had a supply-side bias'. I argued in Arestis and Sawyer (1999) that an industrial strategy needs to be supported by appropriate macroeconomic policies (fiscal and monetary). Chang and Andreoni (2020) specifically mention interest rate policy in this regard working through the effects of interest rate on investment. In the context of the climate emergency a rather different perspective is adopted. I continue to acknowledge the role of fiscal policy in securing high levels of employment. The importance of fiscal policy though is much more about ensuring that the structure of public expenditure, and particularly public investment, is supportive of the climate emergency industrial strategy.In the area of central bank policies, the operations of the central bank should be co-ordinated with the government and its environmental policies. The mandate of a central bank should be formulated along the lines of that of the European Central Bank – that is to 'support the general economic policies in the [European] Union with a view to contributing to the achievement of the objectives of the Union', where it is particularly important is that the general economic policies and objectives strongly feature addressing the climate emergency and environmental degradation.

There have been suggestions that central banks should pay regard to the 'green' credentials of the financial assets in their portfolio. Christine Lagarde, then incoming President of European Central Bank, for example, said that a 'move to a gradual transition to eliminate this type [carbon] of assets' was 'something which needs to be done'.[4] The implementation of such a policy requires a clear definition of what constitutes a 'green investment', particularly in terms of financial assets, and that should come from the general government policies.

The operations of the central bank should be pursued in ways consistent with the broader political and social objectives of the government – in this context, that is it should be a decision of government as to what is regarded as 'green investment' and the degree of 'greenness'. In effect, a central bank being willing to purchase/hold some financial assets but not others shifts the composition of demand for the financial assets, and to that extent the price of 'green' financial assets is somewhat higher than otherwise, and the price of 'brown' financial assets somewhat lower. The effects could be similar those of the disinvest movement (which then raises the question of who decides which type of assets are to be favoured and which not). Dafermos et al. (2020) provides an example from the Bank of England of how the present arrangements can have adverse effects in respect of climate change. '[I]n its present guise [in 2020], the CBPS [Corporate Bond Purchase Scheme] is mis-aligned with the government's climate goals and implicitly creates better financing conditions for carbon-intensive economic activities. The CBPS biases the allocation of capital towards carbon-intensive sectors, while at the same time failing to reflect climate-related financial risks.'

Chang and Andreoni (2020) further consider industrial policy in terms of conflict management. 'It has to be admitted that industrial policy may be most prone to open conflicts, as it tends to be more explicitly selective than other poliices; it inevitably chooses between secvtors, technological, or even individual firms in the same industry. Therefore, conflict management is more important for industrial policy than for other policies.' As Lewney (2020) notes, 'the transition would therefore create both losers and winners. The restructuring impact is made more severe by the fact that the activities that would be phased out tend to be geographically concentrated (regionally and internationally) either as a consequence of geology or because they are subject to economies of scale and so tend to have large plants that are major local employers.' This again emphasises the need for an industrial strategy which includes the development alternative employment in the areas with specialisation in the vulnerable sectors.

A major part of an environmental industrial strategy has to be conceptualising what is to be regarded as 'green investment', 'green consumption patterns' etc. The central purpose is to shift resources into environmentally sustainable sectors and away from the environmentally damaging. This is intended to extend beyond issues of carbon-intensity. It would also seek to take a vertically integrated approach – to avoid encouraging an investment which appears 'green' in relation to one part of the production process but not in other parts of the production process.

5. Roles for a re-structured financial sector

The financial sector has in many respects grown too large, and perhaps paradoxically may well have tended to reduce the rate of growth of GDP, as mentioned above. The climate emergency and environmental degradation requires dramatic shifts in the composition of economic activity, and the funding of those shifts. It is largely not a matter of more investment (and more saving) but rather the composition of investment; and indeed, slower economic growth could well involve lower levels of overall investment. Thus, the focus here is on the role of financial institutions in the allocation of funds in ways consistent with averting the climate emergency, and not in raising higher levels of funds.

Financial institutions can be viewed as intermediaries between savings and investment, and in the simplest form between households as savers and firms, though it has to be recognized that a large element of investment is funded by retained profits and that much lending takes the form of household debt (including mortgages). Commercial banks provide loans through which bank deposits (and thereby money) are created. The central issue is to whom the financial institutions provide finance and funds and on what terms. Credit rationing is a pervasive feature of the behaviour of banks and other financial institutions in the sense that financial institutions have to assess the risks of non-payment and default of loans, and the interest rate charged and the other conditions of any loan will reflect that risk assessment. The credit allocation processes depend on risk assessments which in an uncertain world can only be perceptions of frequency of default etc., rather than based on well-established probability distributions. Financial institutions, as all of us, operate in a world of fundamental uncertainty and path dependence. The assessments of credit risk etc. are inevitably fraught with difficulties in these circumstances, and those assessments have to be influenced by the needs in addressing the climate emergency. The influences can range over policy measures such as selective credit allocation whereby government lays down requirements for the flow of credit. It can also come from the construction and development of the types of financial institution which are designed to foster certain types of funding. The structure of the financial system and the legal framework must be such as to ensure that credit rationing practices do not operate against 'green investment', and more generally environmentally friendly investment.

Policies of selective credit allocation may provide one route through which 'green investment' can be encouraged. Many countries operated policies of selective credit allocation in the first decades after World War II, though such policies have fallen out of favour in recent decades. Selective credit allocation requires banks to allocate finance and funds towards specified priority sectors – at the limit only to those sectors, and at other times requiring a high proportion of lending to the priority sectors. 'Virtually all central banks have engaged in "industrial policy" or "selective targeting". The difference lies in which industries they have promoted. Significantly, the whole tenor of economic development can be fundamentally affected by which of these industries the central bank and associated institutions promote.' (Epstein 2007). Epstein further indicate the wide range of ways in which credit controls have been operated.

A similar route could come from considers a 'differentiation of reserve requirements according to the destination of lending' (Campiglio 2016), and specifically that lending credit creation directed towards low carbon activities would attract lower reserve requirements for the banks – this may be in the form of lower reserves held with the central bank (though many countries do not operate with such reserve requirements) or in the form of capital requirements.

These types of policies face the formidable difficulty of defining and then monitoring those investments which may be considered 'green', low carbon, preserving nature and bio-diversity etc. EU Technical Expert Group on Sustainable Finance (2020) provides an example of seeking to draw up a taxonomy of 'green investments'. It also has to recognized that the assessment of whether an investment should be deemed 'green' has to take into account not just the parts of a production process to which the investment relates but also to other parts of the overall chain of production. It is then highly relevant that the pursuit of selective credit policies be undertaken within the context of a 'climate emergency industrial strategy' which formulates the criteria on which 'green investment' etc is to be judged, and which sets up arrangements to monitor the application of the criteria. There are though weaknesses from labelling bonds as 'green bonds'. 'The current system of green bond labels does not necessarily guarantee a material reduction in carbon emissions. Indeed, these labels would signal emission reductions only if the relevant projects were to transform the activities of the bond issuer radically enough for its carbon emissions to fall.' However, the authors show that 'green bond labels are not associated with falling or even comparatively low carbon emissions at the firm level' (Ehlers, Mojon, and Packer 2020, 31)

Pollin (1998), writing in the context of financial structures and egalitarian policies, noted that 'finance is the conduit for all economic activity in market economies. Because nothing happens unless it is financed, exerting control over the financial system is an efficient way to influence the widest possible range of activity with a set of relatively small and simple policy tools.' (p.163). In respect of financial institutions, he argues that 'bank-based systems are better equipped [than market-based systems] to promote longer time horizons and a stable financial environment. Their structures also create more favourable conditions for activist government interventions, including both traditional macro policies and public credit allocation policies.' (p.164). This line of argument can be broadened out in a number of respects. It has to be recognised that the differences between bank-based and market-based systems are one of degree in that most financial systems have banks and financial markets. Indeed, banks are the major routes through which money is created (and destroyed) in the process of loans being provided. The term 'banks' though covers a wide range of financial institutions including what may be termed commercial banks, some of whose liabilities (bank deposits) are treated as money with close relationships with the central bank. There are also savings banks, investment banks etc., which are deposit accepting and in general regulated. Within the banking sector, the patterns of ownership vary from the private profit-seeking, mutual and co-operatives through to the State owned. Mutual financial institutions have often been more focused on households – for example housing finance (as in the UK's building societies) and credit unions.

Ayadi et al. (2010) and others have drawn the distinction between *Stakeholder Value* (STV) banks and *Shareholder Value* (SHV) banks. They 'conceptualise SHV banks as those whose primary (and almost exclusive) business focus is maximizing shareholder interests, while STV banks in general (and cooperative banks in particular) have a broader focus on the interests of a wider group of stakeholders (notably customer-members in the case of cooperative banks, the regional economy and the society in the case of savings and public banks)' (p. 7).

Block (2014) views a good way forward is the introduction of 'significant competition from financial intermediaries who are not seeking to generate profits. These could take the form of credit unions, community banks, nonprofit loan funds, or banks that are owned by government entities; but the key is that their mission is defined as facilitating economic development in a particular geographical area. With this mission, they have a reason to employ loan officers who develop the skill set needed to provide credit to individuals and firms who fall outside the parameters of the standard lending algorithms' (p. 16). He advocates a 'a combination of governmental supports and grassroots entrepreneurialism to create an expanding network of non-profit financial institutions that would redirect household savings to finance clean energy, growth of small and medium-sized enterprises, and infrastructure' (p.3).

A major form of stakeholder value institutions is mutual and co-operative banking. Groenweld (2015) advocates such retail banking as one which 'demonstrably results in a moderate risk profile and close links with the real economy and local communities' (p. 6). In a similar vein, the argument is put that 'empirical evidence in this study suggests that no radical differences exist between cooperative banks and their peers in terms of performance and efficiency. More important, there are economic, systemic and welfare benefits to be derived from a successful cooperative sector in the banking systems in Europe. A financial system populated by a diversity of ownership and governance structures, and alternative business models, is likely to be more competitive, systemically less risky and conducive to more regional growth than one populated by a single model' (Ayadi et al. 2010, p. vi).

The development of more localised banking can come from regional banking. Klagge and Martin (2005) put a case for regional banking in terms of three advantages. 'First, the presence of a local critical mass of financial institutions and agents – that is of a regionally identifiable, coherent and functioning market – enables local institutions, SMEs, and local investors to exploit the benefits of being in close spatial proximity. ... Second, the existence of regional capital markets specialising in local firms may help to keep capital within the regions, as local investors direct their funds into local companies – and hence into local economic development – rather than investing on the central market. ... Third, in a nationally integrated financial system, the case can be made for a regionally decen-tralized structure on the grounds that it increases the efficiency of allocation of invest-ment between the centre and the regions' (p. 414).

These arguments would echo those of Epstein (2018) when he wrote that 'by reducing the size of "too big to fail" banks, imposing financial transactions taxes, implementing asset-based reserve requirements, establishing "Green Banks" and other initiatives, a restructured finance could help to make the green transition and generate jobs and sustainable growth as well. This is just one – but one very important – example of how restructuring finance can be a much better alternative to roaring banking and bubble finance for job creation and socially useful investment.' (Epstein 2018, 348)

Davis and Cartwright (2019) have considered the suitability of crowdfunding (through small contributions for a large number of sources) for the public sector. They note that crowdfunding is rarely used by the public sector whfor which two main reasons were identidied: lack of knowledge and expertise within public bodies and a concern that higher costs (including administrative) would be involved.

In the project of Davis and Cartwright, six case studies were undertaken with three UK local authories and three NHS bodies to conduct feasibility studies on using crowdfund-ing to finance specified infrastruture project including green energy initiastive, commu-nity regeneration schemes. 'Our research found that investment-based crowdfunding provides a viable and significant opportunity for public bodies seeking additional models of finance whilst also growing local engagement between the public sector and their community. This opportunity is not without its challenges (both real and perceived) ... ' (Davis and Cartwright 2019, 5)

The developments of these alternative (to single bottom line of profits) financial institutions would help to create a more diverse financial system. But more importantly would be the impacts on the ways in which funds are allocated, and the potential for savers to have some influence on the ways in which their savings are deployed. An individual seeking ethical investments can make that choice, and another seeking to support local producers may likewise be able to make that decision.

The roles of State development banks (with names such as Green Investment Bank) are of particular importance. As Griffith-Jones and Cozzi (2016) argue, 'the existence of development banks is justified by the existence of sectors and investment projects that require funding for the future development of the economy, but have high uncertainty as to their future success.' These projects may find difficulty in securing private funding because of the uncertainties involved. The long time horizons, the fundamental uncer-tainty of investment projects, and the path dependence involved are often particularly important with the climate emergency.[5]

There are projects which would be environmentally friendly and socially beneficial but which do not readily generate financial returns. There are activities which need to be de-commodified and taken outside of the market (e.g. provision of universal basic services). Finance Watch (2020) argue for the role of what they term public finance as 'being especially useful for projects that provide public value without being financially bankable themselves', in particular with regard to protection of nature and bio-diversity. It points out that 'the nature of private investment is to focus on financial risks and returns and revenue streams, but **many nature-related projects have no revenue source**. Indeed, nature tends to benefit when there is less economic activity. Conversely, it can be easier to finance businesses that keep their costs low by harming nature' (bold in original). It argues that '**Nature projects are often too small for institutional investors to invest in directly**. They are complex to understand, illiquid and take a long time to mature. . . . . the [finance] industry's reliance on CAPM and other structural factors mean that fund managers have little appetite for alternative or long-term investments.'

It is well recognized that climate change itself impacts on the financial system through two main channels of risks which serve to reduce asset values. One can be described in terms of the physical risks from damage to property, infrastructure, and land. The other set of risks come from 'changes in climate policy, technology, and consumer and market sentiment during the adjustment to a lower-carbon economy.' (Grippa, Schmittmannm, and Suntheim 2019, 26). 'Financial stability concerns arise when asset prices adjust rapidly to reflect unexpected realizations of transition or physical risks, there is some evidence that markets are partly pricing in climate change risks, but asset prices may not fully reflect the extent of potential damage and policy action required to limit global warming to 2°C or less' (Grippa, Schmittmannm, and Suntheim 2019, 27–8). More generally, 'Financial institutions could incur losses on exposures to such firms with business models not built around the economics of low carbon emissions. These firms could see their earnings decline, businesses disrupted, and funding costs increase due to policy action, technological change, and shifts consumer and investor behavior. Risks can materialize especially if the shift to a low-carbon economy is abrupt (as a consequence of prior inaction), poorly designed, or uncoordinated globally. Going forward, a key next step in developing stress tests for transition risks will be to capture "second-round" effects – in which a decline in asset prices leads to fire sales, which further depress asset prices, generating a vicious cycle and an amplifying mechanism for an initial shock.'

There are issues of financial instability arising from the challenge of climate change. There is recognition that there is what has been termed 'transition risks' arising from the re-valuation of carbon-intensive assets arising from shifts to a low-carbon economy.[6] The change in asset values (assumed to be downwards though the valuation of carbon unintensive assets could well rise) have implications for the range of financial institutions and households who own the corresponding assets. This may well be another example of financial markets mis-pricing financial assets – why has the risks involved not been incorporated into the financial asset prices? While it may be relevant for the central banks and others to warn about the likely shifts in asset prices, it is far from clear what actions would follow for monetary policy.

Reduced global demand for fossil fuels is likely to lead to 'stranded assets' with loss of profits on the underlying assets and falls in market valuation of the assets. Mercure et al. (2018) analyse the macroeconomic impact of stranded fossil fuel assets. 'Our analysis suggests that part of the SAFA [stranded fossil fuel assets] would occur as a result of an already ongoing trajectory, irrespective of whether or not new climate policies are adopted; the loss would be amplified if new climate policies to reach the 2° target of the Paris Agreement are adopted and/or if low-cost producers . . . maintain their level of production . . . despite declining demand; the magnitude of the loss from SFFA may amount to a discounted global wealth loss of US$1-4 trillion.'

Warren (2020) argued that many of the largest banks and asset managers have actually *increased* their holdings of fossil fuel assets after the Paris Agreement was adopted, the six largest U.S. bank investors in fossil fuel companies loaned, underwrote, or otherwise financed over 700 USD billion for fossil fuel companies. Wall Street banks are making a quick buck accelerating climate change, all while communities across the country are suffering from the lasting impacts of industrial pollution and the increasingly devasting effects of climate change." (Warren 2020, 2). 'We will not defeat the climate crisis if we have to wait for the financial industry to self-regulate or come forward with piecemeal voluntary commitments. Winning a Green New Deal and achieving 100% clean energy for our global economy … will be near impossible so long as large financial institutions are allowed to freely underwrite investments in dirty fossil fuels.' (Warren 2020).

The clear dangers here are that financial institutions and others continue to own financial assets of carbon-intensive companies, and the value of those financial assets will decline as the underlying assets become stranded. The financial institutions and the owners of the stranded assets then seek bail-outs for their mistaken investment decisions.

6. Concluding comments

The processes of financialisation have involved the growth of the financial sector and its institutions, which from a number of perspectives have become too large (individually and collectively). The climate emergency, environmental damage and loss of bio-diversity will require, inter alia, major re-structure of economic activity and much lower rates of its growth. I have argued for environmental policies to be embedded within what may be termed a climate emergency industrial strategy which seeks to map out 'green scenarios' and which would help to identify sectors and activities which are to be developed. The focus on short-term shareholder value maximisation runs into conflict with a longer term perspective of research and development and investment directed towards sustainability (Brett, Buller, and Lawrence 2020). The direction of funds towards 'green investment' (broadly viewed to include all policies which seek to address environmental degradation, destruction of bio-diversity) will be aided by support for and development of a much more diverse financial system with different types of ownership. It also requires funding arrangements which are capable of supporting activities and investments which do not yield a monetary return, and promote de-commodification of the environment.

Notes

1. I am grateful to the referee for the very helpful comments on the initial draft.
2. I have reviewed the evidence in Sawyer (2016, 2017).
3. For further elaboration on this see Sawyer (2000).
4. Reported in Euractiv 22 September 2019, 'European Central Bank should "gradually eliminate" carbon assets, Lagarde says'
5. See See Spratt, Griffith-Jones, and Ocampo (2013), Griffith-Jones (2016) for more discussion.
6. For example, Breeden (2019), Dafermos, Nikolaidi, and Galanis (2018).

Disclosure statement

No potential conflict of interest was reported by the author(s).

References

Arestis, P., and M. Sawyer. 1999. "The Macroeconomics of Industrial Strategy." In *Industrial Policy in Europe*, edited by K. Cowling, 352–370. London: Routledge.

Ayadi, R., Llewellyn, R. H. Schmidt, E. Arbak, and W. De Groen. 2010. "Investigating Diversity in the Banking Sector in Europe: Key Developments, Performance and Role of Co-operative Banks." *Centre for European Policy Studies*, Brussels. http://ssrn.com/abstract=1677335

Block, F. 2014. "Democratizing Finance." *Politics & Society* 42 (1): 3–28.

Breeden, S. 2019. "Avoiding the Storm: Climate Change and the Financial System." *Speech given at Official Monetary & Financial Institutions Forum, London*, April 15.

Brett, M., A. Buller, and M. Lawrence 2020. "A Blueprint for A Green New Deal." https://www.common-wealth.co.uk

Campiglio, E. 2016. "Beyond Carbon Pricing: The Role of Banking and Monetary Policy in Financing the Transition to a Low-carbon Economy." *Ecological Economics* 121: 220–230.

Chang, H.-J., and A. Andreoni. 2020. "Industrial Policy in the 21st Century." *Development and Change* 51 (2): 324–351.

Clark, E., and K. Hermele 2014. "Financialisation of the Environment: A Literature Survey." *FESSUD Working Paper* no.32.

Crotty, J., and G. Epstein. 2013. "How Big Is Too Big? on the Social Efficiency of the Financial Sector in the United States." In *Capitalism on Trial Explorations in the Tradition of Thomas E. Weisskopf*, edited by J. Wicks-Lim and R. Pollin, 293–310. Cheltenham: Edward Elgar.

Dafermos, Y., D. Gabor, M. Nikolaidi, and F. van Lerven 2020. "Decarbonising the Bank of England Pandemic QE: Perfectly Sensible?" *Policy Briefing*, June. New Economics Foundation.

Dafermos, Y., M. Nikolaidi, and G. Galanis. 2018. "Climate Change, Financial Stability and Monetary Policy." *Ecological Economics* 152: 219–234.

Davis, M., and L. Cartwright 2019. "Financing for Society: Assessing the Suitability of Crowdfunding for the Public Sector." http://eprints.whiterose.ac.uk/145481)

Ehlers, T., B. Mojon, and F. Packer 2020. "Green Bonds and Carbon Emissions: Exploring the Case for a Rating System at the Firm Level." *BIS Quarterly Review*, September, 31–47

Epstein, G. 2005. "Introduction: Financialization and the World Economy." In *Financialization and the World Economy*, edited by G. Epstein, 3-16. Cheltenham and Northampton: Edward Elgar.

Epstein, G. 2007. "Central Banks as Agents of Economic Development." In *Institutional Change and Economic Development*, edited by H.-J. Chang, 95–114. New York and London: United Nations University and Anthem Press.

Epstein, G. 2018. "On the Social Efficiency of Finance." *Development and Change* f49 (2): 330–352.

EU Technical Expert Group on Sustainable Finance. 2020. *Financing a Sustainable European Economy: Taxonomy*. Brussels: European Union.

Finance Watch. 2020. *Nature's Return: Embedding Environmental Goals at the Heart of Economic and Financial Decision-making*. Brussels: Authors Ludovic Suttor-Sorel, Nicolas Hercelin.

Gabbi, G., and E. Ticci 2013. "Implications of Financialisation for Sustainability'." *FESSUD Working Paper*, no. 47.

Gambacorta, L., J. Yang, and K. Tsatsaronis. 2014. "Finance Structure and Growth." *BIS Quarterly Review*, March. 21–35

Graziani, A. 2003. *The Monetary Theory of Production*. Cambridge: Cambridge University Press.

Griffith-Jones, S. 2016. "National Development Banks and Sustainable Infrastructure: The Case of KfW." *GEGI Working Paper* No. 6, July. Boston University, Global Economic Governance Initiative (GEGI).

Griffith-Jones, S., and G. Cozzi. 2016. "The Roles of Development Banks; How They Can Promote Investment, in Europe and Globally." Chapter 5 in *Efficiency, Finance, and Varieties of Industrial Policy: Guiding Resources, Learning, and Technology for Sustained Growth*, edited by A. Noman and J. E. Stiglitz, 131-155. New York: Columbia University Press.

Grippa, P., J. Schmittmannm, and F. Suntheim 2019. "Climate Change and Financial Risk." *Finance and Development*, December.

Groenweld, H. (2015), "European Co-operative Banking Actual and Factual Assessment", http://www.globalcube.net/eacb/medias/publications/eacb_studies/TIUAS Coop_Banking.w.pdf

Hein, E. 2012. *The Macroeconomics of Finance-dominated Capitalism – And Its Crisis*. Cheltenham: Edward Elgar.

HM Government. 2017. Industrial Strategy: Building a Britain fit for the future. UK government. www.gov.uk/beis

IPCC. 2018. *Global Warming of 1.5°C. An IPCC Special Report on the impacts of global warming of 1.5°C above pre-industrial levels and related global greenhouse gas emission pathways, in the context of strengthening the global response to the threat of climate change*. Intergovernmental Panel on Climate Change.

Jackson, T., and P. A. Victor. 2019. "Unraveling the claims for (and against) green growth." *Science* 366 (6468): 950–951

Keynes, J. M. 1936. *The General Theory of Employment, Interest, and Money*. London: Macmillan Publishing Company, UK.

Klagge, B., and R. Martin. 2005. "Decentralized versus Centralized Financial Systems: Is There a Case for Local Capital Markets?" *Journal of Economic Geography* 5 (2005): 387–421.

Lewney, R. 2020. "Environmental Policies to Save the Planet." In *Economic Policies for a Post Neo-liberal World*, edited by P. Arestis and M. Sawyer. Houndmills: Palgrave Macmillan.

Mercure, J.-F., H. Pollitt, J. E. Viñuales, N. R. Edwards, P. B. Holden, U. Chewpreecha, P. Salas, I. Sognnaes, A. Lam, and F. Knobloch. 2018. "Macroeconomic Impact of Stranded Fossil Fuel Assets." *Nature Climate Change* 8: 588–593.

Pollin, R. 1998. "Financial Structures and Egalitarian Economic Policy." In *The Political Economy of Economic Policies*, edited by P. Arestis and M. Sawyer, 162–201. Houndmills: Macmillan.

Pollin, R. 2020."The Industrial Policy Requirements for a Global Climate Stabilization Project." *International Review of Applied Economics*. forthcoming.

Sawyer, M. 2000. "The Theory of Industrial Policy." In *New Challenges to Industrial Policy*, edited by W. Elsner and J. Groenewegen, 23–58. Dordrecht, Netherlands: Kluewer Academic Publishers.

Sawyer, M. 2016. "The Processes of Financialisation and Economic Performance." *Economic and Political Studies* 5, (1). doi:10.1080/20954816.2016.1274523.

Sawyer, M. 2017. "'Financialisation and Economic and Social Performance' in Krzysztof Opolski and Agata Gemzik-Salwach." In *Financialisation and the Economy*, edited by K. Opolski and A. Gemzika-Salwach, 9–25. London: Routledge. 9781138241039.

Spratt, S., S. Griffith-Jones, and J. A. Ocampo. 2013. *Mobilising Investment for Inclusive Green Growth in Low-Income Countries*. Berlin: Deutsche Gessllschaft fur Internationale Zusammenarbeit.

Sullivan, S. 2013. "Banking Nature? the Spectacular Financialisation of Environmental Conservation." *Antipode* 45 (1): 198–217.

Van der Zwan, N. 2014. "State of the Art: Making Sense of Financialization." *Socio-Economic Review* 12: 99–129.

Vercelli, A. 2017. *Crisis and Sustainability. The Delusion of Free Markets*. Houndmills: Palgrave Macmillan.

Warren, E. 2020. "Stop Wall Street from Financing the Climate Crisis." https://medium.com/@teamwarren/stop-wall-street-from-financing-the-climate-crisis-db267e3145c1

UK and other advanced economies productivity and income inequality

Philip Arestis

ABSTRACT

The focus of this contribution is on weak productivity growth, in the UK and other advanced economies, which has been slowing down ever since around 2000, and on increasing income inequality; both of which have been caused by a number of factors, including 'secular stagnation'. Actually, recent evidence clearly suggests that labour productivity and income inequality have been closely and significantly related; this is so since there is a strong relationship between productivity, inequality, economic growth and real wages. Productivity growth is the key determinant of how demand can grow without inflation, thereby reducing inequality of income. The slowdown in productivity growth and increase in inequality have been in evidence in the UK and other advanced economies. Indeed, they have become more pronounced following the Global Financial Crisis. It is the case that although weak productivity growth and increased income inequality predate the Global Financial Crisis, and the Great Recession, they have clearly worsened following them.

1. Introduction

Productivity growth slowdown emerged around 2000 in the UK and in many other countries, and it is not still improving, which suggests that structural as opposed to cyclical factors are behind the slowdown, as Liu, Mian, and Sufi (2019) suggest. It is indeed the case, and as reported at a European Central Bank (ECB) conference on 'Investment and Growth in Advanced Economies, 'the most remarkable fact about economic growth in recent decades is the slowdown in productivity growth that occurred around the year 2000' (ECB, 2017).[1] It is also the case that recent secular stagnation has caused inequality increase and slower productivity. Actually, recent evidence suggests that labour productivity and income inequality have been closely and significantly related (Mishel 2015); this is clearly so since there is a strong relationship between productivity, inequality, economic growth and real wages, as shown below. More so, there are recent contributions that clearly suggest that productivity growth is a significant factor in terms of inequality (Castle and Hendry 2014; Blundell, Crawford, and Jin 2013; Disney, Jin, and Miller 2013; Tenreyro 2018; Haldane 2017, 2018a, 2018b).

Productivity growth is the key determinant of how demand can grow without infla-tion, thereby reducing inequality of income, wealth and opportunity, as shown below. The slowdown in productivity growth and increase in inequality have affected the UK and other advanced economies. Although weak productivity growth and increased income inequality predate the Global Financial Crisis (GFC), and the subsequent Great Recession (GR), they have become more pronounced following them. As Stiglitz (2012) argues, inequality is one of the causes of the slow recovery after the GR. Highly unequal economies recover more slowly since growth mostly benefits the rich, who save more than what they spend. And spending, not savings, enhances economic recovery. Raising incomes for the poor would proportionately increase consumption, which would increase economic growth. And increase in productivity thereby emerges.

We proceed as follows. In section 2 we discuss the state of productivity and inequality in the UK, and in other advanced economies. Section 3 concentrates on the causes of productivity and inequality. Section 4 discusses relevant policy reforms to tackle pro-ductivity and inequality. Finally, section 5 summarises and concludes.

2. The state of productivity and inequality in the UK and other advanced economies

2.1. Productivity

It is important to note that in the UK there has been a large and persistent fall in labour productivity growth and also an increase in inequality. The standard definition of labour productivity is the amount of output produced per unit of labour input, and measured in relation to either the number of workers or the number of hours worked. Over the period of the first quarter of 2008 and the second quarter of 2009 labour productivity (output per worker) fell by 4.3%. That was a weaker productivity performance than in previous post-world war II recessions. Patterson (2012, 13) provided data to show that the drop in productivity and in terms of downturns before the GFC, it was short–lived. Productivity fell during the two previous recessions before the August 2007 GFC, began to rise only after a few quarters;, and with output beginning to recover, productivity regained its peak level very quickly. As far as the GFC and GR, and between the first quarter of 2008 and the third quarter of 2012, productivity, measured as output per worker, fell by 3.2% in absolute terms. Disney, Jin, and Miller (2013) also provided relevant data and showed that labour productivity was 12.3% below its pre-recession trend. The UK Office for Budget Responsibility (OBR) data shows that productivity (output per hour) was 20% lower than the trend prior to GFC up to the end of 2007. And as for July 2018, OBR reports, 'Productivity growth has been 1.5 percentage points lower on average since the crisis than in the pre-crisis period, with three quarters of that fall attributable to the manufacturing and financial sectors, and smaller negative contributions from "Information and Communication Technology" (ICT) and professional services. Together, these four sectors are found to have been responsible for the entire productivity slowdown, despite representing only around a third of total output' (available at: https://obr.uk/box/productivity-growth-long-term/).

Barnett et al. (2014) discuss productivity between the GFC and 2012 to conclude that over these two periods labour productivity growth was 14% below the level of the pre-

GFC growth rate. Unlike previous recessions, productivity did not pick up during the recovery (this is the essence of the 'productivity puzzle'). McCafferty (2018) reports that 'on average since 2010, annual productivity growth has fallen to 0.5% (output per hour) or 0.7% (output per worker), down from the 2.2% (hour) and 1.8% (worker) of pre-crisis' (p. 7). Haldane (2018a) suggests that in the case of UK productivity, 'the problem is a big one by any historical standards' (p. 2). Moreover, and although the present productivity weakness in the UK stands out more starkly, it is in line with that of several other European countries (Patterson 2012, 16). The US productivity growth is by contrast strong (Patterson, op. cit.), and according to ONS (2016) estimates, productivity over the period 2007–2016 was on average 27.3% higher in the US than in the UK. It was also 18% higher in the other six members of the G7, and 35% higher in Germany. In fact, these gaps have been increasing since then (Tenreyro 2018, 5), and 'the UK's productivity slowdown appears to have been larger than in almost any other country' (Haldane 2018a, 2). According to Bloom et al. (2019), the Brexit process has reduced UK productivity by between 2% and 5% since the 2016 referendum vote (see, also, Broadbent et al. 2019). In the Euro Area, and according to relevant data from the ECB, (2006), labour productivity growth has practically halved over the last twenty years or so.

Liu, Mian, and Sufi (2019) propose a theoretical model that shows low long-term interest rates might produce slower productivity growth. The conventional assumption is that low-interest rates expand investment and output; low-interest rates is a reflection of low productivity growth. Liu et al. (op. cit.) theoretical framework suggests that under the assumption that lower interest rates could produce monopolistic tendencies, productivity is reduced. It is the case, though, that lower interest rates and their impact on productivity depend crucially on when lower interest rates emerge. If the decline emerges from a high rate of interest, an increase in productivity occurs. However, if the interest rate reduction occurs at low-interest rates, a reduction in productivity is more likely. This analysis offers an alternative interpretation when consumption is taken on board. Reduction in consumption can lower the equilibrium interest rate to very low levels, and more importantly lower investment and productivity would occur. This, though, is not due to zero lower bound interest rates. Empirical evidence is provided by Liu, Mian, and Sufi (2019), utilizing US relevant data from 1980 onwards and the 10-year Treasury interest rate. Their empirical results support their theoretical propositions.

2.2. Inequality

The UK, as Atkinson (1997) suggests, 'stands out for the sharpness of the rise in recorded income inequality in the 1980 s' (p. 301). Also, according to Turner (2010), 'there has been a sharp rise in income differential between many employees in the financial sector and average incomes across the whole of the economy'. Milanovic (2011) shows that financial liberalisation between the 1980s and 2010, 'the United States and the United Kingdom – and indeed most advanced economies – have become much richer and much more unequal. In 2010, real per capita income in the United States was 65 percent above its 1980s level and in the United Kingdom, 77 percent higher. Over the same period, inequality in the United States increased from about 35 to 40 …. and in the United Kingdom, from 30 to about 37 Gini points. These increases reflect significant adverse movements in income distributions. Overall, between the mid-1980 s and the mid-2000 s, inequality rose in 16 out of 20 rich OECD countries' (p. 8).

Arestis and González Martínez (2016) summarise the Gini coefficients for the 15 most unequal and 15 least unequal countries around the world. These Gini coefficients clearly show that within country inequality around the world has increased substantially; and imply that there is urgency for relevant economic policies around the world to reduce inequality. In fact, and as Goda, Onaran, and Stockhammer (2016) show, the increase in income and wealth inequality was the cause of the euro crisis of 2010. The evidence produced by Atkinson, Piketty, and Saez (2011) shows that the share of US total income going to top income groups had risen dramatically prior to the GFC. The top pre-tax decile income share reached almost 50% by 2007, the highest level on record. The share of an even wealthier group – the top 0.1% – more than quadrupled from 2.6% to 12.3% over the period 1976 to 2007.

Arestis and Karakitsos (2013) demonstrate that an important characteristic of the period 1983 to 2007, in terms of the declining wage and rising profit shares, in the case of the US but also around the world, was the increasing concentration of earnings at the top, especially in the financial sector. Financial-sector relative wages, and the ratio of the wage bill in the financial sector to its full-time equivalent employment share, enjoyed a steep increase over the period mid-1980 s to 2006; Arestis and Karakitsos (op. cit.) suggest that such levels of inequality were one of the main causes of the GFC (see, also, Arestis 2016). Galbraith (2012) also shows that countries with larger financial sectors have more inequality (see, also, Philippon and Reshef 2009).

Hein and Mundt (2012) refer to the G20 developed economies, where it is suggested that since the early 1980s a falling trend of the wage share clearly materialised. In the case of emerging G20 countries, an overall falling trend of the wage share with the exception of India is portrayed. In the European Union, and since the 1980s the welfare state in the EU has failed to reduce inequality as a result of explicit policy decisions aimed at cutting back on benefits. Inequality has in fact increased between the highest and lowest incomes. According to Social Europe (2017), and between 2005 and 2015, the Gini coefficient in the whole of the EU rose from 30.6 to 31 and income disparity increased from 4.7 to 5.2 between the top and bottom 20% of income recipients. Social Europe (op. cit.) suggests that globalisation and migration put pressure on wages thereby becoming factors that led to inequality in the EU. The deterioration in working conditions, increased temporary working, and weakening of collective bargaining were further sources of inequality.

Alvaredo et al. (2017) compare the evolution of inequality in China, US, and France over four decades, utilising data from WID.World (available at: www.wid.world). In 1978 the top 10% of Chinese earned just over a quarter of overall income before tax. That was significantly below the relevant proportions in the US and France. However, by 2015 the top 10% of Chinese earners were paid two-fifths of total income, above the relevant share in France but still below the US. The bottom 50% income share in China was above the US and France in 1978 but by 2015 it was below that of France but still above that of the US. Inequality in China increased substantially more recently. Not just in China, but in nearly all countries in recent decades. Alvaredo et al. (2017) suggest in terms of global inequality, 'the magnitude varies substantially across countries; thereby suggesting different country-specific policies and institutions matter considerably' (p. 408).

3. Causes of weak productivity and high inequality

3.1. Productivity

There has been an increase in labour supply with the labour market more flexible but with less productive workers in the UK.[2] As aggregate demand has fallen, employment has held up thereby productivity has fallen (Arestis and Peinado 2018). Wage demands have been held down, which encouraged companies to hire more staff, but less productive, with lower real wages, which allowed firms to retain more workers, thereby lower labour productivity emerged, especially since 2008. Schneider (2018) analyses the UK productivity by examining the period 2002–2014. Schneider (op. cit.) splits the examined period to pre-crisis 2002–2007 and post-crisis 2007–2014 and shows that the UK productivity slowdown is driven entirely by post-crisis reallocations of less productive workers. This is so since workers shift from high productivity to low productivity sectors, thereby reducing average productivity.

The decline in trade union membership over the last 30 years, which reduces workers' bargaining position, is a further explanation of reduced productivity. In view of trade unions contributing to productivity, it clearly follows that trade union density and productivity are thereby closely related. The International Labour Organisation (ILO) (2015) suggests it is important to strengthen trade unions by increasing their membership and encouraging them to be active; this is a significant factor in tackling low productivity and also inequality. Gosling and Machin (1995) provide evidence in the case of the UK that suggests the decline in unionisation accounted for 15–20% of earnings dispersion in the 1980 s. The reduced role of trade unions is also highlighted by OECD (2011), which shows that in every OECD country, with the exception of Spain, trade union membership was lower in 2008 than in 1980. In the US the overall trade union membership was reduced substantially from 1980 to 2017; it declined from 20.1% in 1983 (the first year for which comparable union data are available) to 11.9% in 2010. In 2014, the union membership rate was 11.1%, down 0.2 percentage point from 2013; in June 2016 it was 10.7%, which is half of what it was in the 1980 s (US Bureau of Labour Statistics, 23 January 2015 and 26 January 2017). It is clear that the US case is an example where labour has little bargaining power. No wonder the labour GDP share is at a post-World-War II low. Mishel (2015) suggests that inequality in the US is central to the low productivity there. This is due to 'changes in labour's share', 'compensation inequality' and 'terms of trade' (p. 2). This is a serious problem for an economy, which is 70% dependent on consumer spending. No wonder that over the period since the GFC and GR, the US economic 'recovery' has been unusually weak. Similar trends prevail in many other countries. There is general agreement that such a decline in the role of trade unions and of collective bargaining coincides with a widening of the pay distribution. The rebirth of labour as an economic force is therefore required. The legal framework of trade unions is another important consideration. Atkinson (2015) provides a relevant example in the case of the UK: 'a succession of laws enacted between 1980 and 1993 that reduced the autonomy of trade unions in the UK and the legitimacy of industrial action'. This clearly implies that 'The end result of the legislation is that unions are considerably weakened in their legal status and protection' (pp. 128–129). Barth, Bryson, and Dale-Olson (2017) support the argument that increases in union density produce substantial increases in productivity and wages. This is supported by their theoretical and empirical evidence based on Norwegian firms over the period 2001 to 2012.

Another explanation relates to the reduction of the capital/labour ratio as a result of the reduction in investment, which fell following the GR, and in view of animal spirits and uncertain economic outlook (Keynes 1936).[3] Uncertainty of future demand, thereby, produces productivity decline. Reduction in the pace of innovation as firms reduce their spending on research and development against a background of uncertainty of future demand is an important factor in this respect. The reduction in investment was significantly larger after the GR than in previous recessions and has been low since then, as argued by Blundell, Crawford, and Jin (2013). Ramsden (2018) suggests that ten years after the GFC, 'cumulative growth in business investment is still around 50 and 30 percentage points below where it was at the equivalent stage of the recoveries seen in the decades after the 1979 and 1990 recessions respectively' (p. 2). Blundell, Crawford, and Jin (2013) argue that reductions in productivity, resulting from a fall in the capital-labour ratio, also contribute to reductions in real wages and hence labour costs; they find it to be the primary driver of productivity falls.

The role of the financial sector is another factor. The GFC and GR produced a situation where banks could not provide the lending required by firms; expansion could not occur, thereby productivity suffered. It is indeed the case that in the aftermath of GR business investment was constrained by companies' inability to borrow as banks shored up their balance sheets. A further related issue is the record-low interest rates, making it easier for firms to finance their loans. This has allowed some highly unproductive firms to avoid going bust. This explains productivity stagnation but only part of it, as the Bank of England (2016) suggests; at most, 3% of the 20% shortfall in productivity growth can only be explained by the record-low interest rates.

Tridico and Parboni (2018) study the case of the UK, and also of several other developed countries in recent years and suggest that all have experienced productivity slowdown. Weak gross domestic product (GDP) and a decline in wage shares are argued to be the major explanatory factors of sluggish productivity. Financialization has also contributed to the productivity slowdown. This is so in view of 'financial investment of non-financial businesses has been rising, and the accumulation of capital goods has been declining' (Stockhammer 2004, 719). This causes a reduction in productivity. Productivity growth is thereby claimed to depend positively on the rate of growth of GDP and the wage share, and negatively on income inequality and financialization; the latter has also contributed to sluggish productivity and increase in inequality. The Tridico and Pariboni (2018) study provides empirical evidence that supports their relevant theoretical propositions. Kim and Loayza (2019) produce empirical results over the period 1985–2015 and for more than 100 countries to conclude that the main determinants of total factor productivity are innovation, education, market efficiency, infrastructure and institutions. The empirical findings of Cette, Fernald, and Mojon (2016), in the case of the US and four EU countries (Germany, France, Italy, and Spain) suggest that productivity growth and real interest rates are linked positively. This is so since, for example, low-interest rates produce flow of funds, which may not be allocated efficiently. This rising misallocation reduces productivity growth. This, it is suggested, is the case across the world and creates concerns on secular stagnation.

3.2. Inequality

It is of vital importance to tackle inequality as our analysis so far has demonstrated (see, also, Arestis and Sawyer 2011). This is in agreement with Keynes (1936) argument that the two outstanding faults of economic policy were the failure to secure full employment and to tackle the inequitable distribution of income. Reducing inequality lowers the productivity-pay gap by raising pay (Mishel 2015), growth and with appropriate economic policies full employment could be achieved. Haldane (2017) suggests that 'Until the crisis, it is difficult to identify a period in the past 50 years when inequality was close to the top of the public policy or academic agenda'. More studies have similar suggestions. A relevant example is the study by Kumhof and Rencière (2010), which suggests, 'Restoring equality by redistributing income from the rich to the poor would not only please the Robin Hoods of the world, but could also save the global economy from another major crisis' (p. 31; see, also, Berg and Ostry 2011; OECD 2011). Kumhof, Rencière, and Winant (2015) go further to argue that such crisis could occur from changes in income distribution and high household debt. A key assumption in terms of this proposition is that the top earners, whose income share increases, provide loans to bottom earners instead of increasing their consumption. Kumhof et al. (op. cit.) present a relevant theoretical framework and US empirical support to explain the emergence of crises over two periods, 1920–1929 and 1983–2008. They suggest that a large increase in the income of high-income households (5% of income distribution) along with a large increase in the household debt emerged; and thereby higher leverage of low- to middle-income households (95% of income distribution) generated financial fragility, which eventually caused the Great Depression of 1929, the GFC of 2007 and the GR of 2008 (see, also, Arestis 2016; McCombie and Spreafico 2015). The slow and uneven recovery in the US and elsewhere since the GFC and GR are due to the existence of inadequate demand; the latter would be stronger if lower income groups received a higher share of income.

Grigoli and Robles (2017) examine non-linear income inequality in relation to economic development, and in the case of 77 countries. An 'inequality overhang' is identified whereby the relationship between economic development and inequality turns negative from positive. This, it is shown, occurs when the Gini coefficient reaches 27%, indicating that the inequality overhang occurs at low levels of income inequality. Ostry (2015) provides empirical evidence that clearly suggests more equality in the income distribution is 'robustly and positively' associated with more and sustainable growth paths. Dabla-Norris et al. (2015) provide evidence, utilising a sample of 159 developed, emerging and developing economies for the period 1980–2012, which suggests that there is an inverse relationship between the disposable income share of the top 20% and economic growth (a 1% increase in the disposable income of the top 20% is associated with a 0.08% decrease in GDP growth in the following five years). A similar increase in the disposable income of the bottom 20% produces a 0.38% higher income growth. This is also the case with the second and third quintiles (the middle class). The empirical results of Dabla-Norris et al. (op. cit.) are in line with the findings of the OECD (2014) study, which utilises a smaller sample of developed countries.

Policy-makers have discussed inequality (see, for example, Bank of England 2012; Carney 2014; Yelen, 2014; Draqhi 2015; ECB. 2016). Most important is the former IMF managing director's (Lagarde 2014) speech, in that 'One of the leading economic stories of our time is increasing income inequality and the dark shadow it casts across the global economy' (p. 11). Lagarde (op. cit.) went on to suggest that 'The facts are familiar. Since 1980, the richest 1 percent increased their share of income in 24 out of 26 countries for which we have data. In the US, the share of income taken home by the top one percent more than doubled since the 1980 s, returning to where it was on the eve of the Great Depression. In the UK, France, and Germany, the share of private capital in national income is now back to levels last seen almost a century ago' (p. 11). Lagarde (2015) summarises the relevant arguments when she suggests that growing income inequality has become a serious problem for economic growth and development, and that 'if you want to see more *durable* growth, you need to generate more *equitable* growth'.

The inequality effects discussed above were greatly affected by a number of factors. An important contributory factor to inequality is deregulation and liberalization of finance in many countries around the world, especially so before the GFC. As a result, financial markets became bigger and more global. Of particular importance on this score was the financial liberalization framework in the US, one of the main causes of the GFC (see Arestis and Karakitsos 2013; Arestis 2016, for further details; also Crouch 2019). Globalisation is another important factor (IMF 2007; Crouch 2019). Goldberg and Pavcnik (2007) provide a great deal of evidence that inequality increased in developing countries as a result of globalisation. This is in view of less-skilled workers, who are relatively abundant in these countries and are not better off in relation to higher skill and educated workers. Atkinson (1997) refers to the demand and supply for skilled and unskilled labour as a possible explanation of the earnings dispersion. When the relative number of skilled workers rises, then a rise in the demand for them emerges, thereby shifting the demand for labour. One explanation for this possibility is international trade liberalisation and increased competition from the countries where unskilled labour is abundant. Not only is the increase in demand for educated workers being driven by globalisation but also by technological changes in terms of information and communication technologies, which have displaced low-skilled workers and created demand for those with better education. Atkinson (2015) summarises the contributory factors to inequality as follows: 'globalisation, technological change (information and communications technology), growth of financial services; changing pay norms, reduced role of trade unions; scaling back of the redistributive tax-and-transfer policy' (p. 82; see, also, Crouch 2019).

Stockhammer (2013), utilising panel estimations of the determinants of the wage share in 71 countries (28 developed and 43 developing and emerging countries) from 1970 to 2007, concludes that globalisation had negative effects; not just in developed but also in developing countries. Stockhammer (op. cit.) also shows that financialisation (see, also, Kohler, Gushanski, and Stockhammer 2019), welfare state retrenchment, decline in the bargaining power of trade unions over time had negative effects on the wage share. In addition, the increased market power of firms in relation to labour had some effect on the wage share of the countries considered; also, changes in technology, which was not one of the main drivers in income distribution.

Furceri and Loungani (2013) suggest two further explanations of the increased inequality: capital account liberalization and lower government budget deficits. Fifty-eight episodes of large-scale capital account reforms are considered in 17 advanced economies to conclude that 'on average, capital account liberalization is followed by a significant and persistent increase in inequality. The Gini coefficient increases by about

1 percent a year after liberalization and by 2 percent after five years' (p. 26; see, also, Furceri and Loungani 2015). Furceri and Loungani (2013) also show that 'Over the past 30 years, there were 173 episodes of fiscal consolidation in our sample of 17 advanced economies. On average across these episodes, policy actions reduced the budget deficit by about 1 percent of GDP. There is clear evidence that the decline in budget deficits was followed by increases in inequality. The Gini coefficient increased by 2 percentage points two years following the fiscal consolidation and by nearly 1 percentage point after eight years' (pp. 26–27).

It is also important to note that rising inequality is affected by secular stagnation (lower rates of economic growth in major economies around the world). Lower economic growth has been around the world for the last 50 years[4]; but especially so since the GFC/GR. Ineffective monetary policy and weak demand, especially investment, are the main factors of such lower economic growth. There are other factors, like hangover from the GFC/GR, and more recently, 'austerity', especially in Europe, has been another cause; hysteresis is of course a further factor.[5] Such lower economic growth is associated with rising inequality and lower productivity, as mentioned above. Very low-interest rates, what is these days termed 'the zero lower bound', is another factor that increases inequality, which prevents a robust economic recovery. Households accumulate a great deal of debt, in view of them being unable to save under very low-interest rates. Households have no hope of wealth accumulation under such circumstances; unlike of course the bond-holders.

It clearly follows from our discussion so far that policy reforms to tackle low productivity and inequality are paramount. We proceed to discussing relevant policies.

4. Policy reforms to tackle low productivity and inequality

There is the argument that low productivity growth is a sign of 'secular stagnation', and also of inequality increases in view of the relationship between productivity and inequality as explained above. Furthermore, and since productivity and wage increases are linked, stagnating productivity would not allow a significant reduction in income inequality. Relevant policy reforms to tackle productivity and inequality are thereby very important and urgently needed. We begin with productivity.

4.1. Productivity

It is the case that in the UK productivity performance is one of the most striking characteristics of the recent economic history of this country (Chadha 2017). Similarly, OECD (2015) highlights productivity growth as one if the biggest challenges since the GFC, with the UK being one of the worst affected. Output per hour worked, grew by around 0.5% per quarter on average in the decade prior to 2008, and has been broadly flat over the past decade, with serious implications. Industrial policies to direct capital to more productive firms, along with new technologies and relevant re-training, not just for staff but also and of equal importance management training, would substantially improve productivity. Equipment and skills need to be a great deal more efficient in the UK to enhance productivity, as reported in the 2017 Skills and Employment Survey, a government funded survey (available at: https://www.gov.uk/government/publications/employer-skills-survey -2017-uk-report). As the report also suggests, it is the case that workers in the UK work very hard but productivity is low. More efficient equipment and skills are thereby greatly needed. Clearly, then, further relevant policies should include boosting skills, expenditure increase on infrastructure and research, and more support for the National Productivity Investment Fund (NPIF). The NPIF was initiated as part of the Autumn Statement 2016 when the UK Government announced its creation worth in total £23bn. NPIF is for investment in areas that are key to boosting productivity: transport, digital communications, research and development (R&D), investing in human capital, and housing. Furthermore, a new 'industrial strategy council' was created in October 2018, chaired by the Bank of England chief economist, Andy Haldane, to tackle the weak productivity in the UK as one of its objectives.[6] Further relevant policies, and from the wage-led strategy point of view, include minimum wage policies, along with legislation that strengthens the status of labour unions and collective bargaining institutions. Bivens (2019) provides evidence of the link between wage-led productivity growth, using US data from 1949 to 2019. Biven's (op. cit.) main conclusion is that 'With no other controls, a 1 percent-point increase in the labour share of income leads to a 0.125 percentage-point increase in average productivity growth over the next two years' (p. 3).

Christine Lagarde, at the G20 meeting of finance ministers and central bank governors (in Fukuoka, Japan, 9 June 2019) suggested that the IMF cut its global growth to 3.3% in 2019, due to slow productivity, and excessive economic inequality; also to negative impact of trade tensions between China and the US, Brexit effects, and corporate debt levels. Relevant policies suggested by Christine Lagarde in her speech are fiscal policy that strikes the right balance between growth, debt sustainability and social objectives; and greater participation of women in the workforce. In all these economic policies, coordination of them is very important not merely within countries but also between them. More recently, Christine Lagarde, as the president-elect of the ECB, and in her address to the European Parliament (4 August 2019), called for changes in Europe's budget-rules and for more and closer coordination of the governments over fiscal policy to stimulate the Euro Area economy (Financial Times, 5 September 2019). The ex-ECB president, Mario Draghi, in an interview with Financial Times (as reported in the Financial Times on 30 September 2019) strongly supported fiscal policy to avoid global slowdown. He also suggested that a fiscal union in the Euro Area is vital. Clearly reforming the EU's 'fiscal rules' is very urgent.

We proceed to examine below policy reforms to tackle inequality. It should be noted at this stage that policies that reduce inequality do have a positive impact on productivity. So the sub-section that follows is also closely related to productivity as well.

4.2. Inequality

Fiscal policy in terms of making taxes fairer, especially so on the financial-sector profits, is very important (see, Korinek and Kreamer 2013). Berg and Ostry (2011) also show that a redistributive tax system is associated with higher and durable economic growth. Raising the minimum wage and indexing it to inflation, and other wage rises, is another important tool to fight inequality. In this sense strong trade unions, collective bargaining and high minimum wages are beneficial. All this would ensure that wage growth catches up with productivity growth and hence consumption and income growth along with inequality reduction. The IMF (2015) study suggests that 'Fiscal policy is a powerful and adaptable tool for achieving distributional objectives. Considering tax and spending programs together enhances the effectiveness of fiscal redistribution'. Thereby, 'improving both distributional outcomes and economic efficiency is possible' (p. 1). In terms of specific guidance on the use of fiscal policy for redistribution, it is suggested that this is a country-specific problem (IMF 2014, 2015; see, also, Dabla-Norris et al. 2015).

Government progressive taxation and public expenditure policies, especially investment in infrastructure, should be able to reduce inequality.[7] It all depends of course on the nature and location of infrastructure. The focus of such policies should be to tax the top more than the rest, and direct social expenditure towards the low-income households. The policies pursued by the UK and the US with an enormous decrease of the progressivity of the income tax since 1980 does not help on this score; indeed this 'explains much of the increase in the very highest earned income' (Piketty 2014, pp. 495–496). Unemployment is another factor, which was significantly lower in the period after the Second World War until the late 1970s; subsequently, it increased substantially, especially in Europe; in the UK, unemployment is at a 40-year low. In terms of reducing unemployment, Atkinson (2015) suggests, 'The government should adopt an explicit target for preventing and reducing unemployment and underpin this ambition by offering guaranteed public employment at the minimum wage to those who seek it' (p. 140). It is also suggested (Atkinson, op. cit.) that an unemployment target of 2% is necessary, with the government acting as 'an employer of last resort', thereby introducing guaranteed public employment. It is also important, as Atkinson (2015) suggests, for the renewal of 'social security for all' in view of the fact that 'One reason for rising inequality in recent decades has been the scaling back of social protection at a time when needs are growing, not shrinking'. It is indeed the case that in the past, the welfare state 'played a major role in reducing inequality'. The welfare state 'is the primary vehicle by which our societies seek to ensure a minimum level of resources for all members' (p. 205).

It is of paramount importance to have in place proper distributional policies, especially so fiscal policies along with wage policies, if a viable growth regime is to emerge and be sustained. However, in order to reduce inequality significantly proper coordination of monetary and fiscal policies along with financial stability, the main focus of monetary policy, would be the best way forward (see, also, Arestis, 2012, 2015, 2016; Arestis 2017, 2018). Fiscal policy should be directed at reducing inequality through appropriate expenditure and progressive tax policies, which should be supported by monetary and financial stability policies. The focus of financial stability should be on reforms to regulate the financial sector and avoid the type of inequality consequences and financial architecture that led to the GFC; for it is the case that such regulation had been neglected prior to the GFC and following it inequality increased further as argued in this contribution (see, also, Arestis 2016). Financial stability policies are necessary to avoid sharp and unsustainable increases in debt-to-income ratios among lower- and middle-income households, thereby containing the leverage ratio and the risks of crises like the GFC. Relevant empirical evidence (Eggertsson 2006) fully supports such coordination of policies. At the end of the day, crises can be avoided if economies are well managed and financial markets are sufficiently regulated.

5. Summary and conclusions

We have discussed in this contribution inequality and productivity in the UK and in other global countries. Relevant policies are clearly required and this conclusion follows from our discussion. A number of economic policy proposals have been suggested to reduce inequality and increase productivity, and thereby weaken 'secular stagnation'. Most important proposal is that fiscal, monetary and financial stability policies should be implemented in a coordinated way to restore full employment, along with focused policies for redistributive goals to reduce inequality and enhance productivity. An interesting question, though, is Crouch's (2019) observation that despite the warnings by the OECD and IMF 'about the economic implications of rising inequalities', is 'their advice likely to be heeded while the prevailing balance of political power remains favourable to more inequality?' (p. 23). This is indeed an interesting question.

Notes

1. Cette, Fernald, and Mojon (2016) also make the point that 'the slowdown in advanced-economy total factor productivity (TFP) growth was broadly underway prior to the crisis' (p. 3).
2. The increase in the UK labour supply has emerged from later retirement, and with fewer workers from the EU, businesses in the UK have sought new sources for labour successfully. Also, the Office of National Statistics (2013) reported that women was a significant source.
3. The doubtful future relationship of the UK with the European Union provides a great deal of uncertainty in terms of the future UK economic prospects (Tenreyro 2018; McCafferty 2018).
4. There was slower growth in the 1970 s, but not in the 1980 s and 1990 s (at least in the UK); also slower growth has emerged since the early 2000 s.
5. Hysteresis is the case of long periods of unemployment, which render workers' skills obsolete. This does not allow them to rejoin the labour market, thereby reducing the number of productive workers.
6. An interesting question posed by Haldane (2018b) is how the strong jobs growth is accompanied by weak wage growth. This is a case in both the UK and other countries. Haldane (op. cit.) suggests this is due to a number of factors, but mainly to weak productivity and workers power.
7. There is the argument closely supported by the proponents of the New Consensus Macroeconomics theoretical framework that fiscal policy is ineffective in view of the Ricardian Equivalence Theorem and the crowding-out argument. As argued in Arestis (2011, 2012, and 2015) and Arestis and González Martinez (2015), such arguments lack convincing theoretical backing and empirical credence. Actually more and more contributors suggest that monetary policy is not the only game in town; fiscal policy is the only instrument left. The ex-ECB President, Mario Draqhi, has repeatedly suggested that EMU governments should get on with fiscal expansion.

Disclosure Statement

No potential conflict of interest was reported by the author(s).

References

Alvaredo, F., L. Chancel, T. Piketty, E. Saez, and G. Zucman. 2017. "Global Inequality Dynamics: New Findings from WID.World." *American Economic Review: Papers and Proceedings* 107 (5): 404–409.

Arestis, P. 2011. "Keynesian Economics and the New Consensus in Macroeconomics." In *A Modern Guide to Keynesian Macroeconomics and Economic Policies*, edited by E. Hein and E. Stockhammer, 88–111. Cheltenham: Edward Elgar.

Arestis, P. 2012. "Fiscal Policy: A Strong Macroeconomic Role." *Review of Keynesian Economics, Inaugural Issue* 1 (1): 93–108.

Arestis, P. 2015. "Coordination of Fiscal with Monetary and Financial Stability Policies Can Better Cure Unemployment." *Review of Keynesian Economics* 3 (2): 233–247.

Arestis, P. 2016. "Main and Contributory Causes of the Recent Financial Crisis and Economic Policy Implications." In *Emerging Economies During and After the Great Recession*, Annual Edition of *International Papers in Political Economy*, edited by P. Arestis and M. C. Sawyer. Basingstoke: Palgrave Macmillan.

Arestis, P. 2017. "Monetary Policy since the Global Financial Crisis." In *Economic Policies since the Global Financial Crisis*, Annual Edition of *International Papers in Political Economy*, edited by P. Arestis and M. C. Sawyer. Basingstoke: Palgrave Macmillan.

Arestis, P. 2018. "Importance of Tackling Income Inequality and Relevant Economic Policies." In *Inequality: Trends, Causes, Consequences, Relevant Policies*, Annual Edition of *International Papers in Political Economy*, edited by P. Arestis and M. C. Sawyer. Basingstoke: Palgrave Macmillan.

Arestis, P., and A. R. González Martinez. 2015. "The Absence of Environmental Issues in the New Consensus Macroeconomics Is Only One of Numerous Criticisms." In *Finance and the Macroeconomics of Environmental Policies*, edited by P. Arestis and M. Sawyer. Basingstoke: Palgrave Macmillan.

Arestis, P., and A. R. González Martinez. 2016. "Income Inequality: Implications and Relevant Economic Policies." *Panoeconomicus* 63 (1): 1–24.

Arestis, P., and E. Karakitsos. 2013. *Financial Stability in the Aftermath of the 'Great Recession'.* Basingstoke: Palgrave Macmillan.

Arestis, P., and P. Peinado. 2018. "Explaining, Restoring Low Productivity Growth in the UK." *Challenge: The Magazine of Economic Affairs* 61 (2): 120–132.

Arestis, P., and M. Sawyer. 2011. "Economic Theory and Policies: New Directions after Neoliberalism." In *New Economics as Mainstream Economics*, edited by P. Arestis and M. C. Sawyer, 1–38. Basingstoke: Palgrave Macmillan.

Atkinson, A. B. 1997. "Bringing Income Distribution in from the Cold." *The Economic Journal* 107 (441): 297–321.

Atkinson, A. B. 2015. *Inequality: What Can Be Done?* Cambridge, MA: Harvard University Press.

Atkinson, A. B., T. Piketty, and E. Saez. 2011. "The Top Incomes in the Long Run of History." *Journal of Economic Literature* 49 (1): 3–71.

Bank of England. 2012. "The Distributional Effects of Asset Purchases." *Bank of England Repor t*, July 12.

Bank of England. 2016. "Understanding and Measuring Finance for Productive Investment." *Bank of England Discussion Paper*, April.

Barnett, A., A. Chiu, J. Franklin, and M. Sebastiá-Barriel. 2014. "The Productivity Puzzle: A Firm-Level Investigation into Employment Behaviour and Resource Allocation over the Crisis." *Bank of England Working Paper No. 495*, April.

Barth, E., A. Bryson, and H. Dale-Olson. 2017. "Union Density, Productivity, and Wages." NIESR Discussion paper No. 481, October 30. National Institute of Economic and Social Research.

Berg, A. G., and J. D. Ostry. 2011. "Inequality and Unsustainable Growth: Two Sides of the Same Coin?." *IMF Staff Discussion Note 11/08*. Washington: International Monetary Fund. http://www.imf.org/external/pubs/ft/sdn/2011/sdn1108.pdf

Bivens, J. 2019. "Looking for Evidence of Wage-Led Productivity Growth." *Working Economics Blog*, December 10. Economic Policy Institute.

Bloom, N., P. Bunn, S. Chen, P. Mizen, P. Smietanka, and G. Thwaites. 2019. "The Impact of Brexit on UK Firms." *Bank of England Staff Working Paper No. 818*, August.

Blundell, R., C. Crawford, and C. Jin. 2013. "What Can Wages and Employment Tell Us about the UK's Productivity Puzzle?" *The Economic Journal* 124: 377–407.

Broadbent, B., F. De Pace, T. Drechsel, R. Harrison, and S. Tenreyro. 2019. "The Brexit Vote, Productivity Growth and Macroeconomic Adjustments in the United Kingdom." *Discussion Paper No. 51*, August. External MPC Unit, Bank of England..

Carney, M. 2014. "Inclusive Capitalism." Speech Given at the Conference on *Inclusive Capitalism*, London, May 27. http://www.inc-cap.com/IC_ESSAY_Book_Introduction_Keynotes.pdf

Castle, J., and D. Hendry. 2014. "T He Real Wage–Productivity Nexus." January 13. https://scholar.google.co.uk/scholar?q=Castle,+J.+and+Hendry,+D.+(2014),+%E2%80%9CThe+Real+Wage%E2%80%93Productivity+Nexus%E2%80%9D,+13+January+2014&hl=en≈sdt=0≈vis=1&oi=scholart

Cette, G., J. Fernald, and B. Mojon. 2016. "The Pre-Great Recession Slowdown in Productivity." *European Economic Review* 88: 3–20.

Chadha, J. S. 2017. "T He Productivity Puzzle." Gresham College Lecture, September 11.

Crouch, C. 2019. "Inequality in Post-Industrial Societies." *Structural Change and Economic Dynamics* 51: 11–23.

Dabla-Norris, E., K. Kochhar, N. Suphaphiphat, F. Ricka, and E. Tsounta. 2015. "Causes and Consequences of Income Inequality: A Global Perspective." *IMF Staff Discussion Note* 15 (13). Available at https://www.imf.org/external/pubs/ft/sdn/2015/sdn1513.pdf

Disney, R., W. Jin, and H. Miller. 2013. "The Productivity Puzzles." In *The IFS Green Budget*, edited by C. Emmerson, P. Johnson, and H. Miller, 53-90. London, UK: Institute of Fiscal Studies.

Draqhi, M. 2015. "The ECB's Recent Monetary Policy Measures: Effectiveness and Challenges." May 14. Washington, DC: Camdessus Lecture, IMF. https://www.ecb.europa.eu/press/key/date/2015/html/sp150514.en.html

Eggertsson, G. B. 2006. "Fiscal Multipliers and Policy Coordination." *Federal Reserve Bank of New York Staff Reports, No. 241*. New York: Federal Reserve Bank of New York.

European Central Bank (ECB). 2006. "Labour Productivity Developments in the Euro Area." Occasional Paper Series, No 53, October.

European Central Bank (ECB). 2016. *Annual Report*. https://www.ecb.europa.eu/pub/pdf/annrep/ar2016en.pdf

European Central Bank (ECB). 2017. "Investment and Growth in Advanced Economies." ECB Forum on Central Banking, Conference Proceedings, June. Sintra: Portugal.

Furceri, D., and P. Loungani. 2013. "Who Let the Gini Out?" *Finance and Development* 50 (4): 25–27.

Furceri, D., and P. Loungani. 2015. "Capital Account Liberalisation and Inequality." *IMF Working Paper 15/243*. Washington D.C.: International Monetary Fund.

Galbraith, J. K. 2012. *Inequality and Instability: A Study of the World Economy Just before the Great Crisis*. Oxford: Oxford University Press.

Goda, T., O. Onaran, and E. Stockhammer. 2016. "Income Inequality and Wealth Concentration in the Recent Crisis." *Development and Change* 48 (1): 3–27.

Goldberg, P., and N. Pavcnik. 2007. "Distributional Effects of Globalisation in Developing Countries." *Journal of Economic Literature* 45 (1): 39–82.

Gosling, A., and S. Machin. 1995. "Trade Unions and the Dispersion of Earnings in UK Establishments 1980-90." *Oxford Bulletin of Economics and Statistics* 57 (2): 167–184.

Grigoli, F., and A. Robles. 2017. "Inequality Overhang." *IMF Working Paper WP/17/76*. Washington D.C.: International Monetary Fund.

Haldane, A. 2017. "Productivity Puzzles." Speech Given at the London School of Economics, March 20. https://www.bankofengland.co.uk/speech/2017/productivity-puzzles

Haldane, A. 2018a. "The UK's Productivity Problem: Hub No Spokes." Speech Given at the Academy of Social Sciences Annual Lecture, London, June 28. https://www.bankofengland.co.uk/speech/2018/andy-haldane-academy-of-social-sciences-annual-lecture-2018

Haldane, A. 2018b. "Pay Power." Speech given at the Acas 'Future of Work' Conference, Congress Centre, London, 10 October. Available at: https://www.bankofengland.co.uk/speeches

Hein, E., and M. Mundt. 2012. "Financialisation and the Requirements and Potentials for Wage-Led Recovery – A Review Focusing on the G20." *Conditions of Work and Employment Series* No. 37.

IMF. 2007. "The Globalisation of Labour." Chapter 5. In *World Economic Outlook*. April. Washington D.C.: International Monetary Fund.

IMF. 2014. "Fiscal Policy and Income Inequality." IMF Policy Paper, January 23. Washington D. C.: International Monetary Fund. http://www.imf.org/external/np/pp/eng/2014/012314.pdf

IMF. 2015. "Harnessing the Power of Fiscal Policy to Mitigate Inequality." IMF Survey Magazine: Policy, IMF Fiscal Affairs Department, September 25. Washington D.C.: International Monetary Fund. http://www.imf.org/external/pubs/ft/survey/so/2015/pol092515a.htm

International Labour Organization (ILO). 2015. *Productivity Improvement and the Role of Trade Unions: A Workers' Education Manual*. edited by Mohammed Mwamadzingo and Paliani Chinguwo. Geneva, Switzerland: International Labour Organization, Bureau for Workers' Activities.

Keynes, J. M. 1936. *The General Theory of Employment, Interest and Money*. London: Macmillan and Company Limited.

Kim, Y. E., and N. V. Loayza. 2019. "Productivity Growth: Patterns and Determinants across the World." *Policy Research Paper 8852*. World Bank Group.

Kohler, K., A. Gushanski, and E. Stockhammer. 2019. "The Impact of Financialisation on the Wage Share: A Theoretical Classification and Empirical Test." *Cambridge Journal of Economics* 43 (4): 937–974.

Korinek, A., and A. Kreamer. 2013. "The Redistributive Effects of Financial Deregulation." *NBER Working Paper 19572*, October. Washington D.C.: National Bureau of Economics Research.

Kumhof, M., and R. Rencière. 2010. "Leveraging Inequality." *Finance and Development* 47 (4): 28–31.

Kumhof, M., R. Rencière, and P. Winant. 2015. "Inequality, Leverage, and Crises." *American Economic Review* 105 (3): 1217–1245.

Lagarde, C. 2014. "Economic Inclusion and Financial Integrity." Speech Given at the Conference on *Inclusive Capitalism*, London, 27 May. Available at: http://www.inc-cap.com/IC_ESSAY_ Book_Introduction_Keynotes.pdf

Lagarde, C. 2015. "Lifting the Small Boats", Address at Grandes Conferences Catholiques Brussels, June 17. http://www.imf.org/external/np/speeches/2015/061715.htm

Liu, E., A. Mian, and A. Sufi 2019. "Low Interest Rates, Market Power and Productivity Growth." *Becker Friedman Institute, Working Paper No. 2019-09*. https://ssrn.com/abstract=3327402

McCafferty, I. 2018. "Changing Times, Changing Norms." Speech Given at the City Lecture OMFIF. London. www.bankofengland.co.uk/speeches

McCombie, J. S. L., and M. R. M. Spreafico. 2015. "Income Inequality and Growth: Problems with the Orthodox Approach." Paper presented at the session, *Analyzing Growth and Inequality in the 21st Century*, at the INET conference, Paris, April.

Milanovic, B. 2011. "More or Less: Income Inequality Has Risen over the past Quarter-Century Instead of Falling as Expected." *Finance and Development* 48 (3): 6–11.

Mishel, L. 2015. "Inequality Is Central to the Productivity-Pay Gap." *Working Economics Blog, Economic Policy Institute* 29 (July): 2015.

OECD. 2011. *Divided We Stand: Why Inequality Keeps Rising*. Paris: OECD Publishing.

OECD. 2014. *Rising Inequality: Youth and Poor Fall Further Behind*. Paris: OECD Publishing.

OECD. 2015. *Economic Surveys: United Kingdom*. Paris: OECD Publishing.

Office for National Statistics (ONS). 2013. "Women in the Labour Market." https://www.ons.gov. uk/employmentandlabourmarket/peopleinwork/employmentandemployeetypes/articles/wome ninthelabourmarket/2013-09-25

Office for National Statistics (ONS). 2016. "Statistical Bulletin." *ONS Statistical Bulletin*, May 31.

Ostry, J. D. 2015. "Inequality and the Duration of Growth." *European Journal of Economics and Economic Policies: Intervention (EJEEP)* 12 (2): 147–157.

Patterson, P. 2012. "The Productivity Conundrum, Explanations and Preliminary Analysis." Office for National Statistics (ONS). http://www.ons.gov.uk/ons/dcp171766_283259.pdf

Philippon, T., and A. Reshef. 2009. "Wages and Human Capital in the U.S. Financial Industry: 1909-2006." *National Bureau of Economic Research Working Paper 14644*.

Piketty, T. 2014. *Capital in the Twenty-First Century*. Cambridge, Massachusetts: Harvard University Press.

Ramsden, D. 2018. "The UK's Productivity Growth Challenge." February 23. Cambridge: Speech Given at Babraham Hall.

Schneider, P. 2018. "Decomposing Differences in Productivity Distributions." *Bank of England Staff Working Paper No. 740*, July.

Social Europe. 2017. "Inequality in Europe." *Social Europe Dossier*. Published by Social Europe in Cooperation with Friedrich-Ebert-Stiftung and Hans Böckler Stiftung.

Stiglitz, J. 2012. *The Price of Inequality: How Today's Divided Society Endangers Our Future*. New York: W.W. Norton.

Stockhammer, E. 2004. "Financialisation and the Slowdown of Accumulation." *Cambridge Journal of Economics* 28 (5): 719–741.

Stockhammer, E. 2013. "Why Have Wage-Shares Fallen? An Analysis of the Determinants of Functional Income Distribution." Chapter 2. In *Wage-Led Growth: An Equitable Strategy for Economic Recovery*, edited by M. Lavoie and E. Stockhammer. Basingstoke: Palgrave Macmillan.

Tenreyro, S. 2018. "The Fall in Productivity Growth: Causes and Implications." Speech Given at the Preston Lecture Theatre, January 15. Queen Mary University of London. www.banofeng land.co.uk/speeches

Tridico, P., and R. Pariboni. 2018. "Inequality, Financialization and Economic Decline." *Journal of Post Keynesian Economics* 41 (2): 236–259.

Turner, A. 2010. "What Do Banks Do? What They Should Do and What Public Policies are Needed to Ensure Best Results for the Real Economy?." http://www.fsa.gov.uk/pubs/speeches/ at_17mar10.pdf/

Yellen, J. L. 2014. "Perspectives on Inequality and Opportunity from the Survey of Consumer Finances." Speech Given at the Conference on Economic Opportunity and Inequality, Federal Reserve Bank of Boston, Boston, Massachusetts, October 17. http://www.federalreserve.gov/ newsevents/speech/yellen20141017a.htm

Financial oversight, the third flawed pillar of the European Union: the missing piece in the Arestis-Sawyer critique of EMU macropolicy design

Gary A. Dymski and Annina Kaltenbrunner ⓘ

ABSTRACT

This paper presents a chronological survey of the 20 academic papers that Malcolm authored or co-authored between 1997 and 2017 on the flawed design – and hence flawed implementation – of the European Monetary Union (EMU)'s macroeconomic policy pillars. We augment his analyses by pointing out a third – complementary – design flaw: the EMU's two-tiered structure of financial regulation and oversight. While this financial pillar aimed at reconciling Europe's historically bank-based financial systems with large European banks' entry into global financial competition, it created a combustible mix when combined with the EMU's macroeconomic-policy pillars. The Global Financial Crisis lit the fire: member-nations, forced to rescue their domestically-chartered too-big-to-fail megabanks, had to adopt austerity policies that both slowed the pace of post-crisis economic growth and eroded support for pro-Union political leaders. Only marginal changes have been made in these policy pillars post-crisis. Consequently, Europe faces a financial bifurcation point: either to continue 'whatever it takes' support for its megabanks, or to rethink both its financial architecture and its macroeconomic and financial policy pillars.

1. Introduction

Malcolm Sawyer has established himself as a global figure in heterodox political economy for several reasons. His writings on macroeconomic theory have synthesized the insights of Keynes and Marx and their interpreters, and have enhanced contemporary understanding of Kalecki's work and legacy (see, for example, Sawyer 1985). Through the years, he has created publishing and research outlets for younger scholars, including this journal; indeed, he capped his long career at Leeds by leading the 2011–16 FESSUD research project, which encompassed researchers at 14 universities in 12 countries.[1]

Malcolm has also secured a place as one of the foremost analysts of Europe's transformation into a monetary union. This paper undertakes a chronological review of the 21 academic papers, written over 20 years, that Malcolm has authored or co-authored on the European Union (EU) and European Monetary Union (EMU, or Eurozone). Even as the Euro was launched, Sawyer, his long-time collaborator Philip Arestis, and other co-

authors, argued that the EMU's two defining macroeconomic-policy pillars were fundamentally flawed: it featured an independent central bank focused solely on inflation
targeting to the exclusion of employment or growth targets; and it placed tight constraints on member-nation fiscal policy, and lacked any plan for Euro-area fiscal policy.
They carried through this critique after the Great Financial Crisis (GFC), as the EMU
entered a sustained period of stagnation.[2]

In this paper, we supplement Malcolm's work on the applied macroeconomics of
Europe by focusing on the links between these two macroeconomic pillars and the
financial policies adopted by the European Union and its member-nations. We show
that in combination with the EMU's approach to monetary and fiscal policy, these
policies created a combustible mix. As the first decade of Eurozone operations unfolded,
the interaction of EU and member-nation financial policies lit the fuse that burned down
this brittle – and hence fragile – construct. On one hand, the EU was financially open to
external capital flows – and in accordance with its law of one market, intra-EU capital
movements were unconstrained. On the other hand, EU member-nations, confronting
the prospect of open competition for their domestic banking markets, facilitated the
creation of national banking champions that could compete on the global stage. This was
the deadly formula that combined to bring financial crisis to all of Europe.

Exposing this third unstable pillar in EU design, and linking it to the two exposed in
Sawyer's work, brings the links between Minsky's financial instability hypothesis and
European policy dilemmas into clearer focus. Sawyer (1999) wrote appreciatively of
Minsky's contention that avoiding a financial market meltdown requires the coordinated
use of central-bank and fiscal policy, and pointed out that the EMU's limits on macroeconomic policy represented an invitation to financial disaster. We show that this third
unstable pillar made the European financial crisis a matter of not if, but when.

In what follows, section 2 reviews the 12 academic papers that Sawyer authored or co-
authored on the EMU before the GFC. Section 3 and 4 complement Malcolm's analysis
with a description of the European Union's flawed two-tiered financial pillar: permitting
open cross-border financial flows inside Europe and across its external borders at the EU
level, while its member-nations retained control of banking regulation and structure. We
show how this approach led many European nations to encourage the growth of 'national
champion' banks that could compete in global financial markets. Section 5 summarizes
Malcolm's papers on the EU and EMU after the GFC. These papers show that the EMU's
failure to rethink its macroeconomic policy pillars are responsible for its stagnant post-
GFC growth. Section 6 shows how the EU's two-tiered financial pillar led EMU nations
into a double-loop of financial crisis whose second loop takes the form of economic
policy failures by member-nations with current-account deficits. Section 7 argues that
European nations in the post-GFC period may have to choose between stabilizing their
locally-focused banks and redoubling their reliance on megabanks tied into global
financial cycles.[3] Section 8 briefly summarizes and concludes.

2. The monetary and fiscal policy design flaws of the European Monetary Union

Malcolm Sawyer's first academic paper on the EU was co-authored in 1997 with Philip
Arestis, as were many of his subsequent papers. Arestis and Sawyer (1997a) – or

A-S (1997a) – critically analyses the proposal for an 'independent European system of central banks' (IESCB). The IESCB proposal represented a plan for the European Monetary Union (EMU, or Eurozone), which would be led by a European Central Bank, which member states could join by meeting macroeconomic convergence criteria involving price inflation (no more than 1.5% per annum), government finance (with government deficit of no more than 3% of GDP, and the government debt/GDP ratio capped at 60%), and interest- and exchange-rate levels maintained close to European averages.[4] The authors argue that this proposal will have 'severe implications' for unemployment, which certainly is not the intent of its authors. This divergence in views is rooted in different ways of understanding how macroeconomies work. Arestis and Sawyer take a Keynesian/Kaleckian approach, which sees output and employment dependent on the adequacy of aggregate demand. The IESCB proposal takes an orthodox approach which incorporates the 'classical dichotomy', wherein output and employment (the real economy) depend solely on supply-side factors, not aggregate demand; and since inflation is a monetary phenomenon, the only task of the central bank is to keep prices stable through its control of the rate of interest – the aim being to achieve a non-accelerating inflation rate of unemployment (NAIRU).

These authors' exploration of the theoretical basis of the convergence criteria reveals that the EMU proposal is consistent with 'new consensus macroeconomics' (which Keynes would have termed the 'Classical' worldview).[5] In lieu of this proposal, which sets up the European Central Bank (ECB) and member-nations' central banks as 'custodians of international capital' (p. 359), the authors propose that the European Monetary Union operates a regional version of Keynes' Bancor proposal (Keynes (1980a), Keynes (1980b)).

In a further analysis, Arestis and Sawyer (1997b) argue that central bank price-inflation targets lead to sub-optimal employment and output, and that the central bank cannot control the price level by manipulating the rate of interest. In turn, A-S (1997c) identifies the elements that can lead to underemployed resources: a shortfall of aggregate demand and/or inflationary pressure due to productivity and balance of trade constraints.

With the launching of the Euro in 1999, Malcolm's writings on Europe turned to the problematic implications of self-imposed EMU constraints on the use of macro-policy instruments. A 1999 paper (A-S 1999) reviewed the EMU from the perspective of Hyman Minsky's writings on financial instability. Minsky's policy dicta can be summarized in the terms 'big bank' and 'big government': responding to the endogenously-generated debt-fuelled downturns in economies with developed financial structures requires a central bank willing and able to stop panics and/or runs with lender-of-last-resort interventions ('big bank'), followed by countercyclical fiscal policies that re-establish aggregate demand ('big government'). Adopting Minsky's analysis, A-S argue that 'The Protocols under which the ECB is established enables, but does not require, the ECB to act as a lender of last resort.' (ibid., page 2); the mandate for this independent body extends only to price stability, not employment. Further, the budget of the European Union (EU) itself is small (limited to 1.5% of member-nations' GDP) and tightly constrained.

The authors then develop a Minskyian theme: because the financial and economic structures of member nations differ, occurrences of instability are likely to be asymmetric across Europe; and because the ECB's mandate focuses on Europe (and European inflation) as a whole, member-states' plunges into instability are more likely to be branded as instances of 'irresponsibility' than as occasions warranting lender-of-last-resort intervention. The authors also highlight the problem of capital mobility. They note that whereas 90% of portfolios have consisted of domestic assets prior to 1999, the combination of 'single market' freedom to move and the uniform currency ensures that 'the amount of funds moving within the euro area will make a quantum leap' (*ibid.*, p. 6). The authors warn against the formation and bursting of asset bubbles, recalling recent East Asian experience.

Two Levy Institute working papers in March 2001 reiterate these points of criticism in light of the Euro's decline against other leading currencies during its first 15 months of existence. While other analysts cited labour-market inflexibility in Europe as the cause of this runoff, A-S (2001) and Arestis (2001) pointed instead to the importance of diverging trends of the Eurozone member-states' core macroeconomic indicators (GDP, inflation, and unemployment, in particular). In light of this evidence, Arestis, McCauley, and Sawyer (2001) refine their 1999 argument about the EMU's missing policy instruments. They highlight 'the separation between monetary policy conducted by the ECB and the constrained fiscal policy operated by national governments' (Arestis, McCauley, and Sawyer 2001, 120). Once the euphoric Euro-launch period is over, long-term investors will worry about whether the EMU can survive a future crisis given its current policy architecture:

> 'What is needed is an expanded institutional setting, allowing the co-ordination of fiscal and monetary policy and large-scale regional transfers, guided by an alternative to the stability and growth pact' (Arestis *et al.*, *ibid.*, p. 7).

While the Euro subsequently recovered some ground, its member-states' economies fizzled – Germany went into recession in 2003, followed by a slowdown in Italy in 2005. Two A-S papers in 2003 pointed out that these problems could be traced back to the Classical foundations of the EMU's Stability and Growth Pact (A-S, 2003a), and to the confused and badly-communicated conduct of EMU monetary policy (A-S, 2003b).

In sum, Malcolm's work on the EMU shows how its limits on macroeconomic policy, given the clash between member-nations' heterogeneity and the EMU's one-size-fits-all mandate, invites economic instability. Financial considerations were implicit in this analysis, but didn't take center-stage. We bring these elements into the analytical light in the following sections.

3. Cross-border capital flows as a disciplining device in EU design

Malcolm Sawyer and co-author Philip Arestis establish that the combination in the EMU of a Europe-wide central bank committed to inflation targeting with severe constraints on member-nations' fiscal policies can only work under rarified conditions: for example, that all external shocks symmetrically affect all member nations; that price and interest-rate movements affect all member nations equally; and that the economic growth engine in every member nation depends only on supply-side forces operationalized by anticipatory price shifts. That is, the EMU's flawed macroeconomic policy design rests on an unachievable idealization of market-based allocation.

The flawed third pillar in EU/EMU design is its unleashing of free capital movements throughout Europe (and between Europe and the rest of the world), together with a purposeful inattention to EU/EMU-wide financial oversight. A two-tiered system was put in place: member-states would remain in control of financial institutions: they would continue to issue financial charters, decide on the form and powers of these financial organizations, and carry out prudential regulation; the movement of capital and provision of financial services would be covered by the principle of the 'European single market': a good or service provided anywhere in the 27 nations of the European Union could be provided everywhere in that 'single market.'

This solution represents a compromise between two historical dynamics in Europe. An older dynamic involved the relationship between national economic development and banking structures across Europe. As scholars including Gerschenkron (1962) and Cameron (1972) have documented, European nations' economic development from the 1800s onward typically involved strong central-state guidance and close cooperation with financiers, whether foreign or domestic. The foundational texts of European economists Hilferding (1919) and Schumpeter (1934), while grounded in different theoretical entry points, both emphasized the central role of financial intermediaries in guiding economic development. A groundbreaking 1981 study conducted for the Joint Economic Committee of the US Congress pointed out the contrast between the roles played in national industrial growth by the US and UK financial sectors, on one side, and West German, French, and Swedish financial institutions, on the other. John Zysman (1984), one of that study's co-authors, formalized these observations in an acclaimed 1984 book: he argued that the credit-based financial systems in France, West Germany, and Japan encouraged a longer-term approach consistent with technology development, in contrast to the short-term-oriented market-based financial systems of the US and UK. This contrast was embedded in the distinction made in the 'varieties of capitalism' literature between 'liberal market economies' and 'coordinated market economies' (Hall and Soskice 2001, page, 8).

While the notion of differing capitalist formations leads to important insights, the notion that Europe's national financial sectors are at the service of government-guided industrial development has to be reconsidered. The same pressures toward financial liberalization that had broken the regulated US banking system in the mid-late 1970s, were at work in the UK and in continental Europe. The UK had its markets' 'big bang' in 1986, resulting in what Philip Augur (2001) termed the 'death of gentlemanly capitalism.' That was only the most dramatic case in a broad western European shift. Consider the case of France: the Chirac government privatized 13 large financial groups in 1986; by 2002, the public-sector accounted for only 5% of all French employment in banking (Abdelal 2006). A French futures market was opened, and securities and foreign exchange markets were liberalized (Cerny 1989). France turned in just over a decade into a 'financial market economy' (Morin 2000, 36), open to foreign capital, with 'a transformation in strategic orientation of the most profound type: a surrender to short-term goals, to accountability for meeting targets, and 'submitting to the imperative of profitability.' (Loulmet and Morin 1999, 14). The French model was no longer bank-based; large, globally active French companies turned increasingly to equity markets to raise money in the 1990s (Levy, 1999). French experience was repeated in other European countries.

These moves to liberalize and to open financial markets to global flows of capital in the name of sectoral efficiency and profitability constituted a newer European dynamic, which rested on a critique of state-led economic growth. States had to be disciplined to permit Europe to compete in open global markets. The disciplining of member states was, in any case, precisely the point of the EMU convergence criteria. And the idea that market forces, not national policy tools, was implicit in the elimination of national currencies and their replacement by a common currency emitted by an EMU central bank that did not have prudential responsibility for European banks, and that lacked a lender-of-last-resort mandate.

This third financial pillar in EU/EMU design did respect the older dynamic of European banking: the diverse national financial eco-systems would continue to be structured and regulated at the nation-state level. However, these structures had to coexist with the newer dynamic: free financial flows in the hands of fund managers and arbitrageurs would punish member-nations whose policy choices were inconsistent with EMU convergence criteria.

The Delors Commission report (1989) sets out the principle of using free cross-border capital flows to discipline member-nation policy makers in so many words. After affirming that 'Greater convergence of economic performance is needed' as is 'more intensive and effective policy coordination' (p. 11), it argues for 'a large degree of freedom for market behavior' (p. 17), so that market forces can discipline states: 'Financial markets, consumers and investors would . . . penalize deviations from commonly agreed budgetary guidelines or wage settlements, and thus exert pressure for sounder policies.' But since access to markets can 'even facilitate the financing of economic imbalances,' and thus 'The constraints imposed by market forces might either be too slow and weak or too sudden and disruptive', member nations would 'have to accept that sharing a common market and a single currency area imposed policy constraints' (p. 20). Installing an independent central bank and establishing 'the full liberalization of capital movements and financial market integration' (*ibid.*, p. 16) before exchange rates are fixed would make it feasible to coordinate monetary policy across all banks and the entirety of the Euro area. So, as in Mundell (1963), making capital completely mobile would lead to welfare-improving equilibria.

As the Delors report shows, the design of the Eurozone was based on the dual premise that capital is scarce and globally mobile, and will be attracted to ports of call where it has fewest constraints on its movements (Dymski 2019). The framework put in place conformed to the policy views of the Classical mainstream view. As Chari and Kehoe (2006) put it:

> 'the practice of macroeconomics by economists have changed significantly—for the better. Macroeconomics is now firmly grounded in the principles of economic theory" [specifically] . . . a "commitment regime . . . [wherein] all policies for today, tomorrow, the day after and so on, are set today and cannot be changed' (p. 6). . . .

> 'We think of commitment as a situation in which at the beginning of time society prescribes a rule for the conduct of monetary policy in all periods. The monetary authority then simply implements the rule. . . . The message . . . is that discretionary policy making has only costs and no benefits. . . . [One possibility] is to delegate policy to an independent authority' (p. 7)

Not surprisingly, these authors have glowing words for the Eurozone: 'Perhaps the most vivid example of both the movement toward independence and the movement toward a rule-based method of policy making is to be found in the charter of the European Central Bank ... [whose] 'primary objective' is ... to 'maintain price stability.' This focus on stability and credibility was reinforced by Issing (2001) and his insistence that a uniform monetary policy across the zone ('one size fits all') would encourage convergence; it was seemingly guaranteed by the fact that the ECB was designed as a virtual duplicate of the Deutsche Bundesbank (Lohmann 1994).

So the European Commission's plan for the ECB forgot Minsky (1986), and forgot that even for the globally-hegemonic United States during the 'golden age of capitalism,' periodic financial-market malfunctions that caused so little damage – as 'credit crunches' – precisely because of timely central-bank interventions (Wojnilower, 1980). The notion that the ECB would lack LLR powers was greeted with incredulity by economists with central-banking experience. As early as 1992, Folkerts-Landau and Garber (1992) warned that the 'narrow' approach being taken would make it necessary 'to slow or even prevent the ongoing development of Community-wide liquid, securitized financial markets' (p. 1). Goodhart (1999) pointed out the impossibility of dispensing with the LLR function. Blinder (1999) critiqued the ECB plan's New Classical emphasis on credibility by asking central bankers about it, and summarizing their answers: 'Respondents think central banks get their credibility the old-fashioned way: They earn it by building a track record, ... not by limiting their discretion via commitment technologies or by entering into incentive-compatible contracts' (pp. 21–22).

Nonetheless, the Euro was launched in January 1999.[6] The ECB responded to critiques of its architecture by establishing several initiatives – such as the Financial Services Action Plan (1999) and the Committee of Wise Men (2001) – that would more closely harmonize European financial markets (Hartmann, Maddaloni, and Manganelli 2003). These did not mollify analysts worried that ECB jurisdiction would be inadequate in a crisis, such as Dominguez (2006) and Schinasi and Teixeira (2006). The latter two IMF economists recommended the 'centralization, or rather the federalization, of financial stability functions', observing that 'given the decentralized banking supervision and financial market surveillance, it may prove difficult to work out responsibilities on an ad-hoc basis in the midst of a crisis' (pp. 21–22).

Clearly, this flawed pillar, like the EU's two macropolicy pillars, rested on an idealized view of capital flows and financial markets. Just as the EMU's macroeconomic structure could only work under rarified general-equilibrium laboratory conditions, its open regulatory financial architecture would work seamlessly only if speculation-free financial-market efficiency obtained both inside and outside the single market. But whether or not deregulated financial markets would yield more allocative efficiency, they would accomplish the goals of the EU architecture of reining in the macrobehavior of EMU member states.

4. The impact of the single market and financial globalization on EMU banking structure and competition

However effective it was in disciplining convergence, the new EU/EMU architecture had problematic implications for banking policy. The twin ideas of opening Europe to overseas capital flows to discipline European economies, and of permitting unchecked capital flows within the single European market, posed two dilemmas for member-nations. The first dilemma arose because nation-states would remain responsible for the safety and soundness of their domestically-chartered financial intermediaries.

On one hand, each national market could be freely entered by other nations' financial-services firms and the products they sold. On the other, domestic central banks – because they would no longer issue domestic currencies – would lose their LLR capacities. These powers were instead vested in a European Central Bank whose formal mandate extended only to inflation targeting. This central-bank design doubled down on the idea that EMU banks and the member-nations that chartered them should avoid excessive risk-taking; further, it made clear that the Eurozone could not withstand a deep financial crisis without setting aside its own rules – as should become evident in the Eurozone crisis

The second dilemma arose because of Europe's historically-rooted heterogeneous financial structures, whose diversity was threatened by the single market. France's big banks had until the 1980s been government-owned tools for achieving national purposes; large German banks typically had cross-shareholding and long-term relationships with large domestic corporations. Some countries favoured arms-length markets for credit and financial-service provision (for example, Great Britain), whereas others favoured 'relationship-based' banking (with Germany as a paradigm case). In some parts of Europe – such as Spain, Italy, and Germany – regional and city-based financial institutions provided core financial services; elsewhere – the Netherlands, Luxembourg, and Switzerland – regional differences were unimportant, and instead international linkages were fundamental.

While these strong home-country advantages had led to dependable income flows for incumbent banks servicing their home markets (Dermine, 1996), EMU banking structure was already under pressure due to foreign entry; in particular, US investment banks were penetrating European markets, dismantling the 'webs of national influence built up over decades' (*The Economist*, 23 June 2001). The arrival of the single market broke the dam: EMU nations, in particular, encouraged their banks to secure their market position through mergers as the Euro era began. *Time International* (22 March 1999) observed: 'Banks within domestic markets are beefing up in preparation for the next stage: a slew of crossborder banking tie-ups between the remaining players.'

Numerous mergers aimed at establishing 'national champion' banks that could stand up to heightened intra-European competition, while also competing globally, followed. For example, in June 1999, Banque Nationale de Paris (BNP), primarily a commercial bank, succeeded in buying Paribas, an investment bank. In March 1999, Italy's two largest banking groups made merger bids–UniCredito Italiano for Banca Commerciale Italiana (BCI), and Sanpaolo IMI for Banca di Roma. In June 1999, Italy's fourth-largest bank, Banca Intesa, merged with the fifth-largest, Banca Commerciale Italiana (BCI). The second-largest Spanish bank, Banco Bilbao Vizcaya, had been created by the merger of two Basque banks in 1988; in January 1999 Banco Santander, the largest Spanish bank, consolidated its lead over the second-largest Spanish bank, Banco Bilbao Vizcaya (created by the 1988 merger of two Basque banks) by merging with BCH, then third-largest. Generally, these mergers permitted branch closings, cost cutting, and increases in market capitalization.

Apart from defensive considerations, scaling up to compete globally was another motivation for large European banks. Not all designated 'national champion' banks sought to become megabanks (see footnote 2), but some did. The two Spanish megabanks mentioned above used acquisitions to take a leading role in Latin American banking markets. Italian banks moved aggressively into emerging European markets. The largest German bank, Deutsche Bank tried but failed to merge with the second largest, Dresdner Bank, in mid-1999, as part of its quest to achieve global scale in investment banking and underwrite megacorporations' 'bulge bracket' issues.[7] Dresdner, after several unsuccessful merger attempts, was finally bought by Allianz, the world's second-largest insurer, in April 2001; this combination represented an effort to create what the *Wall Street Journal* (2 April 2001) termed a 'banking, insurance, and asset-management colossus' – a German Citigroup. ABN Amro, based in the small-market Netherlands, took on substantial leverage so as to expand its consumer banking operations in Asia and Latin America.

This push to create globally-competitive European megabanks came in a period of rapid global innovation in financial practices. US megabanks had taken advantage of the collapse of the thrift-based US mortgage market in the 1980s to take a global lead in securitization, an advantage that was fed by the systematic US capital-account inflows (a consequence of its current-account deficit from 1982 onward). The 1999 repeal of the US Glass-Steagall Act (that had required the institutional separation of commercial and investment banking), along with the increasing availability of liquidity and risk-tolerant investment funds, permitted US megabanks to construct shadow-banking systems and to make and securitize high-risk loans – most notably, subprime mortgage loans. Market-based credit increasingly replaced bank-based credit.

Large European banks also had to turn increasingly to market-based instruments to maintain their larger European customers. Without expanding their deposit networks beyond their domestic borders, they could tap the explosively-growing markets for borrowed funds to buy assets and find loan customers in other countries. These shifts coincided with the post-Euro intensification of trade and financing relations within a consolidating Euro area: now, cross border imbalances were no longer reduced through currency re- and/or devaluations, but instead through compensating capital flows from trade surplus to trade deficit countries.

These competitive and market-opening dynamics, in the early years of the Euro, suggested that many apparent benefits flowed from the EU's embrace of open financial markets. Financial integration in Europe increased, as measured by higher levels of interbank lending and securitization, by narrowed interest-rate differentials, and by the larger deficits being run by the poorer nations:

'Prior to European monetary union, investors would typically have required larger country risk premia to fund such deficits, and the risk of a speculative attack on a debtor's currency would have increased. However, these countries are now largely insulated from such pressures. In effect, claims on other euro-area members are increasingly viewed as good substitutes for claims on domestic parties' (Lane 2006, 55).

While these intra-European capital flows, from surplus to deficit areas, conformed with the expectations of efficient-capital-market theory (Lucas 1990), these were not the only flows of cross-border capital in which European megabanks were involved. US mega-banks and shadow banks were combining innovations in borrowed-money markets with innovations in securitized debt and in derivatives markets to vastly expand their asset bases and revenue flows. Given that globally unbalanced financial flows favoured the US,

European megabanks competing globally had to absorb exchange risk and buy the new asset classes at a distance if they wanted to keep up. The 'go-go' atmosphere of the pre-crisis years took hold, transforming formerly staid European banks beyond recognition. As *Fortune's* Guyon (2000) put it:

> 'The bank may still be called Deutsche, but the center of gravity has clearly moved from the old-line German commercial bankers in Frankfurt to a polyglot team of investment bankers headquartered in London. transform[ing] it into a money machine that has finally brought Deutsche within spitting distance of investment banking's perennial leaders, Goldman Sachs and Morgan Stanley.'

The elements for a severe crisis were falling into place worldwide. In the US, megabanks were at the heart of US subprime lending: the creation, bundling, and selling of loans whose viability depended on sustaining unsustainable price increases in the housing market was refined into a high art by the time housing prices fell (Dymski 2019). The UK had its own plunge into subprime lending and megabank over-expansion. Many European megabanks, eager to compete head-to-head across the landscape of esoteric finance, were pulled along by US and UK megabanks' momentum into speculative position-taking, improbable mergers, and risky cross-border lending, especially for residential and commercial real estate. Tooze (2018, Chapter 3) describes this as the system of 'Transatlantic Finance.'

Since domestic rules on bankruptcy and default blocked the expansion of subprime lending in many EU member-states, European megabanks compensated for this disadvantage by taking positions in securitized loans originated in the US subprime markets. Often, as Lewis (2010, Chapters 2–3) points out, they were gamed by insider Wall Streeters who were a step ahead in riding the bubble even as it collapsed. And then the subprime crisis hit: the US and UK housing bubbles slowed and fell, then Northern Rock's failure led to the collapse of the asset-backed commercial paper market; then came US and UK megabanks' insolvencies. Initially, the subprime crisis hit European banking selectively: Fortis and Commerzbank and Germany's *Landesbanken* failed due in large part to subprime securities holdings. Through mid-2009, the situation stabilized: these failed-bank situations were resolved via fiscal (taxpayer) injections and asset fire sales, and social-welfare provisions seemed to be holding. But by late 2009, austerity macro provisions were imposed throughout most the EMU, and with the change of government in Greece, the scope of the bad debt problem of Greek banks was revealed. Then came a deep macroeconomic recession in the GIPSI nations, and stagnation in most of the rest of Europe.

5. The GFC leads the stability and growth pact into a macroeconomic policy cul-de-sac

This brings us to the second – post-GFC – round of Malcolm Sawyer's work with Philip Arestis on the EMU. Published between 2011 and 2017, these six papers, one co-authored with Giuseppe Fontana, focus on how policy design errors, about which these authors had warned ten years earlier, now limited EMU responses to what they termed the 'Great Recession' (Arestis and Sawyer 2011a, 2011b).[8]

Several of these post-GFC papers show how the EMU's inflexible rules worsened the macroeconomic downturn that followed the financial turmoil of 2008, especially for nations in the EMU's southern and western (GIPSI) periphery.[9] The implementation phase of the EMU converted its convergence criteria into Stability and Growth Pact (SGP) targets (Arestis and Sawyer 2012). The economic slowdown – the Great Recession – then exposed Europe's structural problems (Arestis and Sawyer, 2011a): increased budget deficits, falling GDP levels, and rising debt hit all EMU countries hard, but 'peripheral' countries were hit harder. These countries had been losing competitiveness to Northern EMU nations since the launching of the Euro. The modest relative advances achieved by labour market reforms were swept due to the scale of macroeconomic losses (Sawyer 2015).

The conventional path to diverging unit-labor costs among close trading partners would be currency devaluation for those slipping behind; but this path was closed off – so doing was the very heart of the EMU's policy design. These differentials widened current-account imbalances among EMU nations. And whereas before the crisis period, private-sector capital flows had readily covered peripheral nations' deficits, these flows dried up during the GFC. Arestis and Sawyer (2011a) then show how this led directly to crises of non-compliance with SGP criteria for EMU economies with intra-EMU current-account deficits:

> 'The current account deficits of the south European countries required those countries to borrow heavily from other countries, and in the main from north European banks as well as British and American ones. Because south European countries had much lower interest rates than previously, they rapidly built up their debt. The debts were mainly, though not exclusively, private sector rather than public sector. However, when the Great Recession hit, borrowing was increasingly done by the government.' (p. 8)

The consequences for macroeconomic policy management are stark: repaying these loans depended on bringing borrower (peripheral) nations' macroeconomic parameters back to a glide path consistent with (every nation's) SGP targets; but the only way onto that path while remaining a member of the EMU was through unprecedently draconian budget cuts.

The second point made in these post-GFC papers is that only fundamental reforms in the design architecture of the EMU can block future macroeconomic crises. Arestis and Sawyer (2011b) set out the sequence of flawed and interlocking EMU policies leading to the current impasse.[10] They begin with the convergence criteria, which led directly to the SGP criteria and 'impose a general deflationary bias' (p. 25), strictly limiting fiscal transfers from richer to poorer regions within Europe. They next list the ECB, which is independent and tasked only with meeting European inflation targets. It lacks a LLR mandate, and its 'one size fits all' policy – in the context of national governments that cannot issue their own domestic currency – means that central-bank policies take no explicit account of local variations in employment, growth, and investment. They finish by pointing out that the convergence criteria do not take current-account imbalances within the EMU into account, and thus imbalances can only be handled by bailouts or by deficit countries' wage and price reductions. Since the former are prohibited, the latter represent the only feasible path consistent with maintaining EMU membership. In sum, the crisis is due not to 'the wrong application of the relevant economic policies of some member states' (p. 31) but from design flaws in the EMU.

Arestis, Fontana, and Sawyer (2013) further refine this argument. They point out that the small scale of the European Union budget itself limits meaningful fiscal transfers, and that EU/EMU policy architecture, in closely conforming with 'new consensus macro-economics', undercuts the 'European social model', which rests on 'the universalistic character of welfare provision and a high degree of coordination between economic actors' (p. 30). In effect, the flexible labour markets required to achieve competitiveness necessarily undercut social protection. Arestis and Sawyer (2013) go further, and wonder whether a comprehensive political union – the creation of a United States of Europe – will be necessary to permit the fiscal stimulus – and thus intra-EMU fiscal transfers – needed in the face of adverse macro shocks (such as the GFC). The European Stability Mechanism does not provide an effective substitute for political union. Further, the tepid response to post-crisis plans for a European Monetary Union demonstrates that some member-nations' demands for complete separation between bank oversight and monetary policy can be effectively countered only through adoption of a full-scale European political union.

6. The double loop of Europe's banking crisis and Europe's financial bifurcation point

Malcolm Sawyer's post-crisis essays emphasize the fragility built into the EMU's economic architecture by its two flawed macroeconomic policy pillars, and the asymmetric losses that would result from any downturn. The cyclical downturn associated with the GFC is seen as a case in point. This extended analysis does not examine the implications of the perverse financial dynamics that triggered the GFC: the possibility that structural changes in global finance may now be driving business cycle dynamics.

Yet what launched the EMU and most of the world into macroeconomic downturn was an immense financial crisis – one which involved the collapse of market-based cross-border capital flows within Europe and between Europe and the US, with European megabanks centrally involved and many others (such as Spain's cajas and Germany's landenbanken) deeply – even mortally – scarred.

From the perspective of macroeconomic policy per se, why does the financial trigger of the subsequent downturn and period of stagnation matter? The answer is that linking these two together can explain why Europe's post-GFC period led directly into the Eurozone crisis. For what Arestis and Sawyer leave out of their indictment of EMU macroeconomic policy is that the European sovereign debt crisis arose because the GFC's impact involved a double loop through the balance sheets of large private banks – the very ones whose size and scale had been championed in the launching of the European single market.

Loop one of the crisis involved the cash-flow impact of the crisis itself – the spike in the cost of borrowed funds after the 2007 collapse of European interbank markets, the failure of subprime paper bought by European banks, and so on. Many European banks' cross border loans inside Europe also went bad. Some large banks' asset sizes dwarfed their home nation-states' GDPs; nurtured as national champions in global financial competition, they'd become 'too big to save.'[11] So they were not allowed to fail.

Then, as noted above, the Greek debt crisis was revealed as George Papandreou's government came to power in October 2009. Pressures rose on the ECB due to its hesitant crisis response (Tooze 2018), while European banks' opacity fed market participants' fear and uncertainty. Not only was Greek government debt larger than was previously known, but loan defaults throughout the GIPSI member-nations rose asymmetrically, compared to Northern Europe. French and German banks that had developed loan-customer relations with borrowers in GIPSI nations, including many large banks, now found much of their loan portfolios in default. GIPSI nations' current-account deficits had to be financed.

This led to loop two: large banks took on this debt. Already weakened from loop one, they were in no position to write down more bad debt. However, in the GFC's dire straits, these sovereigns became what Minsky would have called 'Ponzi' units: far from repaying accumulated debts, they required further borrowing even to meet debt-servicing obligations. So the EMU's large private banks holding this debt were forced to finance deficit nations' further borrowing. This was done grudgingly, and under the oversight of the 'troika' – the European Commission, the ECB, and the IMF (Varoufakis 2017, Chapter 2). As Varoufakis put it, this was not a debt crisis – it was both a banking crisis (2017, Chapter 2), and a crisis of the failure of Europe's missing surplus recycling mechanism (Varoufakis 2013).

At the end of the day, the banking situation was not solved – the banks recapitalized; but those who aimed at megabanking status have, for the most part, had their wings clipped. Deutschebank has shrunk in size and ambition. RBS is a shadow of its former self; ABN Amro has survived as a shadow of its former self; Fortis was taken over by BNP Paribas; and so on. Europe's capital markets union and banking union are off to rocky starts due to the unwillingness of nation states to cede ground to one another on the possibility that crises could result in cross-country subsidies.

7. A financial bifurcation point in European banking?

We do not pursue the details of the GFC and its aftermath here. It is sufficient to note that the collapse of leveraged subprime securitization from 2007 onward has compromised global liquidity and forced the use of nation-states' fiscal capacity to prevent collapse. The banks were rescued, but credit remains unavailable for small/medium businesses. To some extent, the shadow banking market has filled in the gaps. US investment megabanks' global dominance is threatened more by emerging IT platforms (Platt, Noonan, and Bullock 2019) than by European megabank rivals, who have withdrawn from US markets (Noonan 2020).

The key point made in section 6 is that the dysfunctional EU financial pillar was responsible for the asymmetric depths that the EMU's macroeconomic pillars forced, in particular, on the peripheral countries of the Eurozone. Whether this interaction among the pillars of the flawed EMU/EU policy architecture becomes an infinite loop depends in part on how European policy makers manage the banking structures whose reform they thought would make them future-proof (section 4), but which the EMU's policy pillars have now revealed as a point of exquisite vulnerability.

EMU member-nations were caught in a particularly vicious whipsaw in the GFC, and now are threatened anew amidst the coronavirus pandemic. EMU rules still do not provide for either fiscal transfers or central-bank liquidity provision in the case of either financial or macroeconomic meltdowns (Hall, Arnold, and Fleming 2020). Plans for consolidated financial supervision, via a European Banking Union managed by the ECB, have not been approved (Fleming and Johnson 2019). Faced with a coronavirus sudden-stop, nation-states are turning to ECB and national-bank lending stop-gaps and to Treasury fixes; but in the absence of further EU/EMU reforms such short-term fixes will compromises market confidence and lead to destabilising negative sovereign-debt loops. The stock of unpayable bank and government obligations can multiply faster than resources can be freed even with extreme austerity policies.

Attacks by suspicious global investors on bankrupt governments and insolvent banks will cease only in one of two circumstances. One would involve shrinking the megabanks and reining them in (along with their penumbra of shadow-banks). Eliminating the need to make good (for global investors' sake) on the obligations of domestically-chartered megabanks whose liabilities approximate the scale of national GDP could recenter attention on how banks can best serve domestic loan customers. The other circumstance would involve ensuring that a willing and able central bank provides lender-of-last-resort interventions as necessary for too-big-to-fail European megabanks. This would not be the ECB as it currently exists; and it would have to be a Europe in which the megabanks in question would indeed be European in scope, providing payments, savings, and investment facilities that serve the EU in its entirety.

This is then the financial bifurcation point. In one direction, a diverse eco-system of European banks, differing among countries and regions, all operating at scales and in activities that do not pose risks larger than their national governments can handle. In the other direction, a small set of homogeneous large European banks, offering similar products and services throughout Europe, operating adventurously in global markets – in head-to-head competition with large Wall Street banks; systemic and even catastrophic failure would be a possibility, one that is viewed as more than offset by the gains accruing to international financial centre status.

In the latter scenario, the European Central Bank would have to accept its role as a backstop against meltdown, since meltdown would bring the entire European financial system with it. The first part of this latter option – the existence of megabanks whose scale dwarfs national income flows – has already happened. What has not yet happened is continent-wide expansion by a small set of European (or non-European) megabanks; it may yet come. Regardless of whether the European Central Bank agrees to backstop the banking system, though, this sequence of events is likely to end in catastrophe.

European banks themselves, caught between Wall Street and the City of London, have lobbied since the GFC, as they always do, for maximum regulatory flexibility and no size or bonus restraints, while searching for new business models. But against the view that their further enabling will permit them to better serve Europe and to compete with overseas competitors is the reality in that only extraordinary measures and luck permitted Europe's megabanks to survive the 2008 crisis. The capacity of European governments to support – and if necessary underwrite – new megabank recombinations and rescues under the current patchwork quilt of national bank/national-sovereign circumstances should not be exaggerated – especially in the current moment.

8. Summary and Conclusions

While many analyses of the Eurozone's flaws have been undertaken, Malcolm Sawyer's work stands out for the depth and breadth of its critique of its macroeconomic policy design, and of the implications of that design. His work, frequently undertaken in collaboration with Philip Arestis, initially diagnosed design flaws in the EMU, then turned to its failure to generate prosperity in Europe, and finally showed how the EU's limitations in confronting financial crisis still require a fundamental rethinking of its policy architecture.

This paper has extended this foundational work by pointing out that the equally flawed financial policy pillar of the EU/EMU has both triggered and deepened Europe's 21st-century crises. Just as the European macroeconomic policy architecture could only work under the unachievable assumptions of the New Consensus Macroeconomic model – a point so vigorously made by Malcolm Sawyer over the years – so too the changes made to European banking so as to achieve the efficiencies available in globally-connected, market-based financial markets would only enhance Europeans' economic and social welfare under conditions that real-world financial systems cannot reach.

Notes

1. See www.fessud.eu.
2. The term 'Global Financial Crisis' (GFC) refers herein to the worldwide economic crisis that began with the 2007–08 subprime crisis and led into the Eurozone crisis from 2009 onward.
3. The term 'megabank' used herein refers to large financial firms operating comprehensive, integrated financial service platforms on a global scale.
4. In a complementary analysis, Arestis, McCauley, and Sawyer (1999) describe the historical evolution of European economic union and monetary integration in great depth.
5. Arestis and Sawyer (2008) investigate the behavioral implications of the formal 'new consensus' models used by the ECB and Bank of England, and demonstrate their dependence on the assumption that the macroeconomy is following a pre-given equilibrium long-run growth path.
6. The European Commission leadership calculated that an imperfect union would generate pressures toward constructing a more thoroughly consistent common regime (Spolaore, 2013).
7. Deutsche Bank had already bought US blue-chip Bankers Trust in mid-1998, leading other European megabanks to try (unsuccessfully) to acquire blue-chip U.S. investment banks. Among the European banks seeking to compete at global scale were Credit Suisse First Boston and Lloyds. Ironically, most large European mergers in this period were under-written by US investment banks.
8. Malcolm Sawyer also published a book (Sawyer 2018) on the prospects of the Euro which synthesizes themes presented in the papers reviewed herein.
9. The term GIPSI refers to Greece, Italy, Portugal, Spain, and Ireland.
10. Sawyer (2013) discusses the EMU's reliance on the supposed supply-side benefits that will flow from 'structural reforms' in labour and product markets.
11. Nine of the 20 countries with the highest bank-liability-to-GDP ratios in the world, in 2008, were EU countries; and among the others in the top bracket were the UK, Switzerland, and Denmark (Demirgüç-Kunt and Huizinga, 2013, Table 2, p. 878).

Disclosure statement

No potential conflict of interest was reported by the author(s).

ORCID

Annina Kaltenbrunner (iD) http://orcid.org/0000-0003-3519-5197

References

Abdelal, R. 2006. "Writing the Rules of Global Finance: France, Europe, and Capital Liberalization." *Review of International Political Economy* 13 (1, February): 1–27.

Arestis, P., and M. Sawyer. 1997b. "The Problematic Nature of Independent Central Banks." In *Money, Financial Institutions and Macroeconomics*, edited by A. Cohen, H. Hagemann, and J. Smithin, 263–279. Boston and London: Kluwer.

Arestis, P. March 2001. "Iris Biefang-Frisancho Mariscal, Andrew Brown, and Malcolm Sawyer, 'The Causes of Euro Instability." Working Paper No. 324, Annandale-on-Hudson, NY: Levy Economics Institute.

Arestis, P., and M. Sawyer. 2002. "New Consensus," New Keynesianism, and the Economics of the 'Third Way'." Working Paper No. 364, Annandale-on-Hudson, NY: Levy Economics Institute. December 2002.

Arestis, P., and M. Sawyer. 2012. "Can the Euro Survive after the European Crisis?" in P. Arestis and M. Sawyer, edited by, The Euro Crisis, International Papers in Political Economy. Basingstoke: Palgrave Macmillan. 1–34.

Arestis, P., G. Fontana, and M. Sawyer. 2013. "The Dysfunctional Nature of the Economic and Monetary Union." In *The EU Economic and Social Model in the Global Crisis: Interdisciplinary Perspectives*, edited by D. Schieck, 23–44. Farnham: Ashgate.

Arestis, P., K. McCauley, and M. Sawyer. February 1999. "From Common Market to EMU: A Historical Perspective of European Economic and Monetary Integration." Working Paper No. 263, Annandale-on-Hudson, NY: Levy Economics Institute.

Arestis, P., K. McCauley, and M. Sawyer. 2001. "'An Alternative Stability Pact for the European Union,' Cambridge." *Journal of Economics* 25: 113–130.

Arestis, P., and M. Sawyer. 1997a. "Unemployment and the Independent European System of Central Banks: Prospects and Some Alternative Arrangements." *American Journal of Economics and Sociology* 56 (3, July): 353–368.

Arestis, P., and M. Sawyer. August 1997c. "Reasserting the Role of Keynesian Policies for the New Millennium." Working Paper No. 207, Annandale-on-Hudson, NY: Levy Economics Institute.

Arestis, P., and M. Sawyer. March 2001. "Will the Euro Bring Economic Crisis to Europe?" *Working Paper No. 322*, Annandale-on-Hudson, NY: Levy Economics Institute.

Arestis, P., and M. Sawyer. July 2003a. "Macroeconomic Policies of the Economic Monetary Union: Theoretical Underpinnings and Challenges." *Working Paper No. 385*, Annandale-on-Hudson, NY: Levy Economics Institute

Arestis, P., and M. Sawyer. 2003b. "Making the Euro Work." *Challenge* 46 (2, March/April): 80–96.

Arestis, P., and M. Sawyer. 2008. "Are the European Central Bank and Bank of England Macroeconomic Models Consistent with the New Consensus in Macroeconomics?" *Ekonomia* 11 (2, Winter): 51–69.

Arestis, P., and M. Sawyer. 2011a. "The Ongoing Euro Crisis." *Challenge* 54 (6, November/December): 6–13.

Arestis, P., and M. Sawyer. 2011b. "The Design Faults of the Economic and Monetary Union." *Journal of Contemporary European Studies* 19 (1): 21–32.

Arestis, P., and M. Sawyer. 2013. "Moving to a United States of Europe?" *Challenge* 56 (3, May/June): 42–52.

Augur, P. 2001. *The Death of Gentlemanly Capitalism: The Rise and Fall of London's Investment Banks*. London: Penguin.

Blinder, A. S. June 1999. "Central Bank Credibility: Why Do We Care? How Do We Build It?" *NBER Working Paper 7161*, Cambridge, MA: National Bureau of Economic Research.

Cameron, R. E. 1972. *Banking and Economic Development: Some Lessons of History*. London: Oxford University Press.

Cerny, P. 1989. "The 'Little Big Bang' in Paris: Financial Deregulation in a *Dirigiste* System." *European Journal of Political Research* 17: 169–192.

Chari, V. V., and P. J. Kehoe. 2006. "Modern Macroeconomics in Practice: How Theory Is Shaping Policy." *Journal of Economic Perspectives* 20 (4, Fall): 3–28.

Committee for the Study of Economic and Monetary Union, Jacques Delors, Chairman (Delors Commission). 1989. *Report on Economic and Monetary Union in the European Community*. Brussels: European Community, April 17.

Dermine, J. 1996. "European Banking Integration, 10 Years after." *European Financial Management* 2 (3): 331–353.

Dominguez, K. M. E. 2006. "The European Central Bank, the Euro, and Global Financial Markets." *Journal of Economic Perspectives* 20 (4, Fall): 67–88.

Dymski, G. 2019. "Post-war International Debt Crises and Their Transformation." In *The Handbook of Globalization*, edited by J. Michie and U. K. Edward Elgar, 103–118. 3rd ed.

Fleming, S., and M. Johnson. 2019. "Eurozone Ministers Divided over Banking Union Negotiations." *Financial Times*, 5 December.

Folkerts-Landau, D., and P. M. Garber. 1992. "The European Central Bank: A Bank or A Monetary Policy Rule." *NBER Working Paper No. 4016*, Cambridge, MA: National Bureau of Economic Research, March.

Gerschenkron, A. 1962. *Economic Backwardness in Historical Perspective*. Cambridge: Harvard University Press.

Goodhart, C. A. E. 1999. "Myths about the Lender of Last Resort." *International Finance* 2 (3): 339–360.

Guyon, J. 2000. "The Emperor and the Investment Bankers: How Deutsche Lost Dresdner." *Fortune* 141 (9, May): 1.

Hall, B., M. Arnold, and S. Fleming. 2020. "Coronavirus: Can the ECB's "Bazooka" Avert a Eurozone Crisis?" *Financial Times*, 23 March

Hall, P. A., and D. Soskice. 2001. "An Introduction to Varieties of Capitalism." In *Varieties of Capitalism*, edited by P. A. Hall and D. Soskice, 21–27. Oxford: Oxford University Press.

Hartmann, P., A. Maddaloni, and S. Manganelli. 2003. "The Euro-Area Financial System: Structure, Integration, and Policy Initiatives." *Oxford Review of Economic Policy* 19 (1): 180–213.

Hilferding, R. 1919. *Finance Capital: A Study in the Latest Phase of Capitalist Development*. London: Routledge.

Issing, O. 2001. "The Single Monetary Policy of the European Central Bank: One Size Fits All." *International Finance* 4 (3): 441–462.

Keynes, J. M. 1980a. "Proposals for an International Clearing Union." (August 1942), *The collected writings of John Maynard Keynes*. Vol. 25, London: Macmillan, pp. 168–195.

Keynes, J. M. 1980b. "Speech to a Meeting of the European Allies, 26 February 1943." *The collected writings of John Maynard Keynes*. Vol. 25, London: Macmillan, pp. 206–215.

Lane, P. R. 2006. "The Real Effects of European Monetary Union." *Journal of Economic Perspectives* 20 (4, Fall): 47–66.

Levy, J. 1999. *Tocqueville's Revenge: State,Society,and Economy in Contemporary France*. Cambridge, MA: Harvard University Press.

Lewis, M. 2010. *The Big Short*. New York: W.W. Norton.

Lohmann, S. 1994. "Designing a Central Bank in a Federal System: The Deutsche Bundesbank, 1957-1992." In *Varieties of Monetary Reform: Lessons and Experiences on the Road to Monetary Union*, edited by P. Siklos, 247–277. Dordrecht: Kluwer Academic Press.

Loulmet, L., and F. Morin. 1999. "The Transformation of the French Model of Capital Holding and Management," Conference on "Corporate Governance in Asia: A Comparative Perspective." Organization for Economic Co-operation and Development, Seoul, 3-5 March

Lucas, R. E., Jr. 1990. "Why Doesn't Capital Flow from Rich to Poor Countries?" *American Economic Review* 80 (2, May): 92–96.

Minsky, H. P. 1986. *Stabilizing the Unstable Economy*. New Haven: Yale University Press.

Morin, F. 2000. "A Transformation in the French Model of Shareholding and Management." *Economy and Society* 29 (1, February): 36–53.

Mundell, R. A. 1963. "Capital Mobility and Stabilization Policy under Fixed and Flexible Exchange Rates." *Canadian Journal of Economic and Political Science* 29 (4): 475–485.

Noonan, L. 2020. "The Rise and Dramatic Fall of European Investment Banks in the US." *Financial Times*, 2 March

Platt, E., L. Noonan, and N. Bullock. 2019. "Morgan Stanley, Goldman and JPMorgan's Grip on Tech IPOs under Threat after Uber." *Financial Times*, 22 May.

Sawyer, M. 1985. *The Economics of Michał Kalecki*. London: Palgrave Macmillan.

Sawyer, M. 1999. "Minsky's Analysis, the European Single Currency, and the Global Financial System." *Working Paper No. 266*, Annandale-on-Hudson, NY: Levy Economics Institute, March.

Sawyer, M. 2013. "Alternative Economic Policies for the Economic and Monetary Union." *Contributions to Political Economy* 32: 11–27.

Sawyer, M. 2015. "Can Prosperity Return to the Economic and Monetary Union?" *Review of Keynesian Economics* 3 (4): 457–470.

Sawyer, M. 2018. *Can the Euro Be Saved?* Cambridge: Polity Press.

Schinasi, G. J., and P. G. Teixeira. 2006. "The Lender of Last Resort in the European Single Financial Market." *IMF Working Paper WP/06/127*, Washington DC: International Monetary Fund, May.

Schumpeter, J. A. 1934. *The Theory of Economic Development: An Inquiry into Profits, Capital, Credit, Interest, and the Business Cycle*. Cambridge, MA: Harvard University Press.

Spolaore, E. 2013. "What Is European Integration Really About? A Political Guide for Economists." *Journal of Economic Perspectives, American Economic Association* 27 (3): 125–144.

Tooze, A. 2018. *Crashed: How a Decade of Financial Crises Changed the World*. New York: Viking.

Varoufakis, Y. 2013. "There Is No Such Thing as a Debt Crisis: The Euro Crisis, Asia's Woes and America's Dilemma in a Global Context." *The coffees of the Secretary-General: Bringing new perspectives to the OECD*. Paris: Organization for Economic Co-operation and Development. 1 March.

Varoufakis, Y. 2017. *Adults in the Room: My Battle with Europe's Deep Establishment*. New York: Vintage.

Wojnilower, A. M. 1980. "The Central Role of Credit Crunches in Recent Financial history." Brookings Papers on Economic Activity, 2/1980.

Zysman, J. 1984. *Governments, Markets, and Growth*. Ithaca, NY: Cornell University Press.

The industrial policy requirements for a global climate stabilization project

Robert Pollin

ABSTRACT
This paper presents an industrial policy approach for advancing a global climate stabilization project. The centerpieces of the project are 1) dramatically improving energy efficiency standards in the stock of buildings, automobiles and public transportation systems, and industrial production processes; and 2) equally dramatically expanding the supply of clean renewable energy sources – primarily solar and wind power – available at competitive prices. Global investment spending in these areas will need to average about 2.5% of global GDP over 2024–2050 to achieve a net zero CO_2 emissions global economy by 2050. The paper works within a policy approach similar to that advanced by Malcolm Sawyer that integrated industrial and macroeconomic policies targeted at achieving full employment.

1. Introduction

What has finally become apparent in recent years is that economic analysis and policy practice can no longer ignore the reality of the global climate crisis in assessing the merits of any given analytic approach or policy proposal. This is for the simple reason that it is now almost universally accepted that we are courting ecological disaster by failing to develop a viable global climate stabilization project. The magnitude of the current crisis was well expressed by the eminent Oxford University climate scientist Raymond Pierrehumbert, who wrote in 2019 'Let's get this on the table right away without mincing words. With regard to the climate crisis, yes, it's time to panic,' (2019, 215)

This same sense of alarm was expressed more formally in the 2018 report of the Intergovernmental Panel on Climate Change (IPCC), the most authoritative global organization This report, titled *Global Warming of 1.5°* emphasized the imperative of limiting the increase in global mean temperatures to 1.5 degrees above pre-industrial levels as opposed to the previous consensus target 2.0 degrees. The IPCC concluded that limiting the global mean temperature increase to 1.5 rather than 2.0 degrees by 2100 will dramatically lower the likely negative consequences of climate change. These include the risks of heat extremes, heavy precipitation, droughts, sea-level rise, biodiversity losses, and corresponding impacts on health, livelihoods, food security, water supply, and human security. The IPCC estimates that to achieve the 1.5 degrees maximum global mean temperature increase target as of 2100, global net CO_2 emissions will have to fall by about 45% as of 2030 and reach net zero emissions by 2050.

Working from these specific stabilization goals set out by the IPCC, it is clear that heterodox economists should be focused on the task of advancing a set of policies that are consistent with achieving these targets as a first priority while concurrently upholding their long-standing egalitarian commitments for full employment and broadly shared well-being. Malcolm Sawyer has been a leader in developing the foundations of this reconstituted project for heterodox macroeconomists. One critical contribution here was his directorship of the major European Commission-sponsored project, Financialization, Economy, Society, and Sustainable Development (FESSUD) that ran from 2011–2016. Among the many innovations that Sawyer introduced in his leadership at FESSUD was to directly integrate 25 research papers studies on various aspects of sustainable development into a broader stream of research on macroeconomic policy and finance.

One of the papers in that stream of FESSUD work was Sawyer's own working paper with his co-author Giuseppe Fontana. This paper was subsequently was published in 2014 under the title 'The Macroeconomics and Financial System Requirements for a Sustainable Future.' The paper develops a macroeconomic analysis along broadly Post Keynesian and Kaleckian lines. In such models, growth is understood to be demand-driven, and the unemployment rate is, in turn, determined by the level of aggregate demand. Within this framework, the Fontana/Sawyer paper shows how to bring aggregate demand in line with requirements of ecological sustainability, while still operating a full employment economy. They emphasize two considerations. The first is to establish a more egalitarian income distribution, through which employment levels will be increased for a given level of overall output. This results because more people are employed at somewhat lower average compensation levels, especially after limiting the extremely high levels of compensation at the top of the pay distribution. The second is to establish financial regulations capable of channeling the available supply of credit, such that a relatively large share of credit is allocated into environmentally beneficial investment projects.

This paper by Fontana and Sawyer, as well as companion works (e.g. Fontana and Sawyer 2016) provide a useful framework for developing macroeconomic models that reconcile the twin imperatives of full employment and ecological sustainability. But, to date, Sawyer and his co-authors have not applied their general approach and conclusions to the specific challenges faced by climate change.

As a starting point for such climate-specific macro analyses, we must begin with recognizing the single most critical requirement for advancing a viable climate stabilization growth path. That is to cut the consumption of oil, coal and natural gas dramatically and without delay. The reason this is the single most critical matter at hand is because producing and consuming energy from these fossil fuel sources is responsible for generating about 70% of the greenhouse gas emissions that are causing climate change. Carbon dioxide (CO_2) emissions from burning coal, oil and natural gas alone produce about 66% of all greenhouse gas emissions, while another 2% is caused mainly by methane leakages during extraction.

But in conjunction with driving the fossil fuel industry down to a zero or near-zero level of activity, other categories of economic activity will need to grow massively as an equally critical component of the climate stabilization project. These are the investments associated with the production and distribution of clean energy. More specifically, clean energy investments are those that: 1) dramatically improve energy efficiency standards in the stock of buildings, private vehicles and public transportation systems, and industrial production processes; and 2) equally dramatically expand the supply of clean renewable energy sources – primarily solar and wind power – available at competitive prices to all sectors and in all regions of the globe. Building a clean energy infrastructure with these specifications will be the foundation for enabling economies to avoid facing significant energy input supply constraints as the climate stabilization project proceeds.

Considered in this way, the climate stabilization project needs to be understood as falling primarily within the realm of industrial policy much more than within macro-economic policy, which, as recognized by Fontana and Sawyer, is primarily concerned with aggregate, as opposed to industrial-level, activity. That is, the climate stabilization project is, first and foremost, about how to massively expand the *clean energy industry* while closing down the global *fossil fuel industry*.

With this recognition in mind, it happens that many earlier contributions by Sawyer that focused on the analytics of industrial policy provide valuable guides for our present concerns on climate stabilization. This includes papers in 1992, 1994 and 2000. The 1994 paper, 'Industrial Strategy and Employment in Europe,' is especially useful because of its focus on the channels through which industrial policies and macro policies for full employment can interact most effectively.

The remainder of the paper is structured as follows. In Section 2, I briefly review Sawyer's contributions in the area of industrial policy, especially on the relationship that he describes between industrial and macro policies. Section 3 then lays out the basics of what I consider to be a climate stabilization project capable of achieving the goal of net zero CO_2 emissions by 2050. Section 4 asks 'what is clean energy,' to clarify what should be the specific sub-industry targets within the overall industrial policy framework. Section 5 presents the key industrial policy tools that will be needed to successfully advance the clean energy investment project. Sections 6 and 7 focus on the financing requirements to support the clean energy investments. Section 8 offers some brief concluding remarks, considering again Malcolm Sawyer's contributions in the areas of industrial policy and a Post-Keynesian approach to macroeconomics.

2. Sawyer on Industrial and Macro Policies

Sawyer begins his 1994 paper by establishing a distinction between the terms 'industrial strategy' and "industrial policy, in the following way:

> Industrial policy is often used to mean little more than competition policy (e.g. policies on monopoly, mergers, and restrictive practices), whereas industrial strategy involves a wide range of policy instruments and objectives. Indeed the key element of an industrial strategy is intentions rather than instruments: for example, low interest rates used to promote investment and industrial development would count as an instrument of industrial strategy. The essence of an industrial strategy is that the government seeks to pursue a range of economic and industrial policies which are consistent with the overall strategy ....The general notion of an industrial strategy can be linked with that of a developmental state – a state organized and concerned to promote economic and industrial development (1994, p. 177).

In his subsequent work on the topic, such as his 2000 paper, Sawyer does not emphasize this distinction in terminology per se, and indeed reverts back to using the term 'industrial policy' in referring to the developmental state role of government in industry. But, of course, the main point here is not which particular terminology Sawyer chooses to emphasize at any given time, but rather, conceptually, how he defines the role of government in establishing an effective overall industrial policy approach.

In his 1994 paper, Sawyer goes on to emphasize the complimentary relationship between industrial strategy and the macroeconomic policy aim of achieving and maintaining full employment. His first point is that industrial strategy is critical for easing supply-side constraints that will arise in conjunction with rising capacity utilization. But he then also notes that:

> A high level of demand is required not only for reasons of employment creation but also to underpin confidence necessary for the industrial strategy to operate. Investment would be enhanced by high levels of capacity utilization, and the changes in employment patterns arising from industrial development are less painful for those involved when demand is high (1994, 178)

Sawyer's observation here is important. This is because, just as Sawyer and Fontana demonstrate in their 2016 paper, a growing economy creates a positive environment to support private investment. This starts with rising overall incomes, which in turn yield a relatively high and rising level of aggregate demand, including, considering our particular concerns, demand for energy resources.

Of course, we understand that relatively high economic growth rates by no means themselves guarantee anything in terms of stimulating private investment. Among other factors, it has long been recognized that under neoliberal hegemony for roughly the past 40 years, the gains from economic growth in virtually all countries have persistently favored the rich, and financial market investors in particular. This corresponding rise in inequality has encouraged financial speculation, by putting relatively more money in the hands of high-income people with the capacity to engage in financial asset trading. It has also contributed to diminishing aggregate demand, by delivering a smaller share the overall economic pie to flow to the non-wealthy, who have much higher marginal propensities to consume than the wealthy. Nevertheless, the prospects for reversing inequality and advancing positive program for productive investments will be far less daunting when the overall economy is growing than if the economy experiences intensifying struggles over relative shares within a slowly growing or even contracting level of aggregate output.

3. Some Basics of a Global Climate Stabilization

The core feature of the climate stabilization project needs to be a worldwide program to invest about 2.5% of global GDP every year to raise energy efficiency standards and expand clean renewable energy supply. Through this investment program, as I describe in some detail below, it becomes realistic to drive down global CO_2 emissions to zero by 2050 while also supporting rising living standards and expanding job opportunities. It is critical to recognize that, within this framework, a more rapid economic growth rate will also accelerate the rate at which clean energy supplants fossil fuels, since higher levels of

GDP will correspondingly mean that a higher level of investment funds are channeled into clean energy projects.

The consumption of oil, coal, and natural gas will then be falling to zero over this same 30-year time period. The rate of decline can begin at a relatively modest 3.5% in the initial years of the transition program, but will then increase every year in percentage terms, as the base level of fossil fuel supply contracts to zero as of 2050. Of course, both the privately owned fossil fuel companies, such as Exxon-Mobil and Chevron, and equally, the publicly owned companies such as Saudi Aramco and Gazprom in Russia, have massive self-interests at stake in preventing reductions in fossil fuel consumption as well as enormous political power. These powerful vested interests will simply have to be defeated.

The investments aimed at dramatically raising energy efficiency standards and expanding the supply of clean renewable energy will also generate tens of millions of new jobs in all regions of the world. This is because, in general, building a green economy entails more labor intensive activities than maintaining the world's current fossil fuel-based energy infrastructure. At the same time, unavoidably, workers and communities whose livelihoods depend on the fossil fuel industry will lose out in the clean energy transition. Unless strong policies are advanced to support these workers, they will face layoffs, falling incomes, and declining public-sector budgets to support schools, health clinics, and public safety. It follows that the global clean energy investment project must also commit to providing generous transitional support for workers and communities tied to the fossil fuel industry. These labor market considerations are critical features of the global clean energy investment project. But, for reasons of space, I will not consider them further here.[1]

4. What is Clean Energy?

There are large differences in the emissions levels resulting through burning oil, coal, and natural gas, respectively, with natural gas generating about 40% fewer emissions for a given amount of energy produced than coal and 15% less than oil. It is therefore widely argued that natural gas can be a 'bridge fuel' to a clean energy future, through switching from coal to natural gas to produce electricity. Such claims do not withstand scrutiny. At best, an implausibly large 50% global fuel switch to natural gas would reduce emissions by only 8%. But even this calculation does not take account of the leakage of methane gas into the atmosphere that results through extracting natural gas through fracking. Recent research finds that when more than about 5% of the gas extracted leaks into the atmosphere through fracking, the impact eliminates any environmental benefit from burning natural gas relative to coal. Various studies have reported a wide range of estimates as to what leakage rates have actually been in the United States, as fracking operations have grown rapidly. A recent survey paper puts that range as between 0.18% and 11.7% for different specific sites in North Dakota, Utah, Colorado, Louisiana, Texas, Arkansas, and Pennsylvania. It would be reasonable to assume that if fracking expands on a large scale in regions outside the U.S., it is likely that leakage rates will fall closer to the higher-end figures of 12%, at least until serious controls can be established. This then would diminish, if not eliminate altogether, any emission-reduction benefits from a coal-to-natural gas fuel switch.[2]

Some analysts also consider 'clean energy' to include nuclear power and carbon capture and sequestration (CCS) technologies. Nuclear power does generate electricity without producing CO_2 emissions. But it also creates major environmental and public safety concerns, which have only intensified since the March 2011 meltdown at the Fukushima Daiichi power plant in Japan. Similarly, CCS presents hazards. These technologies aim to capture emitted carbon and transport it, usually through pipelines, to subsurface geological formations, where it would be stored permanently. But such technologies have not been proven at a commercial scale. The dangers of carbon leakages from flawed transportation and storage systems will only increase to the extent that CCS technologies are commercialized and operating under an incentive structure in which maintaining safety standards will reduce profits. As such, the most cautious clean energy transition program requires investments in technologies that are well understood, already operating at large-scale, and, without question, safe.

Thus, the first critical project within the global green growth project is to dramatically raise energy efficiency levels. Energy efficiency entails using less energy to achieve the same, or even higher, levels of energy services from the adoption of improved technologies and practices. Examples include insulating buildings much more effectively to stabilize indoor temperatures; driving more fuel-efficient cars or, better yet, relying increasingly on well-functioning public transportation systems; and reducing the amount of energy that is wasted both through generating and transmitting electricity and through operating industrial machinery.

Expanding energy efficiency investments support rising living standards because raising energy efficiency standards, by definition, saves money for energy consumers. A major 2010 study by the U.S. Academy of Sciences found, for the U.S. economy, that 'energy efficient technologies … exist today, or are expected to be developed in the normal course of business, that could potentially save 30% of the energy used in the U.S. economy while also saving money.' Similarly, a McKinsey and Company study focused on developing countries found that, using existing technologies only, energy efficiency investments could generate savings in energy costs in the range of 10% of total GDP, for all low- and middle-income countries. In her 2015 book, *Energy Revolution: The Physics and Promise of Efficient Technology*, the Harvard University physicist Mara Prentiss argues, further, that such estimates understate the realistic savings potential of energy efficiency investments.[3]

Raising energy efficiency levels will generate 'rebound effects' – i.e. energy consumption increases resulting from lower energy costs. But such rebound effects are likely to be modest within the current context of a global project focused on reducing CO_2 emissions and stabilizing the climate. Among other factors, energy consumption levels in advanced economies are close to saturation points in the use of home appliances and lighting – i.e. we are not likely to clean dishes much more frequently because we have a more efficient dishwasher. The evidence shows that consumers in advanced economies are likely to heat and cool their homes as well as drive their cars more when they have access to more efficient equipment. But these increased consumption levels are usually modest. Average rebound effects are likely to be significantly larger in developing economies. But it is critical that all energy efficiency gains will be accompanied by complimentary policies (as discussed below), including setting a price on carbon emissions to discourage fossil fuel consumption. Most significantly, expanding the supply of clean renewable energy will allow for higher levels of energy consumption without leading to increases in CO_2 emissions. It is important to recognize, finally, that different countries presently operate at widely varying levels of energy efficiency. For example, Germany presently operates at an efficiency level of roughly 50% higher than that of the United States. Brazil is at more than twice the efficiency level of South Korea and nearly three times that of South Africa. There is no evidence that large rebound effects have emerged as a result of these high-efficiency standards in Germany and Brazil relative to those of United States, South Korea, and South Africa.

As for renewable energy, the International Renewable Energy Agency (IRENA) estimated in 2019 that, in all regions of the world, average costs of generating electricity with most clean, renewable energy sources – wind, hydro, geothermal, and low-emissions bioenergy – were roughly at parity with fossil fuels.[4] This is without even factoring in the environmental costs of burning oil, coal and natural gas. Solar energy costs were some-what higher on average. But, according to IRENA, as a global weighted average, solar photovoltaic costs have fallen by over 70% between 2010 and 2017. Average solar photovoltaic costs are likely to also fall to parity with fossil fuels as an electricity source within 5 years. Adnan Z. Amin, Director-General of IRENA summarizes the global cost trajectory as follows: 'By 2020, all mainstream renewable power generation technologies can be expected to provide average costs at the lower end of the fossil-fuel cost range. In addition, several solar PV and wind power projects will provide some of the lowest-cost electricity from any source,' p. 5).[5]

5. The Industrial Policy Framework

Depending on specific conditions within each country, industrial policies will be needed to promote technical innovations and, even more broadly, adaptations of existing clean energy technologies. Again depending on circumstances, governments will need to deploy a combination of industrial policy instruments, including research and development support, preferential tax treatment for clean energy investments, and government procurement policies. Clean energy industrial policies will also need to include regulations of both fossil fuel and clean energy prices as well as emission standards.

One major policy intervention that can facilitate the creation of a vibrant clean energy market will be for governments to themselves become both large-scale investors in energy efficiency and purchasers of clean renewable energy. An important comparable historical experience was the development of the internet within the U.S. military, beginning in the 1940 s. In the process of bringing the internet to a commercial scale, the U.S. military provided a guaranteed market for 35 years, which enabled the technology to incubate while private investors gradually developed effective commercialization strategies.[6]

But guaranteeing stable prices with the private-sector purchases of clean renewables is also critical here. Such policies are termed *feed-in tariffs*. Specifically, these are contracts that require utility companies to purchase electricity from private renewable energy generators at prices fixed by long-term contracts. Feed-in tariffs were first implemented in the United States in the 1970 s, and a number of state and local programs is currently operational in the United States today. However, the impact of feed-in tariffs have been much more significant outside of the United States, especially in Germany, Italy, France, Spain, and Canada. A 2009 study by the U.S. Department of Energy found that these policies in Europe 'resulted in quick and substantial renewable energy capacity expansion.'[7]

This basic result has been affirmed through more recent research, including that by Milanes-Montero et al. (2018), which showed how feed-in tariffs 'have had a significant positive influence on the economic profitability' on solar PV companies in Europe. The key factor in the success of these European programs is straightforward: the guaranteed prices for renewable energy were set to adequately reflect the costs of producing the energy along with a profit for the energy provider. This then encouraged private renewable energy investors by providing a stable long-term market environment.[8]

Feed-in tariffs have also had some successes as a policy tool in Africa. The African Development Bank reports as follows:

> Public investment is critical in bridging the gap between public demonstration of new technologies and mature deployment. Feed-in tariffs are a prominent example of such subsidies. These tariffs are a policy mechanism that offers compensation to renewable energy producers, based on the difference between the cost of electricity generation of each technology and the market price of electricity generation that, in the case of RETs, is usually lower. In Kenya, for example, feed-in tariffs led to the high level of uptake of solar PV. As of 2011, 7 African countries used feed-in tariff policies (African Development Bank Group 2013, 143).

Another important set of policies are those that aim to directly reduce fossil fuel consumption. These include carbon caps and carbon taxes. Carbon taxes also fall within this broad set of measures. In principle at least, a carbon cap establishes a firm limit on the allowable level of emissions for major polluting entities, such as utilities. Such measures will also raise the prices of oil, coal, and natural gas by limiting their supply. A carbon tax, on the other hand, will directly raise fossil fuel prices to consumers, and aim to reduce fossil fuel consumption through the resulting price signals. Either approach can be effective as long as the cap is strict enough, or tax rate high enough, to significantly reduce fossil fuel consumption and as long as exemptions are minimal to none. Raising the prices for fossil fuels will also, of course, create increased incentives for both energy efficiency and clean renewable investments, as well as a source of revenue to help finance these investments. We return to this point below.

However, significant problems are also associated with both approaches. Establishing a carbon cap or tax will have negative distributional consequences that will need to be addressed in the policy design. All else equal, increasing the price of fossil fuels would affect lower-income households more than affluent households, since gasoline, home-heating fuels, and electricity absorb a higher share of lower-income households' consumption. An effective solution to this problem is to rebate to lower-income households a significant share of the revenues generated either by the cap or tax to offset the increased costs of fossil-fuel energy.[9]

Renewable energy portfolio standards for utilities and energy efficiency standards for buildings and transportation vehicles are similar in their intent to a carbon cap. That is, renewable portfolio standards set a minimum standard that utilities must achieve in generating electricity from renewable energy sources. Energy efficiency standards for automobiles set minimum miles-per-gallon levels (or comparable measures) that a given auto fleet must achieve to be in compliance with the law. Comparable efficiency standards can also be established for buildings in terms of allowable levels of energy consumption for a given building size.

However, a major problem that has emerged with carbon caps as well as renewable and efficiency standards has been with enforcement. As a major case in point, when these cap programs are combined with a carbon permit option – as in 'cap-and-trade' policies – the enforcement of a hard cap becomes difficult to sustain or even monitor, thereby weakening the impact of the policy.[10]

6. Sources of Aggregate Financing

In a 30-year clean energy investment cycle between 2021–2050 that I have developed in detail elsewhere (see, e.g. Pollin 2020), I assume that large-scale investments do not begin in earnest until 2024, thereby establishing the actual investment cycle to proceed over the 27-year period 2024–2050. This investment project is designed within the framework of the International Energy Agency's global energy consumption model. This model assumes, among other things, that global GDP proceeds at an average annual rate of 3.4% between 2021–2040. I extend that same growth trajectory to 2050.

Within this framework, the global GDP in 2024 will be at 104 USD trillion. Total energy efficiency and renewable energy investments in 2024 will, therefore, need to total to about 2.6 USD trillion. As an average figure over 2024–2050, global clean energy investments will need to be sustained at an average level of 4.5 USD trillion over the 27-yer investment cycle. This equals 2.5% of the average global GDP over 2024–2050.

There are two separate, but interrelated policy considerations with respect to financial requirements for this specific global clean energy investment project. The first is: where will the funding come from to support approximately 2.6 USD trillion in new clean energy investments in 2024 and 4.5 USD trillion as an annual average over 2024–2050. The second issue is: how can these funds be most effectively channeled into the full range of specific projects that will need to advance every year in order to build a net zero global economy? We consider these issues in turn.

On the first question of aggregate financing, in principle, it should not be especially challenging to solve this problem. To begin with, as of 2018, Credit Suisse estimates that the total value of global financial assets was 317 USD trillion.[11] The 2.6 USD trillion that I am proposing to channel into clean energy investments as of 2024 amounts to 0.8% of this total financial asset pool.

Still, it is important to anchor the discussion in specific proposals. Therefore, for purposes of illustration, I propose four large-scale funding sources to support public investments in clean energy. Other approaches could also be viable. These four funding sources are: 1) a carbon tax, in which 75% of revenues are rebated back to the public but 25% are channeled into clean energy investment projects; 2) transferring funds out of military budgets from all countries, but primarily the U.S.; 3) A Green Bond lending program, initiated by both the U.S. Federal Reserve and the European Central Bank; and 4) Eliminating all existing fossil fuel subsidies and channeling 25% of the funds into clean energy investments. Strong cases can be made for each of these funding measures. But each proposal does also have vulnerabilities, including around political feasibility. The most sensible approach is, therefore, to combine the measures into a single package that minimizes their respective weaknesses as standalone measures. Table 1 presents this set of combined proposals in summary form, focused on Year 1 of the investment cycle, i.e. 2024.

Table 1. Major funding sources for global clean energy investments.

Investment Level for 2024 – Year 1 of investment cycle: $2.6 trillion in public and private investments, at 2.5 percent of GDP
Clean Energy Investment Areas
● *Clean Renewable Energy: $2.1 trillion*
● Wind, solar, geothermal, small-scale hydro, low-emissions bioenergy
● *Energy Efficiency: $500 billion*
● Buildings, transportation, industrial equipment, grid and battery storage upgrades
Public Sources of Funds: $1.3 trillion
● *Carbon Tax Revenues: $160 billion*
● *25% of revenues from tax; 75% returned to consumers as a rebate*
● *Transfers from Military Budgets: $90 billion*
● *5% of global military spending*
● *Green Bond Purchases by Federal Reserve and European Central Bank: $200 billion*
● *1.6% of Federal Reserve Wall Street bailout support during financial crisis*
● *Transfers of 25% of Fossil Fuel Subsidies: $750 billion*
● *Total fossil fuel subsidies = $3 trillion*
● *75% of funds for lower clean energy prices or direct income transfers for lower-income households*
Private Sources of Funds: $1.3 trillion
● **Policies for Incentivizing Private Investors**
● *Government procurement*
● *Regulations*
● *Carbon caps and taxes*
● *Renewable energy portfolio standards for utilities*
● *Energy efficiency standards for buildings and transportation vehicles*
● *Investment Subsidies*
● *Feed-in tariffs*
● *Low-cost financing through development banks and green banks*

6.1. Carbon tax with rebates

As noted above, carbon taxes have the merit of impacting climate policy through two channels – they raise fossil fuel prices and thereby discourage consumption while also generating a new source of government revenue. At least part of the carbon tax revenue can then be channeled into supporting the clean energy investment project. But the carbon tax will hit low- and middle-income people disproportionately, since they spend a larger fraction of their income on electricity, transportation and home-heating fuel. An equal-shares rebate, as proposed by Boyce (2019), is the simplest way to ensure that the full impact of the tax will be equalizing across all population cohorts.

Consider, therefore, the following tax-and-rebate program. Focusing, again, on 2024, the first year of the full-scale investment program, we begin with a tax at a low rate of 20 USD per ton of carbon. Given current global CO_2 emissions levels, that would generate about 625 USD billion in revenue. If we use only 25% of this revenue to finance clean energy investments, that amounts to roughly 160 USD billion for investment projects. The 75% of the total revenue that is rebated to the public in equal shares would then amount to 465 USD billion. This amounts to about 60 USD for every person on the planet, or 240 USD for a family of four.[12]

6.2. Transferring funds out of military budgets

Global military spending in 2018 was at 1.8 USD trillion.[13] The U.S. military budget, at about 700 USD billion, accounted for nearly 40% of the global total. There are solid logical and ethical grounds for transferring substantial shares of each country's total military budget to supporting climate stabilization, if we take at face value the idea that military spending is fundamentally aimed at achieving greater security for the citizens of each country. But to remain within the realm of political feasibility, let us assume that 5% of global military spending will transfer into supporting climate security. That would amount to 90 USD billion.

6.3. Green bond funding by Federal Reserve and European Central Bank

It was demonstrated during the 2007–2009 global financial crisis and subsequent Great Recession that the Federal Reserve is able to supply basically unlimited bailout funds to private financial markets during crises. The extensive 2017 study, *The Costs of The Crisis*, by Better Markets concludes that the Federal Reserve committed approximately 12.2 USD trillion to stop the crash of the financial system, stabilize the economy and try to spur economic growth. I would propose 100 USD billion in Green Bond financing supplied by the Fed. This would amount to a minuscule 0.8% of the Fed's 2007–2009 bailout operations during the crisis. The Fed's funding support could be injected into the global economy through straightforward channels. That is, various public entities, such as the World Bank, could issue long-term zero interest rate Green Bonds. The Fed would purchase these bonds. The various public entities issuing these bonds would then have the funds to pursue the full range of projects that will fall under the rubric of the global clean energy project.

   This framework has not yet been introduced into policy discussions at the Federal Reserve. But they are becoming a central area of focus at the European Central Bank. Thus, the *Financial Times* reported on 12/2/19 that the recently installed ECB President Christine Lagarde is moving quickly on the matter. The *Financial Times* reports that:

> Christine Lagarde . . . is pushing to include climate change considerations in a review the central bank is due to hold into the way it conducts monetary policy. Until now, the expectation was for a review into purely monetary matters, such as whether the inflation target should be revised. An explicit focus on climate change policy would be a huge move. Because the central bank is by far the biggest influence on financial conditions in the market, it can make a significant difference to investment decisions that determine how Europe's climate transition goes.[14]

The *Financial Times* article makes clear that the specific channels through which the ECB would intervene to support clean energy financing will require substantial fleshing out. The type of approach I have sketched for a Federal Reserve intervention would seem like a relatively straightforward and modest form of intervention. I, therefore, propose that the ECB undertakes Green Bond purchases at the same level of the Federal Reserve, i.e. at 100 USD billion as of 2024, and growing over time to support clean energy investments continuing at an average rate of 2.5% of global GDP per year.

6.4. Eliminating fossil fuel subsidies and channeling 25% of funds to clean energy investments

One recent estimate of direct fossil fuel subsidies to consumers – measured as the difference between supply and consumer prices to purchase fossil fuel energy – is about 3 USD trillion globally as of 2015, or about 0.4% of global GDP.[15] Channeling these funds, in full, into supporting public clean energy investments would therefore more than pay for the 2.6 USD trillion estimate for total clean energy investments as of 2024. This 3 USD trillion would also represent more than double the amount necessary to cover a global public investment level of 1.3 USD trillion. However, such fossil fuel subsidies are largely used as a form of general support for all energy consumers. Lower- and middle-income households are therefore major beneficiaries of these subsidies, along with, of course, the fossil fuel energy suppliers. Therefore, in terms of the global income distribution, eliminating these subsidies altogether would likely have a significant regressive impact, comparable to establishing a carbon tax without an accompanying rebate program. As such, to continue to provide support for lower-income households, most of the funds that are now being channeled to these households through fossil fuel subsidies should be redirected into either supporting lower consumer prices for clean energy or to provide direct income transfers for lower-income households.

Given that we will have raised 440 USD billion from the carbon tax, military spending transfers and central bank Green Bond programs, we could then assume that 25% of the 3 USD trillion received as fossil fuel subsidies be transferred into the clean energy investment fund. That would amount to 750 USD billion. With these funds, we will have reached the total of 1.3 USD trillion in public investment funds necessary to attain the total of public and private investment spending of 2.6 USD trillion as of 2024.

7. Delivering Finance at the Project Level

Both general purpose development banks as well as special-purpose green development banks are already significantly engaged in financing clean energy investments. It will be crucial to build from these efforts to achieve the necessary level of financing for clean energy investments.

7.1. Germany's KfW Bank

The case of Germany is instructive, since it has been the most successful large-advanced economy to date in developing its clean energy economy. The publicly owned development bank in Germany, KfW, has been critical to this success. Griffith-Jones (2016) considers KfW's impact on Germany's overall green transformation, including renewable energy as well as energy efficiency investments. She finds that KfW has underwritten roughly one-third of all financing for green investments in Germany. KfW has thus been instrumental in moving policy ideas into effective investment projects, with respect to both energy efficiency and clean renewables. KfW has also been highly active in financing green investment projects elsewhere in Europe and in developing countries. As Griffith-Jones writes:

> KfW plays a key role, domestically and internationally, in supporting energy revolution, through funding major investments in renewable energy and in energy efficiency. In the national German case, this was to a large extent implemented within a clear institutional and policy framework, namely the renewable energy law, through strong policy measures, such as feed in tariffs (FITs) and reverse competitive auctions, which made investment in renewables commercially attractive. A similar modus operandi existed for energy efficiency ... The combination of clear government policies and associated development bank targets has produced very positive results in green infrastructure in Germany, which can be replicated in emerging and developing countries (2016, 4).[16]

Griffith-Jones also describes the financing terms offered by KfW in all of their areas of active lending. These include long-term loans and below-market interest rates, 100% disbursement rates, up to 3-years holidays in making repayments, and repayment bonuses of up to 17.5%.

7.2. Green banks

Special purpose green development banks have also become increasingly active in recent years. A OECD study defines a green investment bank as 'A publicly capitalized entity established specifically to facilitate private investment into domestic low-carbon and climate-resistant infrastructure and other green sectors such as water and waste management,' (p. 15). These special purpose banks have been established at the national level in Australia, Japan, Malaysia, Switzerland, and the UK. Within the US, the states of California, Connecticut, Hawaii, New Jersey, New York and Rhode Island have created green banks. The OECD study describes the banks as having 'diverse rationales and goals, including meeting ambitious emissions targets, mobilizing private capital, lowering the cost of capital, lowering energy costs, developing green technology markets, supporting local community development and creating jobs,' (p. 15). The OECD study does not provide systematic evidence as to the scale at which these institutions are currently providing investment financing. But considering other references, it is reasonable to assume that, in general, their scale of operations is much smaller than KfW.[17] This raises the question as to whether the necessary level of financing can be achieved without the full backing of large-scale national or regional entities, such as the equivalent of KfW.

7.3. Emerging trends in developing countries

Within developing economies there has been a general movement in the aftermath of the 2007–2009 global financial crisis away from the predominant neoliberal financial market policy framework that prevailed prior to the crisis. This trend has included the formation or expansion of development banks. For example, Grabel (2018) describes the emergence of the Development Bank of Latin America and the New Development Bank as potentially significant new sources of subsidized long-term financing for developing economies, including in the area of green energy investments.

Recent studies by the World Bank (Hussain 2013) and African Development Bank Group (2013) examine specific financial models for advancing green investments in developing countries. Both studies consider financing arrangements through which concessionary public financing can be mobilized to encourage, as opposed to crowd out, private investments, thereby creating viable public/private partnerships with clean energy investment projects. The World Bank study in particular, which focuses on renewable energy investments, emphasizes that the long-term funding for these investments has been limited by the range of risks private investors face while working with still relatively unfamiliar technologies. These risks include uncertainty over the reliability of the technology within any given project and shifts in the relevant regulatory environment. The World Bank proposes a series of financing techniques for reducing these risks for private investors. Yet the overall point remains that the public financing interventions – whether they are implemented through formal development banks or otherwise –

will need to absorb a disproportionate share of these risks in order for the financing levels to reach scale rapidly enough.

With respect to financing clean energy investments in developing countries, in particular, it is also critical that the benefits of these investments be shared fully by the society's least-advantaged groups. Spratt, Griffith-Jones, and Ocampo (2013) emphasize this consideration in their 2013 study *Mobilizing Investment for Inclusive Green Growth in Low-Income Countries*. This would mean, as important examples, expanding access to electricity and providing clean energy for electricity and other needs at affordable prices.[18] To accomplish these ends, Spratt et al. emphasize that it is not realistic to expect clean energy investments to consistently generate profits for private businesses at rates comparable to mature investment areas, including fossil fuel energy. The requirement that the financing terms for clean energy investments be affordable for borrowers – that is, not always yielding high returns for lenders – reinforces the centrality of public investment banks with clear social criteria guiding their financing strategies.

8. Conclusion

In light of the most recent assessments within mainstream climate science, as reflected in the 2018 IPCC report, it is reasonable at this historic juncture to invoke Margaret Thatcher's famous dictum, i.e. 'there is no alternative' to developing a global climate stabilization project that is realistically capable of reaching the IPCC's stated goal of CO_2 emissions at net zero by 2050. This paper outlines an approach for achieving that goal. The paper tries to demonstrate that, most fundamentally, this climate stabilization project needs to focus, first and foremost, on industrial policies for advancing a global clean energy investment project. Within a framework of effective industrial policies, I try to argue that it is realistic to allow that, on a global basis, clean energy investments can be sustained at a rate of about 2.5 percent of global GDP over the period 2024–2050. With clean energy investments advancing at that rate over this 27-year investment cycle, it then becomes realistic, in turn, to allow that the global economy can operate with zero net CO_2 emissions by 2050.

Any such global-scale climate stabilization project will further need to operate effectively in conjunction with macroeconomic policies capable of maintaining a supportive environment for both public and private clean energy investments. As Malcolm Sawyer and Giuseppe Fontana illustrate in their pioneering 2016 paper that integrates ecological concerns within a Post-Keynesian macro framework, a slow-growing economy will present major difficulties for sustaining a high-level of aggregate investment activity. Investments in the specific area of clean energy investments will be subject to the same challenges as all other investments within any such slow-growing macro environment.

In addition to his 2016 paper with Fontana, Malcolm Sawyer has made valuable contributions to the economics literature in the area of industrial policy/industrial strategy. The present paper operates within a framework similar to that set out by Sawyer in his 1994 paper that insightfully integrated industrial policy/industrial strategy within a macroeconomic policy program for full employment. Of course, we still have much to learn about how to most effectively integrate industrial and macro policy for achieving a viable climate stabilization path.

Further efforts to deepen our understanding of these matters will continue to benefit from Malcolm Sawyer's contributions in both the areas of industrial and macro policy.

Notes

1. See e.g. Pollin et al. (2015) and Pollin (2015).
   See e.g. Pollin et al. (2015) and Pollin (2015).
2. Alvarez et al. (2012), Romm (2014), Howarth (2015), and Peischl et al. (2016).
   Alvarez et al. (2012), Romm (2014), Howarth (2015), and Peischl et al. (2016).
3. NAS (National Academy of Sciences), NAE (National Academy of Engineering), and NRC (National Research Council) (2010), McKinsey & Company (2010), and Prentiss (2015).
   NAS (National Academy of Sciences), NAE (National Academy of Engineering), and NRC (National Research Council) (2010), McKinsey & Company (2010), and Prentiss (2015).
4. http://about.bnef.com/services/renewable-energy/.
   http://about.bnef.com/services/renewable-energy/.
5. The figures I am citing from the 2018 IRENA study are for 'Levelized Costs of Electricity,' (LCOE). LCOE costs include: levelized capital costs; fixed operations and maintenance; variable operations and maintenance, including fuel costs; transmission; and the capacity factor for the equipment in use. IRENA reports LCOE figures on a national, regional, and global basis.
   The figures I am citing from the 2018 IRENA study are for 'Levelized Costs of Electricity,' (LCOE). LCOE costs include: levelized capital costs; fixed operations and maintenance; variable operations and maintenance, including fuel costs; transmission; and the capacity factor for the equipment in use. IRENA reports LCOE figures on a national, regional, and global basis.
6. See Ruttan (2006).
   See Ruttan (2006).
7. Cory, Couture, and Kreycik (2009, 2).
   Cory, Couture, and Kreycik (2009, 2).
8. Cointe and Nadal (2018) emphasize this point, and contrast it with the official EU aim of liberalizing renewable energy markets.
   Cointe and Nadal (2018) emphasize this point, and contrast it with the official EU aim of liberalizing renewable energy markets.
9. See Boyce (2019) for an effective solution to the distributional problem, via what he terms 'carbon dividends.' Azad and Chakraborty (2019) expand on the idea of an egalitarian carbon dividend program to the global economy.
   See Boyce (2019) for an effective solution to the distributional problem, via what he terms 'carbon dividends.' Azad and Chakraborty (2019) expand on the idea of an egalitarian carbon dividend program to the global economy.
10. See, e.g., Teeter and Sandberg (2017). There is also the problem of the caps, or renewable portfolio standards, being established in law but then ignored in policy implementation. This has been the experience, for example, in New York State. See Pollin, Garrett-Peltier, and Wicks-Lim (2017), pp. 79–80.
    See, e.g., Teeter and Sandberg (2017). There is also the problem of the caps, or renewable portfolio standards, being established in law but then ignored in policy implementation. This has been the experience, for example, in New York State. See Pollin, Garrett-Peltier, and Wicks-Lim (2017), pp. 79–80.
11. file:///C:/Users/RPollin/Downloads/global-wealth-report-2018-en.pdf.
    file:///C:/Users/RPollin/Downloads/global-wealth-report-2018-en.pdf.
12. Azad and Chakraborty (2019) develop a more complex rebate structure, that rewards residents of countries according to the emissions levels of each country.
    Azad and Chakraborty (2019) develop a more complex rebate structure, that rewards residents of countries according to the emissions levels of each country.
13. https://www.sipri.org/media/press-release/2019/world-military-expenditure-grows-18-trillion-2018.
    https://www.sipri.org/media/press-release/2019/world-military-expenditure-grows-18-trillion-2018.

14. https://www.ft.com/content/89f5f412-12bc-11ea-a225-db2f231cfeae.
https://www.ft.com/content/89f5f412-12bc-11ea-a225-db2f231cfeae.

15. Coady et al. (2017). This study distinguishes direct fossil fuel subsidies – what it terms 'pre-tax' subsidies – and 'post-tax' subsidies. They define post-tax subsidies as including global warming damages, air pollution damages, and vehicle externalities, including congestion, accidents and road damage. They estimate post-tax subsidies as amounting to roughly 6% of global GDP. These are valuable calculations. But for the purposes of this discussion on financing, the standard, and much more narrowly defined measure of pre-tax subsidies are more directly relevant.

Coady et al. (2017). This study distinguishes direct fossil fuel subsidies – what it terms 'pre-tax' subsidies – and 'post-tax' subsidies. They define post-tax subsidies as including global warming damages, air pollution damages, and vehicle externalities, including congestion, accidents and road damage. They estimate post-tax subsidies as amounting to roughly 6% of global GDP. These are valuable calculations. But for the purposes of this discussion on financing, the standard, and much more narrowly defined measure of pre-tax subsidies are more directly relevant.

16. Griffith-Jones's conclusions are fully in line with those of other researchers. For example, the overview of the IEA's 2013 *Energy Efficiency Market Report* concluded that "Germany is a world leader in energy efficiency. Germany's state-owned development bank, KfW, plays a crucial role by providing loans and subsidies for investment in energy efficiency measures in buildings and industry, which have leveraged significant private funds (IEA 2013, p. 149).

Griffith-Jones's conclusions are fully in line with those of other researchers. For example, the overview of the IEA's 2013 *Energy Efficiency Market Report* concluded that "Germany is a world leader in energy efficiency. Germany's state-owned development bank, KfW, plays a crucial role by providing loans and subsidies for investment in energy efficiency measures in buildings and industry, which have leveraged significant private funds (IEA 2013, p. 149).

17. See, for example, Pollin, Garrett-Peltier, and Wicks-Lim (2017) for a discussion of the New York State green bank and related public financing initiatives within New York State. See also Pollin et al. (2014) for a discussion of green banks within the U.S. economy, and as one element within a broader framework of measures to support clean energy investments.

See, for example, Pollin, Garrett-Peltier, and Wicks-Lim (2017) for a discussion of the New York State green bank and related public financing initiatives within New York State. See also Pollin et al. (2014) for a discussion of green banks within the U.S. economy, and as one element within a broader framework of measures to support clean energy investments.

18. As one specific policy proposal, Azad and Chakraborty (2018) develop a program for rapidly advancing the expansion of renewable energy supply in India. The proposal includes a carbon tax, with the revenues from the tax being channeled into clean renewable energy investments that will then supply free electricity to low-income communities, many of which still have no access to electricity.

As one specific policy proposal, Azad and Chakraborty (2018) develop a program for rapidly advancing the expansion of renewable energy supply in India. The proposal includes a carbon tax, with the revenues from the tax being channeled into clean renewable energy investments that will then supply free electricity to low-income communities, many of which still have no access to electricity.

Disclosure statement

No potential conflict of interest was reported by the author(s).

References

African Development Bank Group. 2013. *African Development Report 2012: Towards Green Growth in Africa*. Tunis: African Development Bank Group. https://www.afdb.org/fileadmin/uploads/afdb/Documents/Publications/African_Development_Report_2012.pdf

Alvarez, R. A., S. W. Pacala, J. J. Winebrake, W. L. Chameides, and S. P. Hamburg. 2012. "Greater Focus Needed on Methane Leakage from Natural Gas Infrastructure." *Proceedings of the National Academy of Sciences of the United States of America* 109 (17): 6435–6440. doi:10.1073/pnas.1202407109.

Azad, R., and S. Chakraborty. 2019. "Balancing the Climate Injustice: A Proposal for A Differential Global Carbon Tax." In *Handbook of Green Economics*, edited by S. Acar and E. Yeldan, 117–134. London: Elsevier Press.

Azad, R., and S. Chakraborty (2018) "Green Growth and the Right to Energy in India." PERI Working Paper # 477. Political Economy Research Institute. https://www.peri.umass.edu/publication/item/1137-green-growth-and-the-right-to-energy-in-india

Boyce, J. 2019. *The Case for Carbon Dividends*. Cambridge, UK: Polity Press.

Coady, D., I. Party, L. Sears, and B. Shang. 2017. "How Large are Global Fossil Fuel Subsidies?" *World Development* 91: 11–27. doi:10.1016/j.worlddev.2016.10.004.

Cointe, B., and A. Nadal. 2018. *Feed-In Tariffs in the European Union*. Palgrave Macmillan.

Cory, K., T. Couture, and C. Kreycik. 2009. *Feed-in Tariff Policy: Design, Implementation, and RPS Policy Interactions*. Golden, CO: National Renewable Energy Laboratory. http://www.nrel.gov/docs/fy09osti/45549.pdf

Fontana, G., and M. Sawyer. 2016. "Towards Post-Keynesian Ecological Macroeconomics." *Ecological Economics* 21: 186–195. doi:10.1016/j.ecolecon.2015.03.017.

Grabel, I. 2018. *When Things Don't Fall Apart: Global Financial Governance and Development Finance in an Age of Productive Incoherence*. Cambridge, MA: MIT Press. doi:10.7551/mitpress/11073.001.0001.

Griffith-Jones, S. 2016. "National Development Banks and Sustainable Infrastructure: The Case of KfW." Global Economic Governance Initiative (GEGI), GEGI Working Paper #6, July. Boston University.

Howarth, R. 2015. "Methane Emissions and Climatic Warming Risk from Hydraulic Fracturing and Shale Gas Development: Implications for Policy." *EECT*, 45. https://www.dovepress.com/methane-emissions-and-climatic-warming-risk-from-hydraulic-fracturing–peer-reviewed-article-EECT

Hussain, M. Z. 2013. *Financing Renewable Energy: Options for Developing Financing Instruments Using Public Funds*. Washington DC: World Bank and Climate Investment Funds. http://documents.worldbank.org/curated/en/196071468331818432/pdf/765560WP0Finan00Box374373B00PUBLIC0.pdf

IEA (International Energy Agency). 2013. *Energy Efficiency Market Report 2013: Market Trends and Medium Term Prospects*. Paris: OECD/IEA

IPCC (Intergovernmental Panel on Climate Change). 2018. *Global Warming of 1.5°C*, https://www.ipcc.ch/sr15/

IRENA (International Renewable Energy Agency). 2019. *"Renewable Power Generation Costs in 2018."* https://www.irena.org/publications/2019/May/Renewable-power-generation-costs-in-2018

McKinsey & Company. 2010. *Energy Efficiency: A Compelling Global Resource*. New York, NY.

Milanes-Montero, P., A. -F. Alberto, and P. -C. Esteban. (2018) "Assessment of the Influence of Feed-In Tariffs on the Profitability of European Photovoltaic Companies". Sustainability, September.

NAS (National Academy of Sciences), NAE (National Academy of Engineering), and NRC (National Research Council). 2010. *Real Prospects for Energy Efficiency in the United States*. Washington, DC: National Academies Press.

OECD (Organization of Economic Cooperation and Development) 2017. *"Green Investment Banks: Innovative Public Financial Institutions Scaling-Up Private, Low-Carbon Investment Policy Reform."* OECD Environmental Policy Paper. https://www.oecd-ilibrary.org/docserver/e3c2526c-en.pdf?

expires=1582944313&id=id&accname=guest&checksum=70C2AA0A4FE27F94592C9C
BFAC259FF3

Peischl, J., A. Karion, C. Sweeney, E. A. Kort, M. L. Smith, A. R. Brandt, T. Yeskoo, et al. 2016. "Quantifying Atmospheric Methane Emissions from Oil and Natural Gas Production in the Bakken Shale Region of North Dakota." *Journal of Geophysical Research: Atmospheres* 121 (10): 6101–6111. https://agupubs.onlinelibrary.wiley.com/doi/full/10.1002/2015JD024631

Pierrehumbert, R. 2019. "There Is No Plan B for Dealing with the Climate Crisis." *Bulletin of the Atomic Scientists* 75 (5): 215–221. doi:10.1080/00963402.2019.1654255.

Pollin, R. 2015. *Greening the Global Economy*. Cambridge, MA: MIT Press.

Pollin, R. 2020. "An Industrial Policy Framework to Advance a Global Green New Deal." In *The Oxford Handbook of Industrial Policy*, edited by A. Oqubay. Oxford, UK: Oxford University Press.

Pollin, R., H. Garrett-Peltier, J. Heintz, and B. Hendricks. 2014. *Green Growth: A U.S. Program for Controlling Climate Change and Expanding Job Opportunities*. Washington, DC: Center for American Progress. https://cdn.americanprogress.org/wp-content/uploads/2014/09/PERI.pdf

Pollin, R., H. Garrett-Peltier, J. Heintz, and S. Chakraborty. 2015. *Global Green Growth: Clean Energy Industrial Investments and Expanding Job Opportunities*. United Nations Industrial Development Organization and Global Green Growth Institute. http://gggi.org/wp-content/uploads/2015/06/GGGI-VOL-I_WEB.pdf

Pollin, R., H. Garrett-Peltier, and J. Wicks-Lim. 2017. *Clean Energy Investments for New York State: An Economic Framework for Promoting Climate Stabilizing and Expanding Good Job Opportunities*. Amherst, MA: Political Economy Research Institute. https://www.peri.umass.edu/publication/item/1026-clean-energy-investments-for-new-york-state-an-economic-framework-for-promoting-climate-stabilization-and-expanding-good-job-opportunities

Prentiss, M. 2015. *Energy Revolution: The Physics and the Promise of Efficient Technology*. Cambridge, MA: Harvard University Press.

Romm, J. 2014. "Methane Leaks Wipe Out Any Climate Benefit of Fracking, Satellite Observations Confirm." *Think Progress*, https://thinkprogress.org/methane-leaks-wipe-out-any-climate-benefit-of-fracking-satellite-observations-confirm-2ac26dd30381/;

Ruttan, V. W. 2006. *Is War Necessary for Economic Growth? Military Procurement and Technology Development.New*. New York, NY: Oxford University Press.

Sawyer, M. 1994. "Industrial Strategy and Employment in Europe." In *Unemployment in Europe*, edited by J. Michie and J. Smith, 177–187. London: Harcourt, Brace.

Sawyer, M. 2000. "The Theoretical Analysis of Industrial Policy." In *Industrial Policies after 2000*, edited by W. Elsner and J. Groenewegen, 23–57. London: Kluwer Academic Publishers.

Spratt, S., S. Griffith-Jones, and J. Antonio. 2013. *Mobilising Investment for Inclusive Green Growth in Low-Income Countries*. Berlin: Deutsche Gesellschaft fur Internationale Zusammenarbeit. http://www.stephanygj.net/papers/MobilisingInvestmentforInclusiveGreenGrowth2013.pdf

Teeter, P., and J. Sandberg. 2017. "Constraining or Enabling Green Capability Development? How Policy Uncertainty Affects Organizational Responses to Flexible Environmental Regulations." *British Journal of Management* 28 (4, October): 649–665. doi:10.1111/1467-8551.12188.

Challenges to neo-liberalism in the United States

Samuel Rosenberg

ABSTRACT

With the significant changes in the economy and the society occurring under neo-liberalism as a base, this paper analyzes the extent to which government policies, including federal, state, and local, and labor activism since the Great Recession constitute challenges to neo-liberalism in the United States. It investigates the legacy of neo-liberalism including the ineffective federal governmental response to the COVID-19 economic and health crisis, and the emerging discourse within the Democratic Party calling for a major reorientation of government policy away from neo-liberalism. It concludes by discussing the effects of the neo-liberal agenda on economic well-being and evaluating whether the neo-liberal agenda has been successful in its own terms.

1. Introduction

The United States has often been presented as the paragon of neo-liberalism among major capitalist societies by supporters and opponents of neo-liberalism alike. From the late 1970s, the beginning of neo-liberalism in the United States, until the United States-generated global economic and financial crisis of 2008 that was accurate. However, the Great Recession of 2008 served as the catalyst for more critical assessments of the 'free market' and deregulatory government policies, crucial components of neo-liberalism as practiced in the United States. And the inadequate preparedness of the United States to deal with the health and economic crisis caused by the novel corona virus raises serious questions as to the continued viability of neo-liberalism.

The policies of the Obama administration and of the Trump administration, prior to the onset of COVID-19, represent partial, but different, challenges to neo-liberalism. The Obama administration's macroeconomic policy, including a focus on 'deficit reduction' and its international trade policy emphasizing 'freer trade' were broadly consistent with neo-liberalism. Yet, there was increased direct government intervention in the economy, greater regulation overall, and a labor and social policy designed to somewhat protect workers from the negative effects of the competitive forces of the 'free labor market.'

Unlike the Obama administration and often in direct contradiction to its policies, the Trump administration pursued deregulation domestically. Many of the regulations imposed during the Obama administration were rolled back and the safety net was cut back. All this is consistent with neo-liberalism.

While taxes were sharply reduced on corporations and the wealthy, consistent with a neo-liberal fiscal policy, the ensuing federal budget deficits, quite large for an expansion phase of a business cycle, were not. Concluding that neo-liberal globalization no longer served the interests of the United States and wanting to change the 'rules of the game,' rules created by the United States when it was the dominant power in the world, the Trump administration has continually interfered in international trade flows to attempt to achieve its goals.

The ideological turn against regulation at the federal level has led to an increase in labor regulation in some, but not all, states and some localities, a challenge to neo-liberalism. Furthermore, there has been a growing assertiveness of workers, both unionized and nonunionized alike, against the degradation of labor standards and the overall increase in inequality resulting from neo-liberalism. In addition, Joseph Biden, the nominee of the Democratic Party to run against Donald Trump in the 2020 presidential election, has advanced bold policy proposals representing a strong rebuke of the consequences of neo-liberalism.

The article is organized as follows. Section 2, serving as a background for the sections to follow, discusses the nature of neo-liberalism and important changes in the economy and the society occurring under neo-liberalism. Section 3 analyzes the extent to which federal governmental policies from the Great Recession onward represent challenges to neo-liberalism. Section 4 focuses on state and local labor and social policies since the Great Recession and analyzes the extent to which they, too, represent challenges to neo-liberalism. Section 5 investigates the upsurge in labor activism, a direct challenge to the labor conditions created under neo-liberalism. Section 6 treats the legacy of neo-liberalism including the economic and health crisis, and the changing political discourse in the Democratic Party calling for a major reorientation of governmental policy away from neo-liberalism, and the ineffective ongoing federal governmental response to the consequences of COVID-19. Section 7 concludes by discussing the effects of neo-liberalism on economic well-being and evaluates whether the neo-liberal agenda has been successful in its own terms.[1]

2. The nature of neo-liberalism and specific institutional changes in the United States under neo-liberalism

2.1. The nature of neo-liberalism

The essence of neo-liberalism is that individual decision-making coordinated via free competitive markets, both domestically and internationally, will generally result in optimal economic outcomes. Goods and services will be produced efficiently, and income will be distributed based on the productive contributions of factors of production such as labor and capital. There will be strong economic growth, technological progress will be fostered and the economy will gravitate toward full employment. As such, free competitive markets and market relations in general should play a greater role in regulating economic activity. The government and other institutions, such as unions, should play a diminished role as they primarily interfere with the functioning of free competitive markets.

Ideologically, the government is attacked as the primary cause of the economic problems facing a society. Programmatically, the attack on the government often translates into tax reductions for the wealthy and business, and into weakening social protection policies benefitting workers, in general, and the poor, in particular. Calls for a balanced government budget often accompany tax reductions. Assuming the tax cuts do not generate additional income or even pay for themselves, then a balanced government budget is synonymous with a shrinkage in the size of the government, quite consistent with neo-liberalism. Assuming the economy is at full employment, with a smaller government there is less risk of government expenditures 'crowding out' private sector investment spending.

In addition to calling for a shrinkage in the size of the government, proponents of neo-liberalism push for a reorientation of government policy away from cushioning people from adverse changes in demand and supply to exposing people more fully to the competitive forces of the labor market. For example, reducing the real value of the minimum wage relative to the real value of the average wage advances the neo-liberal agenda by making wage determination, particularly at the lower end of the labor market, more market-oriented. Weakening union power at the bargaining table results in wages and working conditions being more market determined. Reducing transfer payments and imposing work requirements for eligibility for such payments expose people more to market forces. On balance, all of these policies reduce the government presence in the income distribution process and expand the role of the market, quite consistent with the neo-liberal agenda.

2.1.1. Deregulatory government policy

The neo-liberal direction in United States government policy began in the late 1970s during the Carter administration and came to fruition in the 1980s during the Reagan administration. Seeking to reverse the economic decline of the 1970s, represented by stagflation, raise the average rate of profit and improve overall productive efficiency in the economy, the Reagan administration with support from a cohesive business community pursued a policy agenda designed to restructure the economy by freeing up market forces. This market-based conservative strategy was closely aligned with an ongoing corporate restructuring with United States-based corporations trying to become more lower-cost producers to meet increased domestic and international competition. The Reagan administration argued that individual decision-making coordinated via free markets will 'generally result in the most appropriate distribution of our economic resources' (Council of Economic Advisers 1982, 78). Their attack on the federal government led to a weakening of the minimal social protection policies benefitting workers, in general, and the poor, in particular, and to a reduction in tax and regulatory burdens faced by businesses along with lower taxes on those with very high incomes.[2]

The bottom of the wage structure was effectively lowered as the federal minimum wage remained fixed throughout the 8 years of the Reagan administration. While the federal minimum wage had hovered around 50% of average hourly earnings in private nonagricultural industries during the 1950s and 1960s, and 45% of average hourly wages in the 1970s, by 1985 it had declined to about 39% of average hourly wages (Mishel, Bernstein, and Allegretto 2007, pp, 190–192).

The social wage was also reduced. As did the Carter administration, the Reagan administration increased the effective rate of taxation of unemployment insurance (UI). Furthermore, they shortened the effective duration of these benefits. Changes at the state level made it more difficult to qualify for these benefits. This was occurring at the same time that rates of unemployment were at their highest levels, up to that time, since the Great Depression. The fraction of unemployed workers receiving unemployment benefits was lower in 1981–82 than in any other postwar recession. In 1982, only 45% of the jobless were receiving unemployment compensation in contrast to 1975, during the 1975–76 recession, when 78% of the unemployed were receiving unemployment benefits (Burtless 1983, 226).

While the Carter administration had unsuccessfully tried to expand Aid to Families with Dependent Children (AFDC), also known as welfare, initially, by the end of its term, they were calling for cuts in outlays for AFDC. The Reagan administration program changes limited the number of people eligible for assistance and the amounts they received. Welfare caseloads declined. As many as 500,000 families previously receiving AFDC may have lost eligibility as a result of these policy changes (Pierson 1994, 118).

The National Labor Relations Board (NLRB) during the Reagan administration rolled back many of the gains made by the labor movement. The anti-union perspective of the NLRB made it more difficult for union organizing efforts to succeed and for unions to gain their goals at the bargaining table. According to William Gould, who headed the NLRB during the Clinton administration, the NLRB in the early years of the Reagan administration 'was more relentless in reversing precedent than any board even under Eisenhower, Nixon or Ford' (Tasini 1988).

Deregulation of industries where unions were strong increased competition in those industries resulting in more difficult collective bargaining environments. In 1978, during the Carter administration, the U.S. Congress passed the Airline Deregulation Act. The telecommunications industry was deregulated as a result of the divestiture of AT&T demanded by the Reagan administration's Department of Justice. The railroad and trucking industries were also deregulated during this period.

While government policy at the start of the neo-liberal era made conditions more difficult for workers and the poor, changes in tax policy benefitted the wealthy, more than the rest of the population, and business. The Economic Recovery Tax Act of 1981 and the Tax Reform Act of 1986 lowered individual income tax rates for all tax payers with the top marginal tax rate initially falling from 70% to 50% and eventually to 28% for the highest income taxpayers. The maximum capital gains tax rates were also reduced as were corporate income tax rates.

The tax cuts were part of a long-term plan to reshape the role of the federal government. By being deprived of revenue, the government would be forced to reduce non-defense government expenditures, particularly social spending.

In addition to the industries discussed earlier that were deregulated, the Reagan administration continued the deregulation of the finance industry begun under the Carter administration. The U.S. financial system was in the process of being transformed into a completely market-based system, with all the instabilities that implied, another key institutional change under neo-liberalism. In 1980, during the Carter administration, the Depository Institutions and Monetary Control Act was passed setting a time period for the complete elimination of Regulation Q which set limits on the interest rates that banks were allowed to pay on different types of accounts. Regulation Q had been part of the Banking Act of 1933, also known as the Glass-Steagall Act. In 1982, President Reagan signed the Garn-St. Germain Depository Act which deregulated the savings and loan (thrifts) industry.

While the Clinton administration, led by centrist Democrats, did deviate from the Reagan policies in some respects, the emergent neo-liberal framework strongly influenced their policy program. While the minimum wage was increased contrary to the Reagan program, in 1999 toward the end of Clinton's second term in office, the minimum wage was still just 39% of average hourly earnings in private nonagricultural industries, the same level as in 1985 during the middle of the Reagan years (Council of Economic Advisers 2001, 330). The Clinton administration set an enabling environment for states to ease eligibility requirements for UI and raise benefit levels. Nevertheless, less than 38% of unemployed workers received UI in 1999, the same percentage as in 1990 (Wenger, 2001, p. 12). However, the unemployment rate was lower in 1999 than in 1990. The Clinton administration did expand the Earned Income Tax Credit, a wage subsidy for the working poor, by making more families eligible and improving the payments to recipients. The income tax rate was increased on the highest earners with the effective top marginal tax rate rising to 39.6%. Businesses also faced higher corporate tax rates.

On balance, however, the Clinton administration allowed its macroeconomic policy to be constrained by the large federal budget deficits of the previous decade. The Clinton administration carried out the Reagan attack on AFDC to its conclusion. In 1996, with the passage of the Personal Responsibility and Work Opportunity Act, the 'entitlement' to welfare was ended and a strong work requirement was imposed. Clinton's labor policy was the least pro-union of any Democratic administration in the post-World War II era, reflecting the decline in the economic and political strength of the labor movement, a legacy of the Reagan era. And President Clinton actively supported Congressional approval of the North American Free Trade Agreement (NAFTA), signed by President George H.W. Bush, anathema to the labor movement which feared the export of unionized jobs to Mexico and the downward pressure this would place on wages in the United States. Finally, the Clinton administration continued with the deregulation of the finance industry. The 1994 Riegel-Neal Interstate Banking and Breaching Efficiency Act removed restrictions on interstate banking branches. The Federal Reserve, in 1996, reinterpreted the Glass-Steagall Act allowing commercial banks to gain a higher percentage of their revenues from activities typically reserved for investment banks. The Glass-Steagall Act was eventually repealed by the 1999 Gramm-Leach-Bliley Act, thereby allowing the creation of financial mega-firms able to pursue banking, investment, and insurance activities inside one company. Finally, the Commodity Futures Modernization Act of 2000 exempted financial derivatives, including credit default swaps, from government regulation (Brine and Poovey 2017; Smith 2016). The consequences of financial deregulation would become apparent in the events leading up to the Great Recession of 2008.

2.1.2. Growing financialization of the economy

Financial deregulation facilitated and was a response to the dramatic growth in the financial sector beginning around 1980, the start of neo-liberalism in the United States. The financial services sector represented just 2.8% of GDP in 1950. Its contribution to GDP increased slowly representing 4.9% of GDP in 1980. From that point onward, the financial sector grew much more rapidly contributing 8.3% to GDP just prior to the Great Recession (Greenwood and Scharfstein 2013, 3).

From 1980 onward, total financial assets, including stocks, bonds, derivatives, and mutual funds shares, grew much more rapidly than GDP. In 1980, their value was approximately five times the GDP. By 2007, their value was approximately 10 times the value of GDP. The value of financial assets also increased much more rapidly than the value of physical assets such as plant and equipment, residential structures, and land (Greenwood and Scharfstein 2013, 4).

With the onset of financial sector deregulation in the early 1980s, financial sector firms gained larger shares of total corporate profits. From the late 1960s to the early 1980s, financial corporations earned between 10% and 15% of total corporate profits. Thereafter, profits of financial sector firms rose much more rapidly than profits of non-financial corporations, peaking at 40% of total corporate profits at the beginning of the twenty-first century. They then fell back to 20% of total corporate profits just prior to the beginning of the Great Recession (Mazzucato 2018, 139).

The rapid growth of finance impacted the non-financial economy. Two important factors driving the growth of the financial sector were asset management fees and residential loan origination fees. As total financial assets increased in value, so did fees from traditional asset-management including mutual funds, pension funds, and exchange-traded funds, and from non-traditional asset management including hedge funds, private equity funds, and venture capital funds. Overall, from 1980 to 2007, the growth in total asset management fees accounted for more than one-third of the total growth in the output of the financial sector (Greenwood and Scharfstein 2013, 11).

The growth of the credit intermediation industry accounted for approximately one-fourth of the growth in the financial sector. Household credit grew more dramatically than did corporate credit, and its most rapid growth occurred during the housing boom of 2000–2006, just prior to the Great Recession. Overall, the value of corporate credit grew from 31% of the value of GDP in 1980 to 50% of the value of GDP in 2007, while the value of household credit, primarily mortgages, increased from 48% of the value of GDP in 1980 to 99% of the value of GDP in 2007. A significant share of the growth in mortgage loans was securitized, packaged into asset-backed securities held by 'shadow-banking' entities (Greenwood and Scharfstein 2013, 21). This easy access to household credit helped maintain the growth of aggregate demand prior to the Great Recession as worker compensation stagnated.

Not only was there an increased financialization of the economy under neo-liberalism, but there was also an ideological shift among executives of financial and non-financial corporations. The primary goal of a public corporation was now to 'maximize share-holder value' and the primary stakeholders were the shareholders. To raise equity prices, firms would engage in financial engineering. Funds that might have been utilized for investment in new plant and equipment, technological innovation and job creation were instead utilized for share buybacks. In fact, for the 248 companies in the Standard and Poor's 500 Index in March 2014 that were publically listed in 1981, stock repurchases as a portion of net income were less than 5% in 1981. Stock repurchases as a portion of net income peaked in the 2006–2008 period at 68%. In fact in 2007, on the eve of the Great Recession, for these corporations the value of share buybacks plus dividends exceeded total net income (Lazonick 2015, p. 160.)

Private equity funds were a creation of neo-liberalism. As late as 1990, private equity funds were managing a few assets and earning minimal management fees. Yet, by 2007, private equity funds were managing approximately 854 USD billion of assets (Greenwood and Scharfstein 2013, 10). Public corporations were often taken private by private equity funds. Their goal was to generate returns by first taking substantial dividends from their newly created company and eventually to sell their interests in the company. However, with the easy availability of credit, post-buyout firms were often downsized and highly leveraged facing a risk of bankruptcy if economic conditions deteriorated (Davis et al. 2020).

In short, with the increased financialization of the economy, U.S. corporations would be less able to cope with cash flow problems during the Great Recession.

2.1.3. Growing monopoly power and monopsony power

Neo-liberal ideology posits that 'free markets' are self-regulating and lead to efficient production in the public interest. Furthermore, assuming 'contested markets,' concentrated corporate power is favorably viewed as fostering consumer well-being by efficiently producing and distributing goods and services at low prices. Thus, as neo-liberalism evolved in the United States, industry concentration increased and monopoly power strengthened.

At the beginning of the neo-liberal era in the United States, from 1975 to 1995, average industry concentration did decline as firms faced increased domestic and international competition, and large conglomerates were broken up by anti-trust enforcement. But, the decline in average industry concentration was short-lived. From 1995 to 2015, more than 75% of industries experienced an increase in concentration. However, while industries were more concentrated in 2015 than in 1995, they were still less concentrated than in 1975, prior to the onset of neo-liberalism (Kahle and Stulz 2017).

Yet industry concentration measures, by themselves, do not necessarily equate to measures of the extent of competition in an industry or the extent of market power held by industry-leading firms. However, at the same time as industries have become more concentrated, firm turnover at the top of their industries has been falling. Ranking firms by profits and market value, Philippon (2019, pp. 52–53) finds that the likelihood of firm turnover among industry leaders has sharply declined. Over three-year periods in the mid-1990s, the likelihood of firm replacement at the top of a given industry was 45%. More recently, by 2012, this figure had fallen to less than 30%, the lowest it had been in the neo-liberal era.

Labor, in general, did not benefit from the increase in market power attained by industry-leading firms. From the mid-1990s onward, labor's share of value-added declined and the overall profit share of value-added grew sharply. Those industries that experienced the largest increase in concentration also experienced the sharpest declines in labor's share of output and the sharpest increase in profit margins. During the mid-1990s, corporate profits as a share of the Gross Domestic Product was 5%. By 2017, this figure had increased to 10%, higher than at any other time in the post-World War II period, with the exception of a few years at the beginning of the 2010s (Philippon 2019, pp. 56–57; Barkai 2016).

Thus, firms with monopoly power also gained monopsony power under neo-liberalism. Statistical analyses of wage determination in manufacturing provide indirect evidence of employer monopsony power. Actual employer behavior further strengthens the case for increased employer monopsony power. In manufacturing, average employer concentration by county increased between 1977 and 2009. Wages paid were lower as employer concentration increased and workers had fewer options in their local labor market. The negative relationship between employer concentration and wages was twice as strong in the second half of the time period as neo-liberalism continued to evolve. However, where workers were unionized, increased employer concentration in a local labor market did not lead to lower wages (Benmelech, Bergman, and Hyunseob 2018).[3] Yet, as will become apparent, deunionization was a defining characteristic of neo-liberalism.

Actual employer behavior, as represented by anti-poaching agreements and non-compete agreements provide further evidence of growing employer monopsony power under neo-liberalism. These practices restrict employees' labor market mobility thereby limiting their ability to bargain for higher wages or seek higher wages elsewhere. No-poaching clauses, typically, but not only, used by major franchisees became more prevalent in recent years. The share of major franchisees with no-poaching clauses rose from 35.6% in 1996 to 53.3% in 2016. Non-compete agreements signed either with the current employer or the previous employer affected approximately 25% of the workforce (Krueger and Posner 2018).

2.1.4. Deunionization, employer domination, and the push for greater 'flexibility'

A declining union presence enables employer monopsony power. The union membership rate has fallen by close to half since 1983, the first year for which there are comparable data. In 2019, only 10.3% of wage and salary workers were union members in contrast to 20.1% in 1983. Deunionization has occurred primarily in the private sector where only 6.2% of private-sector workers were unionized in 2019, a sharp decline from the 16.8% figure in 1983.[4] Union density in the public sector has stayed relatively constant since 1983. In 2019, 33.6% of public sector workers were union members (U.S. Department of Labor, Bureau of Labor Statistics 2020a).

The organizing environment for labor unions deteriorated under neo-liberalism, as employers strengthened their opposition to union organizing efforts.[5] Typically, workers interested in unionization petition the NLRB to hold a representation election. If a majority of workers vote for a union, that union is deemed the duly chosen bargaining agent for the relevant workers. The number of representation elections held has sharply declined in the neo-liberal era. During the 1970s, prior to the onset of neo-liberalism, there were approximately 8,000 union representation elections held annually. That number fell to 4,000 during the 1980s, 3,000 during the 1990s, 2,000 during the 2000s, and 1,400 for the 2011–2018 time period. In fact, the ratio of NLRB representation elections to private sector employment has declined steadily during the neo-liberalism era and is now the lowest since the NLRB began holding representation elections in 1936 (Schmitt and Zipperer 2009; Zuckerman 2019).

Not only are fewer union representation elections being held, the share of elections where the NLRB found that union supporters were illegally fired during an election campaign has sharply risen. During the 1951–1975 time period, in less than 10% of union representation elections were union supporters illegally fired. That figure rose slightly to 12% during the 1976–1980 time period. However, under neo-liberalism, employers have been much more likely to illegally fire union supporters during a campaign for union representation. Dividing the 1981–2007 time period into five-year intervals, the share of union representation elections where union supporters were illegally fired ranged from 16% to 31% (Schmitt and Zipperer 2009). More recently, for the 2016–2017 time period, approximately 30% of union representation elections were characterized by illegal firings. And employers have been increasing their spending on 'union avoidance' consultants to increase their likelihood of defeating unions in representation elections (McNicholas et al. 2019).

Declining union density is often, but not always, a sign of the weakening bargaining power of labor. The inability to strike is another. As neo-liberalism has evolved, there has been a sharp decline in major work stoppages involving 1,000 workers or more. During the 1980s, there were approximately 83 major work stoppages, involving 1,000 workers or more annually, in contrast to 351 work stoppages in the 1970s, a period when labor was still relatively strong. In the 1990s, major work stoppages became even less common with only 35 annually. They continued to decline in the twenty-first century where from 2000 to 2009 they averaged 20 annually and from 2010 to 2018 there were only 14 annually (U.S. Department of Labor, Bureau of Labor Statistics 2020b).

With their increased relative bargaining power, firms sought greater 'flexibility' in labor compensation and labor usage in order to increase profits. Employers not facing unions or facing weak unions have wide latitude in crafting compensation policies. During the neo-liberal era, pay inequality has increased and wage bargaining patterns developed prior to neo-liberalism have been eroded. Differences in compensation practices across firms have been the primary driver of this increased inequality. From 1977 to 2009, more than 80% of the growth in pay inequality among similar workers was accounted for by increased pay inequality across establishments reflecting differences in firm compensation policies and practices (Freeman 2016; Barth et al. 2016).

Furthermore, American companies have also tried to increase the utilization of a 'just in time' cheaper workforce. During the 1980s, there was a growth in alternative working arrangements such as temporary help agency work, on-call work, contract company work, and independent contracting (Rosenberg and Lapidus 1999). There is conflicting evidence regarding whether alternative working arrangements became more prevalent since then. The U.S. Department of Labor, Bureau of Labor Statistics (2018) found that in 2017 10.1% of workers had, as their primary job, a position considered to be an alternative working arrangement in contrast to 9.9% in 1995.

Yet, tax filings reporting all jobs held by individuals not merely their primary job point to a slow growth in the share of tax filers reporting self-employment income, particularly those with self-employment income and no business expenses. These people are providing labor services, for example as an actual independent contractor or a misclassified independent contractor. Overall, from 2000 to 2014, the share of individuals filing reporting Schedule C income (self-employment income) rose from 8.5% to 11.0% (Jackson, Looney, and Ramnath 2017).

Not only have employers gained 'flexibility' through the increased usage of workers in alternative employment arrangements, but they have also gained 'flexibility' in the scheduling of work. The increasing prevalence of varying and unpredictable work schedules accounted for the substantial increase in earnings instability from the 2004-2007 time period to 2008–2012, even among those continuously employed. For prime-age hourly workers, the cumulative probability of reporting varying work hours increased from 36% over the 2004–2007 time period to 46% between 2008 and 2012 (Finnigan 2018).

Thus, as employers gained increased 'flexibility,' workers came to be seen more as 'on-demand' workers. In addition, firms increasingly subcontracted work. Employment in professional and business services is a crude measure of domestic outsourcing. The share of employees in this industry nearly doubled from 7.3% in 1970 to 13.9% in 2015. Half of this growth was accounted for by workers in professional occupations and a half was accounted for by workers in low-wage non-professional occupations (Bernhardt et al. 2016).

2.1.5. Growing globalization and a more integrated world economy

In addition to institutional changes in the domestic U.S. economy under neo-liberalism prior to the Great Recession, there were also structural changes in the world economy that affected the United States. Economic activity became more globalized and the world economy became more economically integrated.

First, foreign direct investment increased in importance. In 1980, the market value of the stock of foreign direct investment in the United States was equivalent in size to less than 5% of United States GDP. By 2007, the market value of the stock of inward foreign direct investment was equivalent in size to approximately 25% of United States GDP. Similarly, the market value of the stock of United States firms' direct foreign investment abroad also grew in size much more rapidly than the United States GDP. In 1980, the market value of the stock of foreign direct investment of United States firms abroad was equivalent in size to somewhat less than 10% of United States GDP. By 2007, the market value of foreign direct investment of United States firms abroad was equivalent in size to approximately 35% of United States GDP (Clausing 2019, 140).

Second, the United States became more open to international trade. The sum of the value of exports plus imports increased faster than the United States GDP. In 1980, the value at constant prices of exports plus imports was equivalent in size to 12.3% of United States real GDP. By 2007, they were equivalent in size to 28.6% of United States real GDP (University of Pennsylvania 2020). Of particular note was the substantial increase in imports of labor-intensive manufacturing goods from less-developed countries, particularly China. Imports from China between 1990 and 2007 accounted for approximately 25% of the decline in United States manufacturing employment during that period, and lowered wages and reduced the labor force participation rate of affected workers (Autor, Dorn, and Hanson 2013).

Not surprisingly, those most negatively affected by job losses attributable to trade, particularly with China, lived in voting precincts which heavily favoring Donald Trump in the 2016 United States presidential election. President Trump ran on a platform promising trade protection and the renegotiation of trade agreements.

3. Federal governmental policy from the Great Recession to just prior to the novel corona virus: a challenge to neo-liberalism?

Even the most ardent 'market fundamentalists' would allow for limited government intervention to protect property rights, provide public goods and resolve such market imperfections as monopoly power which do not foster economic well-being. Yet, there can also be significant government interventions which, on the surface, appear to be inconsistent with neo-liberalism but ultimately serve to maintain or strengthen neo-liberalism.

The Great Recession was an economic crisis rooted in neo-liberalism. There was an increased financialization of the economy along with the deregulation of the financial sector. In fact, Bernanke, Geithner, and Paulson (2019) maintain that the economic damage of the Great Recession illustrates the 'costs of running a financial system with weak oversight' (p. 212). In addition, there were asset bubbles in housing and financial services, and excess personal debt due to stagnating real earnings and a decline in employer-provided health benefits. With profits rising relative to wages and increasing inequality in household incomes, there was a growing volume of funds seeking investment opportunities and a perceived shortage of profitable, productive investment opportunities leading to asset bubbles in real estate and financial services. The collapse of these asset bubbles led to the Great Recession.

In 2008, the Bush administration made large capital investments in more than 200 banks and guaranteed the assets of the most troubled banks. These policies were designed to save neo-liberalism not to replace it.

The Obama administration was willing to address, at least to a degree, the imbalance of power between labor and management, improve the social wage, restructure firms and industries in crisis, strengthen the regulation of the financial system and overall utilize government action to rebuild the economy. They were willing to push back against the 'market fundamentalism' of neo-liberalism.[6]

The NLRB, during the Obama administration, issued several rulings favorable to union organizing and collective bargaining including shortening the time between the filing of a petition for a union representation election and the holding of such an election. In so doing, it lessened the time employers would have to try to dissuade their workers from voting for union representation.[7] In its Browning-Ferris decision, the NLRB confronted the collective bargaining issues raised by the growth of subcontracting and of staffing agencies. It expanded the types of business arrangements in which companies and their subcontractors or staffing agencies would be considered employers of workers with both being liable for pay and safety violations and both being expected to engage in collective bargaining where workers were unionized.

Furthermore, the U.S. Department of Labor issued rulings expanding the number of workers eligible for overtime pay and limiting the ability of employers to misclassify workers as independent contractors rather than employees. Employers, at times, classified some low-paid employees as managers or supervisors thereby making them ineligible for overtime pay. The U.S. Department of Labor issued a ruling raising the minimum annual earning required before a worker would be able to be considered a manager or supervisor.[8] Employers, both inside and outside of the 'gig economy' were, at times, misclassifying workers as independent contractors rather than employees to evade workplace protection laws that typically cover employees but not independent contractors. The U.S. Department of Labor issued guidance that implied that many independent contractors should more correctly be classified as employees.

The Obama administration strengthened the social safety net and its tax and transfer payment policies diminished inequality. Its singular achievement was the Patient Protection and Affordable Care Act (ACA). Funded by a tax on very high-income individuals, providing subsidies to low-income individuals, and providing financial incentives to states to expand Medicaid, a publically provided health program for low-income individuals, it gradually enabled more Americans to access health insurance and medical benefits. Between October 2013 and early 2016, the uninsured rate for adults aged 16–64 fell from 20.3% to 11.5% (Uberoi, Finegold, and Gee 2016).[9] Though based on market-principles, thereby consistent with a neo-liberal approach to the provision of health care, the ACA increased the regulation of private health insurance companies.

The Obama administration expanded tax credits for low-income working families, including the Earned Income Tax Credit and the Child Tax Credit. Together with the increase in tax rates for the highest-earning individuals, the administration's tax and transfer payment policies served to reduce inequality more than any previous administration in the neo-liberal period. However, overall inequality still grew as its tax and transfer payment policies were too weak to counter tendencies toward deepening inequality (Congressional Budget Office 2019; Porter 2020).

Contrary to a neo-liberal approach to industrial policy, the Obama administration actively invested in the automotive industry in order to save General Motors (GM) and Chrysler.[10] 'Free market' economists strongly criticized such government intervention arguing that the normal process of market capitalism was being disrupted since companies should rise and fall on their own (Ikenson 2011). Believing that the liquidation of GM and Chrysler would cause at least one million jobs to be lost in the automotive industry and in businesses dependent on the automotive industry, the Obama administration provided 80 USD billion from the Troubled Asset Relief Program (TARP) to GM and Chrysler and oversaw the structured bankruptcy proceedings of both firms. As the price for accepting government funds, these two companies were forced to restructure and lower their cost structures. More than a dozen assembly plants were closed, brands were discontinued, managerial ranks were reduced, executive pay was cut and the United Auto Workers (UAW) were forced to provide wage and work rule concessions. The government intervention achieved its goal, and GM and Chrysler quickly returned to profitability.

Given that the minimally, ineffectively regulated financial system was at the heart of the Great Recession, the Obama administration called for increased, more effective regulation. The U.S. Congress passed the Dodd-Frank Wall Street Reform and Consumer Protection Act of 2010. This legislation created the Consumer Finance Protection Bureau, improved the oversight of financial institutions, increased capital and liquidity requirements for financial institutions, restricted proprietary trading by banks (the 'Volcker Rule'), increased regulation of the derivatives market, and created new procedures for dealing with failing financial institutions. Overall, the share of the U.S. financial system facing leverage requirements increased from 41% at the end of 2007 to 92% at the end of 2017 (Bernanke, Geithner, and Paulson 2019, 210). On balance, the Dodd-Frank Act helped to stabilize the financial system. This legislation did represent a move away from a neo-liberal approach to the financial system.

Some policies of the Obama administration represented partial challenges to neo-liberalism. Others, including macroeconomic policy and international trade policy, did not.

Its macroeconomic policy was initially not in line with neo-liberalism but rather consistent with Keynesianism. Eventually, however, the Obama administration's focus turned toward reducing the federal budget deficit, more in line with neo-liberalism. The American Reconstruction and Recovery Act of 2009 included approximately 800 USD billion in economic stimulus divided between increased spending on unemployment insurance and infrastructure, decreased taxes and increased aid to state and local governments. The federal budget deficit grew from 3.1% of GDP in fiscal year 2008, the last year of the Bush administration, to 9.8% of GDP in fiscal year 2009, the first year of the Obama administration, the largest federal budget deficit relative to GDP since World War II, up to that point. The federal government would, initially, be a primary driver of economic growth rather than waiting for market forces to eventually lead to economic recovery.

However, the Obama administration underestimated the severity of the Great Recession and the economic stimulus was less than needed for a strong economic recovery. The Republicans, claiming to have principled objections to federal budget deficits, took over control of the U.S. House of Representatives in 2011. The Budget Control Act of 2011, passed with bipartisan support, required substantial spending reductions over the next 10 years. While President Obama was not satisfied with this legislation, in signing it he stated that this compromise does make a serious down payment on the 'deficit reduction we need' (Applebaum 2011). This premature shift to tighter fiscal policy consistent with neo-liberalism made the economic recovery slower than it needed to be. And the federal budget deficit relative to GDP continued to decline for most of the time President Obama was in office, falling to 3.2% of GDP in fiscal year 2016.

The Obama administration's international trade policy was in line with neo-liberal globalization. The administration wanted to constrain China and write the 21st century rules for international trade. Central to their effort was the Trans-Pacific Partnership (TPP), a 'freer' trade agreement covering 12 countries, not including China. The TPP reduced or eliminated tariffs and other trade barriers on a wide variety of products, liberalized trade in services, opened up foreign markets to private investment and enabled foreign investors to sue host countries if their policies interfered with private profitability, provided guidelines for e-commerce, protected intellectual property and placed restrictions on state-owned enterprises. It also included labor and environmental standards. The TPP was not ratified by the U.S. Congress while President Obama was in office (McBride and Chatzky 2019).

Domestically and internationally, the Trump administration has tried to undo Obama administration initiatives. With the exception of macroeconomic policy, the Trump administration's domestic policy program is perfectly aligned with neo-liberalism. Internationally, however, the Trump administration rejects global neo-liberalism and favors an 'America-first' regime.

The Trump administration's labor policies are designed to strengthen management and weaken labor. By its rulings, it has moved in the direction of deregulating employer–employee relations. In its first 2 years, the Trump NLRB overturned precedent in more than a dozen cases, and each time they did so their decision favored employers. They made it more difficult for unions to win representation elections by giving employers more say in the determination of the election unit, lengthening the time between the petitioning for an election and the holding of such election, and making it more difficult for union organizers to communicate with workers. It has undermined collective bargaining by broadening management's right to make unilateral changes in a contract (McNicholas, Poydock, and Rhinehart 2019).

The Trump administration rejected Obama-era guidance on joint-employers, independent contractors, and overtime pay. It effectively overturned the Browning–Ferris decision by limiting the types of business arrangements where companies and their subcontractors or staffing agencies would be considered joint-employers of workers, with both thereby being liable for pay and safety violations. The Trump NLRB and the U.S. Department of Labor ruled that workers in the 'gig economy' are independent contractors not employees. (However, these rulings do not preclude states and localities from passing their own legislation regarding the classification of workers.) While the Trump administration's Department of Labor did issue a ruling expanding access to overtime pay, many fewer workers were eligible for overtime pay than under the Obama ruling.

In addition, the Trump administration weakened the safety net and their tax and transfer payment policies increased inequality. Most importantly, the Trump administration first weakened the ACA by eliminating the individual mandate to purchase health insurance. In June 2020, in the middle of the pandemic, the Trump administration asked the U.S. Supreme Court to completely overturn the ACA. If this were to occur, more than 20 million people would lose health coverage and millions more with pre-existing conditions would lose the protections of the ACA. Taking a hard-line politically conservative view of the country's safety net system, President Trump directed federal agencies to develop plans to restrict eligibility for and reduce expenditures on such programs as food stamps, Medicaid, and Temporary Assistance to Needy Families, also known as 'welfare.' These plans were to include satisfying increased work requirements to be eligible for these programs.

The Tax Cuts and Jobs Act of 2017 represented a major overhaul of the tax system. Even after close to four decades of increasing income inequality, this law made the tax system more regressive. It substantially reduced the corporate income tax rate and lowered individual income tax rates. It has been estimated that more than half of the benefits will go to the top 10% of income earners. At issue is the extent to which the corporate income tax cuts will be passed through to workers in the form of higher wages or better benefits. In countries where there are strong unions, the pass through has been substantial but this will likely not be the case in the U.S. with its low union density. Rather, corporate shareholders will be the primary beneficiaries of the corporate tax cuts. And while this piece of legislation included incentives for multinationals to repatriate profits held abroad, it did not lead to increased investment in new plant and equipment but rather increased stock buybacks and dividend payments (Slemrod 2018; Clausing 2020).

The Trump administration would like to deregulate the financial system. Not being able to repeal Dodd-Frank, the Trump administration has instead worked through the Federal Reserve and the regulatory agencies to achieve its goals. In October 2019, the Federal Reserve Board of Governors issued a set of rollbacks to the post-Great Recession financial regulations. They reduced bank capital requirements, weakened bank stress testing and weakened bank obligations to provide resolution plans in the event of bankruptcy. In voting against these changes, Lael Brainard stated that these rules along with other changes underway "weaken core safeguards against the vulnerabilities that caused so much damage in the crisis" (Brainard 2019). In addition, banks are now more able to engage in proprietary trading. Furthermore, the Consumer Finance Protection Bureau has also been weakened. Many consumer and investor protections have been rolled back.

While much of its domestic policy agenda is quite consistent with neo-liberalism, its macroeconomic policy and its approach to international economic relations is not. The Tax Cuts and Jobs Act of 2017 was passed at a time when the economy was expanding and unemployment was falling, typically not a time to pass tax legislation that would substantially reduce tax revenues and increase the federal budget deficit. Interestingly, the Republicans who objected to Obama-era stimulus spending on the grounds that it would increase the federal budget deficit had no problem voting for this legislation. And the federal budget deficit has increased relative to GDP throughout the time President Trump has been in office rising from 3.5% of GDP in fiscal year 2017 to 4.6% of GDP in fiscal year 2019.

Internationally, the Trump administration claimed that global neo-liberalism no longer serves the interests of the U.S. Critical of the operations of multinational governing bodies such as the World Trade Organization (WTO), the Trump administration has blocked appointments to the Appellate Body of the WTO crippling its ability to settle trade disputes. With its 'America First' approach to international economic relations, the Trump administration withdrew the U.S. from the TPP, preferring to confront China directly through bilateral negotiations.

They believe that countries such as China have benefitted disproportionately from global neo-liberalism while not abiding by the rules. China has developed its industrial base while the U.S. industrial base has sharply deteriorated. China has gained access to the latest technology while stealing the intellectual property of U.S. firms. China has become strategically placed within the global supply chain giving it the potential to restrict the U.S. from needed supplies. China has run large balance of trade surpluses, including with the U.S., gaining it substantial foreign exchange reserves able to serve as a counter against U.S. financial power.

In its bilateral negotiations with China, with other countries, and with the European Union, the Trump administration has used tariffs more aggressively than any other previous administration in the post-World War II period. In regards to China, it imposed new tariffs on more than 360 USD billion of Chines goods. Regarding other countries, it imposed new duties on such products as steel, aluminum, and washing machines.

As of now, the increased utilization of tariffs has protected some U.S. firms producing finished goods while raising production costs for U.S. firms that depend on imported inputs to their production processes. An agreement was reached with China agreeing to purchase an additional 200 USD billion of U.S. products by the end of 2021, remove barriers for U.S. firms doing business in China and place additional protections on the intellectual property of U.S. firms. An agreement was reached with Mexico and Canada replacing the North American Free Trade Agreement (NAFTA) with the U.S., Mexico, Canada Agreement (USMCA). It included stronger labor and environmental standards, better enforcement mechanisms, and stronger local content rules regarding automobile production.

4. State and local governmental policy from the Great Recession onward: a challenge to neo-liberalism?

As with federal governmental policy, there are contradictory tendencies in state and local governmental policies. Some states and localities governed by progressive Democrats implemented more progressive labor and social policies than did the Obama administration. Such 'intervention in the market from below' does represent a challenge to 'market fundamentalism.' At the same time, other states governed by conservative Republicans rolled back worker protection policies and blocked progressive policies passed at the local level. In these states, the 'market fundamentalism' of neo-liberalism is as strong as ever.

State governments and localities, when allowed by their state governments, can set minimum wages higher than the federal minimum wage, which is 7.25 USD per hour and has not increased since July 2009. It has fallen in real terms and as of 2018 was just 32% of average hourly wages. In recent years, the 'Fight for 15 USD and a Union' led by the Service Employees International Union (SEIU) and other progressive groups took hold in many parts of the United States. As of May 2020, 29 states and the District of Columbia had minimum wages higher than the federal minimum wage. And 46 cities and other localities had minimum wages higher than their state levels (Economic Policy Institute 2019).

In addition to intervening in the 'free market' and mandating minimum wages higher than the federal minimum wage, some states and localities have mandated paid sick days. There is no federal legislation requiring that firms provide paid sick days for their employees. While higher-paid workers were typically provided paid sick days by their employers, lower-paid workers were not. Currently, 11 states, the District of Columbia and 35 other localities have paid sick day laws. As a result, the share of low-paid workers with paid sick days increased from 18% in 2012 when the first state law was passed to 30% currently while the share of higher-paid workers with this benefit rose from 86% in 2012 to 93% currently (Gould 2020; Gould and Schieder 2017).

Similarly, it is at the state and local levels where there is a movement for fair workweek laws. These laws represent a response to employer 'flexibility' and labor being treated as a totally variable cost. There is no federal legislation regulating the scheduling of work. Some cities, including San Francisco, Seattle, New York City, Chicago, and Philadelphia along with the state of Oregon have passed laws designed to increase the predictability and stability of work schedules (Schneider and Harknett 2019). Eight states and the District of Columbia have legislated mandatory pay during a work shift for reporting to work even when the employee is sent home due to a lack of work (Golden 2015).

Not only has there been a movement for fair and humane work scheduling, but worker rights activists have also pushed for state legislation to limit the misclassification of workers as independent contractors rather than employees. In 2019, California became the first state to pass legislation aimed at app-based companies such as Uber and Lyft. By making it very difficult for a worker to satisfy the legal requirements to be classified as an independent contractor, it effectively requires that almost all workers be classified as employees.

While some states have passed legislation raising labor standards, other states controlled by conservative Republicans have done the exact opposite. Rather these states have passed legislation completely in line with a neo-liberal approach to the labor market. They have pushed back against improved labor standards and have passed legislation preempting the ability of cities and other localities to pass ordinances improving labor standards. In recent years, preemption laws have increased. As of 2017, 25 states had passed laws taking away the ability of local authorities to pass minimum wage ordinances. Local paid leave legislation has been preempted in 20 states and local fair scheduling legislation has been preempted in 9 states (Von Wilpert 2017).

The rights of labor unions in both the private and public sectors have been restricted in some states controlled by conservative Republicans. 'Right to Work' laws are state laws making it illegal to require workers to join or financially support a union as a condition of employment. Thus, workers can be 'free riders,' receiving the benefits of union representation without paying dues or fees to the union. Currently, 27 states have 'right to work' laws, 5 of which, including Michigan, formerly a bastion of unionism, passed these laws since 2012.

5. Upsurge in labor activism: a challenge to neo-liberalism?

During the Great Recession, workers gave concessions to their employers anticipating sharing in the gains once the economy recovered. Yet, as the economic expansion continued and the unemployment rate fell to below 4%, real compensation did not improve as expected. After close to 40 years of neo-liberalism, employers still maintained the upper hand.

Dissatisfaction with the results of neo-liberalism led to more workers participating in major strikes. Overall, in 2018 and 2019, 455,400 workers annually were involved in major work stoppages, the largest 2-year pooled average since 1983 and 1984, at the beginning of the neo-liberal period. Similarly, in the 2018 and 2019 time period, 3.1% of union members were involved in a major work stoppage each year on average, a level not seen since 1983 (Shierholz and Poydock 2020, 3–4). While this may represent a significant turning point, the number of workers involved in major work stoppages is still well below the levels of the post-World War II period through 1979, prior to the onset of neo-liberalism.

The major work stoppages that did occur involved many workers. In 2018–2019, there was an average of 20,000 workers involved in each major work stoppage. This was the highest two-year average on record dating back to 1947 (Shierholz and Poydock 2020, 3–4).

Public sector workers, particularly teachers, were involved in many of the strikes involving more than 20,000 workers. These included public school teachers in West Virginia, North Carolina, Kentucky, Oregon, Los Angeles, and Chicago. While not verbalized in this way, they were striking against neo-liberalism, against the privatization of public education, the chronic underfunding of public education, and the conditions faced by students in poverty. Public sector unions, particularly teachers, often broadened the notion of collective bargaining to encompass 'Bargaining for the Common Good.' This included a shared commitment to the preservation of public services and using the public sector as a tool for building a more just society for all (Sneiderman and McCartin 2018).

Strikes involving at least 20,000 workers also occurred in the private sector. They included job actions at General Motors, AT&T, and the New England grocery chain Stop & Shop. The six-week strike at General Motors was the longest national strike in the automotive industry in decades. Lying behind these and other strikes was a sense of unfairness and economic anxiety, the results of close to 40 years of neo-liberalism.

Comprehensive data does not exist on strikes encompassing less than 1,000 workers or lasting for less than one shift. However, more recently, there have been a range of job actions by 'essential' but often low-paid workers forced to work during the COVID-19 pandemic. They include grocery store and fast-food staff, delivery workers, warehouse employees, and other food-service-related occupations. Some are 'standard' employees while others are 'gig workers.' The strikes and other job actions were typically short and often included only a few workers. They occurred at such companies as McDonald's, Amazon, and its subsidiary Whole Foods, Walmart, Target, FedEx, Starbucks, and Instacart. The issues in dispute revolved around worker safety, including the provision of appropriate personal protective equipment (PPE), paid sick leave, and hazard pay. The job actions target conditions generated by neo-liberalism. And the COVID-19 pandemic has helped to radicalize workers. 'Essential,' front-line, low-paid workers have more leverage than ever. The question is whether they will use it.

6. The legacy of neo-liberalism and the challenges going forward

The legacy of neo-liberalism is apparent in the inadequate preparation of the United States for the COVID-19 pandemic, the nature of the Trump administration's policy response to the health and economic crisis, and the changing political discourse within the Democratic Party, particularly the push for strong, effective government intervention in the 'free-market' to create a fairer, more well-functioning economy.

The "free-market logic of neo-liberalism resulted in the United States finding itself short of PPE for front-line health workers. Rather than utilize the Defense Production Act to require firms to produce the necessary materials, the Trump administration gave states and localities little choice but to compete against each other in the 'free market' for PPE.

The public policy response to the sharp recession resulting from the pandemic also reflected the legacy of neo-liberalism. First, the tattered safety net was addressed, at least temporarily, with the Families First Corona Virus Response Act and the Corona Virus Aid, Relief and Economic Security Act (CARES). There is no federal legislation requiring paid sick leave. Now, those employed at firms with less than 500 employees are able to get up to 80 hours of corona-virus-related paid sick leave. The wage replacement rate of unemployment insurance was increased and more workers became eligible for unemployment benefits at least for a period of time. Those already covered by the unemployment insurance system were eligible to receive their normal unemployment benefit plus 600 USD per week through the end of July 2020. The wage replacement ratio for many low-wage workers exceeded 100%, at least until the end of July 2020. Eligibility for unemployment benefits was extended to 'gig workers,' independent contractors and the self-employed through the end of 2020. Previously, they were not covered by the unemployment insurance system.

In addition, low- and middle-income people received a one-time stimulus check of up to 1,200 USD based on income. Overall, the unprecedented expansion of U.S. pandemic aid helped millions of people avoid poverty even in the face of the highest unemployment rates since the Great Depression (DeParle 2020).

Second, money was allocated for the Paycheck Protection Program, with the aim of keeping employment levels stable at small businesses. 'Forgiveable loans' were granted to firms able to maintain their payrolls. Along the same lines, there were bailouts of firms in particularly hard-hit industries, such as airlines. In the case of airlines, government aid was predicated on not instituting layoffs until 1 October 2020.

Third, reflecting their relative political and economic power under neo-liberalism, the very wealthy and large corporations received substantial benefits from the corona virus relief programs. The CARES Act tax reforms primarily benefit those making very high incomes. In addition, the Federal Reserve is bailing out wealthy investors by purchasing a wide variety of real estate mortgages, student loan securities, and even high-risk corporate 'junk' bonds. Furthermore, in the CARES Act itself, a substantial amount of money was dedicated to helping large corporations. And the Federal Reserve has several trillion dollars at its disposal to lend to large corporations.

Ultimately, the Federal Reserve is benefitting Wall Street and the wealthy and hoping that the benefits of doing so will 'trickle down' to the rest of the economy. Yet, there is another approach for dealing with the economic and social dislocation caused by the pandemic as well as the economic and social problems arising from 40 years of neo-liberalism. During the Democratic Party primaries to choose a presidential candidate to run in the 2020 elections, several of the leading candidates, including Bernie Sanders and Elizabeth Warren, advanced bold policy proposals representing a strong rejection of neo-liberalism and succeeded in moving the political discourse inside the Democratic Party to the 'left.' And prior to the pandemic, in February 2020, the U.S. House of Representatives, with a majority of Democratic members, passed the Protecting the Right to Organize Act (PRO), amending the National Labor Relations Act (NLRA) and related labor laws to redress the imbalance of power between labor and management arising from 40 years of neo-liberalism. This legislation is now blocked in the Republican-dominated U.S. Senate.

Support for the PRO Act was included in the policy proposals presented by the Biden-Sanders Unity Task Force (2020) in July 2020. The task force's goal was to develop a policy program for building a stronger, fairer economy, with the federal government playing an essential role in redressing the problems caused by neo-liberalism. The federal government would be responsible for creating jobs, raising wages, strengthening the safety net, reducing overall inequality, diminishing monopoly and monopsony power, controlling the excesses of financialization, and creating a trade policy benefitting both the United States and its trading partners. If Joseph Biden is elected President and the Biden-Sanders Unity Task Force policy proposals were implemented, this would represent a very strong challenge to neo-liberalism, a step toward the dismantling of neo-liberalism.

7. Conclusion

While there have been and there are currently challenges to neo-liberalism, on balance the neo-liberal agenda has strongly influenced the political economy of the United States over the past four decades. The resilience of neo-liberalism raises two questions. First, has the neo-liberal agenda been successful in its own terms? Second, what have been the effects of neo-liberalism on economic well-being?

In its own terms, the neo-liberal agenda has been only partially successful. During the Reagan administration, federal governmental policies in line with neo-liberalism did help to resolve the stagflation of the 1970s, albeit on terms completely aligned with the interests of the business community. By doing so, the federal government helped to create conditions for the economic growth which followed. While the Great Recession exposed fundamental contradictions within the neo-liberal model, until that time the economic expansions were long, the recessions short and relatively mild, and the rate of inflation remained low. Macroeconomic stability seemed to have achieved. Following the Great Recession, the United States experienced its longest economic expansion in the post-World War II period and the rate of unemployment fell to a historic low. This expansion was cut short by the economic and political response to the COVID-19 pandemic.

However, the increasing importance of firms with monopoly and/or monopsony power is not consistent with the essence of neo-liberalism which emphasizes individual decision-making coordinated via free competitive markets. Furthermore, with such concentrated product market and workplace power, it cannot be argued, as neo-liberalism postulates, that income is being distributed primary on the basis of the productive contributions of factors of production such as labor and capital. Rather unequal power plays a central role in the income distribution process in the United States. Profit margins widened in the United States, a goal of the neo-liberal agenda. However, this did not lead to strong net private business investment as the neo-liberal agenda predicted would occur. Rather, with the exception of the end of the 20th century and the beginning of the 21st century when net private business investment was strong, driven by improvements in and the growing use of information technology, investment spending has not been robust (Gordon 2016, 587). The weak business investment performance under neo-liberalism is, at least, partly due to the increase in industry concentration and the utilization of internally generated funds for dividend payments and share buybacks rather than for net private business investment spending (Philippon 2019; Gutierrez and Philippon 2018; Guitierrez and Philippon 2017). Using growth rates of total factor productivity as a proxy for a more direct measure of innovation, with the exception of the 1994–2004 time period, annualized growth rates of total factor produc-tivity have been very weak under neo-liberalism.[11] And overall business dynamism has declined in the neo-liberal era (Gordon 2016, pp. 575, 584–585). The United States economy has not become more efficient under neo-liberalism.

While the United States economy has not become more dynamic or efficient, it has become more unequal under neo-liberalism. The benefits of economic growth have not been widely dispersed throughout the population.

Wealth is more unequally distributed than income. Since 1989, the first year for which there is comparable data, the wealthiest 1% of families in the United States have gained an increasing share of wealth held by families. They held 25% of all wealth in 1989. By 2019, their share of wealth had grown to 34%. Those wealth holders in the 90th-99th percentiles increased their share of wealth slightly from 36% in 1989 to 38% in 2019. Overall, the top 10% of wealth holders held 72% of total wealth in 2019. On the other hand, the share of wealth accounted for by those in the 50th to 90th percentiles fell from 35% in 1989 to 27% in 2019. Those families below the 50th percentile by wealth holding had virtually no wealth, accounting for just 2% of total wealth in 2019 (Bricker et al. 2020).

As with wealth, income has also become more unequally distributed under neo-liberalism. The richest 5% of households in the United States received 23.0% of all house-hold income in 2019, in contrast to 16.5% in 1979. Overall, the share of income going to the top fifth of households was 51.9% in 2019, in contrast to 44.1% in 1979. The rise in the top fifth's share of income occurred almost entirely within the share of income accounted for by the top 5% of households. The share of household income received fell in each of the other four quintiles from 1979 to 2019. Overall, the Gini coefficient of household income rose from .404 in 1979 to .484 in 2019 (Semega et al. 2020, Table A-4, p. 38).

Not only has there been increased economic inequality, for many households the neo-liberal era has represented a period of relatively stagnant living standards. Leaving aside the issue of income mobility, households at similar points in the bottom three income quintiles of the income distribution in 1979 and 2019 would have experienced relatively stagnant standards of living over this 40 year period. Within the bottom quintile, mean household real income in 2019 was only 16% higher than in 1979. Similarly, within the second quintile mean real household income rose by only 24% over this period and within the third quintile the growth in real mean household income was just 27%. In each of these quintiles, this translates into an annual growth of real mean household income of well below 1%. And within the fourth income quintile, real mean household income rose by only 40% over this 40 year period (Semega et al. 2020, Table A-4, pp. 38, 42).

Not only have living standards stagnated for many under neo-liberalism, the poverty rate has remained relatively constant over 40 years. While it fluctuates with overall economic conditions, the poverty rate fell to a record low of 10.5% in 2019. However, this represented only a minimal improvement compared to the poverty rate in 1979 at the start of the neo-liberal era. At that time, the poverty rate was 11.7% (Semega et al. 2020, Table B-5, p. 61).

Similarly, in 2019, many adults were living 'very close to the edge' financially. In 2019, close to 30% of adults were either unable to pay all of their monthly bills or would not be financially able to handle an unexpected 400 USD expense. Furthermore, 25% of adults skipped needed medical care, such as a doctor or dentist visit, since they were unable to afford the cost of such visit (Board of Governors of the Federal Reserve System 2020).

While many were left behind under neo-liberalism, there were winners as well. These included people at the upper reaches of the income distribution. Those households in the top 5% of the income distribution saw their mean income rise by approximately 108% from 1979 to 2019 while those in the highest quintile overall experienced, on average, a mean increase in household income of close to 80% (Semega et al. 2020, Table A-4, pp. 38, 42).

Increasing economic inequality together with stagnating living standards for many has led to a growing sense of unease and unfairness. Within the political process, this manifests itself both as support for right-wing 'populists' such as Donald Trump and for progressive candidates calling for an end to neo-liberalism though not in such terms. Ultimately, the choice to embark on neo-liberalism was a political choice backed by a coalition of groups, including a cohesive business community anticipating benefitting from such a political-economic framework, and an ideology supporting it. The replacement of neo-liberalism by a more well-functioning, fairer political-economic system will also be a political choice backed by a coalition of groups tired of neo-liberalism and an ideology supporting such a change. And if this were to occur, it would represent the ultimate challenge to neo-liberalism, a political-economic regime whose time is up.

Notes

1. In his review article on Thomas Piketty's 'Capital in the 21st Century', Sawyer (2015) argues for focusing on country-specific aspects and the roles of institutions and policies in trying to understand trends in, for example, inequality rather than searching for general forces at work. This paper is very much in line with Sawyer's preferred research direction.
2. See Rosenberg (2003) for a detailed discussion of these policy changes. See Kotz (2015) for a detailed discussion of business support for neo-liberalism in the United States.
3. Azar, Marinescu, and Steinbaum (2017) and Dube et al. (2020) provide further evidence of the existence of monopsony.
4. Union density rates and the share of private-sector workers covered by collective bargaining agreements are very similar in the United States.
5. See Freeman (1988) and Greenhouse (2019) for discussion of management's increased desire for a union-free environment.

6. The Obama administration was also concerned about the increase in industry concentration and growing monopsony power (Council of Economic Advisers 2016a, 2016b).
7. However, the primary legislative goal of the labor movement, the Employee Free Choice Act, did not gain Congressional approval due to intense business opposition.
8. A Federal Judge in Texas blocked the implementation of this ruling expanding eligibility for overtime pay.
9. All Americans 65 years of age and older are eligible for Medicare.
10. See Wade (2012) for a discussion of the decline of the neo-liberal consensus regarding industrial policy.
11. Public funding was crucial for fostering innovations in information technology, not the image of the government advanced by supporters of neo-liberalism.

Disclosure statement

No potential conflict of interest was reported by the author(s).

References

Applebaum, B. 2011. "Spending Cuts Seen as Step Not as Cure," *New York Times*, August 2.

Autor, D., D. Dorn, and G. Hanson. 2013. "The China Syndrome: Local Labor Market Effects of Import Competition in the United States." *American Economic Review* 103 (6): 2121–2168. doi:10.1257/aer.103.6.2121.

Azar, J., I. Marinescu, and M. Steinbaum. 2017. "Labor Market Concentration." Unpublished paper. University of Pennsylvania.

Barkai, S. 2016. "Declining Labor and Capital Shares." Unpublished paper. London Business School.

Barth, E., A. Bryson, J. C. Davis, and R. Freeman. 2016. "It's Where You Work: Increase in the Dispersion of Earnings across Establishments and Individuals in the United States." *Journal of Labor Economics* 34 (S2): S67–S97. doi:10.1086/684045.

Benmelech, E., N. Bergman, and K. Hyunseob. 2018. "Strong Employers and Weak Employees: How Does Employer Concentration Affect Wages?" Working Paper 24307. Cambridge, MA: National Bureau of Economic Research.

Bernanke, B. S., T. Geithner, and H. M. Paulson Jr. 2019. *Firefighting: The Financial Crisis and Its Lessons*. New York: Penguin Books.

Bernhardt, A., R. Batt, S. Houseman, and E. Appelbaum 2016. "Domestic Outsourcing in the United States: A Research Agenda to Assess Trends and Effects on Job Quality," Working Paper 16-253. Kalamazoo, MI: W.E. Upjohn Institute for Employment Research.

Biden-Sanders Unity Task Force. 2020. "Biden-Sanders Unity Task Force Recommendations." Joe. biden.com, July.

Board of Governors of the Federal Reserve System. 2020. "Report on the Economic Well-Being of U.S. Households in 2019." Featuring Supplemental Data from April 2020. https://www.feder alreserve.gov/publications/default.htm

Brainard, L. 2019. *Press Release: Statement by Governor Lael Brainard*. Washington, DC: Federal Reserve Board. October 10.

Bricker, J., S. Goodman, K. B. Moore, and A. H. Volz 2020. "Wealth and Income Concentration in the SCF, 1989-2019," FEDS Notes. Washington, DC: Board of Governors of the Federal Reserve System, September 28. 10.17016/2380-7172.2795.

Brine, K. R., and M. Poovey. 2017. *Finance in America: An Unfinished Story*. Chicago: University of Chicago Press.

Burtless, G. 1983. "Why Is Insured Unemployment so Low?" *Brookings Papers on Economic Activity* (1): 225–249. doi:10.2307/2534356.

Clausing, K. 2019. *Open: The Progressive Case for Free Trade, Immigration, and Global Capital.* Cambridge, MA: Harvard University Press.

Clausing., K. 2020. "Fixing Five Flaws of the Tax Cuts and Jobs Act." Unpublished Paper. Reed College.

Congressional Budget Office. 2019. "The Distribution of Household Income, 2016". www.cbo.gov/publication/

Council of Economic Advisers. 1982. *Economic Report of the President.* Washington, DC: U.S. Government Printing Office.

Council of Economic Advisers. 2001. *Economic Report of the President.* Washington, DC: U.S. Government Printing Office.

Council of Economic Advisers. 2016a. "Benefits of Competition and Indicators of Market Power." Issue Brief. May.

Council of Economic Advisers. 2016b. "Labor Market Monopsony: Trends, Consequences and Policy Responses." Issue Brief. October.

Davis, S. J., J. C. Haltiwanger, K. Handley, B. Lipsius, J. Lerner, and J. Miranda. 2020. *The Economic Effects of Private Equity Buyouts.* Working Paper 26371. Cambridge, MA: National Bureau of Economic Research.

DeParle, J. 2020. "Buoyed by Virus Aid, Millions Avoided Poverty," *New York Times.* June 22, pp. A1, A7

Dube, A., J. Jacobs, S. Naidu, and S. Suri. 2020. "Monopsony in Online Labor Markets." *American Economic Review Insights* 2 (1): 33–46. doi:10.1257/aeri.20180150.

Economic Policy Institute. 2019. *State of Working America Data Library.* Minimum Wage. www.epi.org

Finnigan, R. 2018. "Varying Weekly Work Hours and Earnings Instability in the Great Recession." *Social Science Research* 74: 96–107. doi:10.1016/j.ssresearch.2018.05.005.

Freeman, R. 2016. *A Tale of Two Clones.* Washington, DC: Third Way.

Freeman, R. B. 1988. "Contraction and Expansion: The Divergence of Private Sector and Public Sector Unionism in the United States." *Journal of Economic Perspectives* 2 (2): 63–88. doi:10.1257/jep.2.2.63.

Golden, L. 2015. *Irregular Work Scheduling and Its Consequences.* Washington DC: Economic Policy Institute.

Gordon, R. J. 2016. *The Rise and Fall of American Growth: The U.S. Standard of Living since the Civil War.* Princeton, NJ: Princeton University Press.

Gould, E. 2020. *Lack of Paid Sick Days and Large Numbers of Uninsured Increase Risks of Spreading the Coronavirus.* Washington, DC: Economic Policy Institute.

Gould, E., and J. Schieder. 2017. *Work Sick or Lose Pay?: The High Cost of Being Sick When You Don't Get Paid Sick Days.* Washington, DC: Economic Policy Institute.

Greenhouse, S. 2019. *Beaten Down, Worked Up: The Past, Present and Future of American Labor.* New York: Alfred A. Knopf.

Greenwood, R., and D. Scharfstein. 2013. "The Growth of Finance." *Journal of Economic Perspectives* 27 (2): 3–28. doi:10.1257/jep.27.2.3.

Guitierrez, G., and T. Philippon. 2017. "Investment-less Growth: An Empirical Investigation." *Brookings Papers on Economic Activity* (Fall): 89–169. doi:10.1353/eca.2017.0013.

Gutierrez, G., and T. Philippon. 2018. "Ownership, Concentration, and Investment." *AEA Papers and Proceedings* 108 (May): 432–437. doi:10.1257/pandp.20181010.

Ikenson, D. J. 2011. *Lasting Implications of the General Motors Bailout.* Washington DC: Cato Institute.

Jackson, E., A. Looney, and S. Ramnath. 2017. *The Rise of Alternative Work Arrangements: Evidence and Implications for Tax Filing and Benefit Coverage.* Working Paper 114. Washington, DC: U.S. Office of Tax Analysis.

Kahle, K. M., and R. M. Stulz. 2017. "Is the US Public Corporation in Trouble?" *Journal of Economic Perspectives* 31 (3): 67–88. doi:10.1257/jep.31.3.67.

Kotz, D. M. 2015. *The Rise and Fall of Neoliberal Capitalism.* Cambridge, MA: Harvard University Press.

Krueger, A. B., and E. A. Posner. 2018. "A Proposal for Protecting Low-Income Workers from Monopsony and Collusion." Policy Proposal 2018–05 The Hamilton Project. Washington DC: Brookings Institution.

Lazonick, W. 2015. "Labor in the 21st Century: The Top 0.1% And the Disappearing Middle Class." In *Inequality, Uncertainty, and Opportunity: The Varied and Growing Role of Finance in Labor Relations*, edited by C. E. Weller, 143–195. Urbana-Champaign, IL: Labor and Employment Relations Association.

Mazzucato, M. 2018. *The Value of Everything: Making and Taking in the Global Economy*. New York: Public Affairs.

McBride, J., and A. Chatzky. 2019. *What Is the Trans-Pacific Partnership?* New York: Council on Foreign Relations.

McNicholas, C., M. Poydock, J. Wolfe, B. Zipperer, G. Lafer, and L. Loustaunau. 2019. *U.S. Employers are Charged with Violating Federal Law in 41.5% Of All Union Election Campaigns*. Washington, DC: Economic Policy Institute.

McNicholas, C., M. Poydock, and L. Rhinehart. 2019. *Unprecedented: The Trump NLRB's Attack on Worker Rights*. Washington, DC: Economic Policy Institute.

Mishel, L., J. Bernstein, and S. Allegretto. 2007. *The State of Working America: 2006/2007*. Ithaca, NY: Cornell University Press.

Philippon, T. 2019. *The Great Reversal: How America Gave up on Free Markets*. Cambridge, MA: Harvard University Press.

Pierson, P. 1994. *Dismantling the Welfare State? Reagan, Thatcher and the Politics of Retrenchment*. Cambridge: Cambridge University Press.

Porter, E. 2020. "Could Economic Downturn Produce Lasting Change?" *New York Times*, June 26, p. B4.

Rosenberg, S., and J. Lapidus. 1999. "Contingent and Non-Standard Work in the United States: Towards a More Poorly Compensated, Insecure Workforce." In *Global Trends in Flexible Labor*, edited by A. Felstead and N. Jewson, 62–83. London: Macmillan.

Rosenberg, S. 2003. *American Economic Development since 1945: Growth Decline and Rejuvenation*. New York: Palgrave Macmillan.

Sawyer, M. 2015. "Confronting Inequality: Review Article on Thomas Piketty on 'Capital in the 21st Century'." *International Review of Applied Economics* 29 (6): 878–889. doi:10.1080/02692171.2015.1065227.

Schmitt, J., and B. Zipperer. 2009. *Dropping the Ax: Illegal Firings during Union Election Campaigns, 1951-2007*. Washington, DC: Center for Economic and Policy Research.

Schneider, D., and K. Harknett 2019. "Unstable and Unpredictable Work Schedules are an Occupational Hazard." www.wipsociology.org/2019/06/17

Semega, J., M. Kollar, E. A. Shrider, and J. F. Creamer. 2020. *Income and Poverty in the United States: 2019. Current Population Reports, P60-270*. Washington, DC: U.S. Government Printing Office.

Shierholz, H., and M. Poydock. 2020. *Continued Surge in Strike Activity Signals Worker Dissatisfaction with Wage Growth*. Washington, DC: Economic Policy institute.

Slemrod, J. 2018. "Is This Tax Reform or Just Confusion?" *Journal of Economic Perspectives* 32 (4): 73–95. doi:10.1257/jep.32.7.73.

Smith, B. 2016. "The Resilience of American Neoliberalism." In *The Crisis and Renewal of American Capitalism: A Civilizational Approach*, edited by L. Cossu-Beaumont, J.-H. Coste, and J.-B. Velut, 66–84. London: Routledge.

Sneiderman, M., and J. A. McCartin. 2018. "Bargaining for the Common Good: An Emerging Tool for Rebuilding Worker Power." In *No One Size Fits All: Worker Organization, Policy, and Movement in a New Economic Age*, edited by J. Fine, L. Burnham, K. Griffith, M. Ji, V. Narro, and S. Pitts, 219–233. Urbana-Champaign, IL: Labor and Employment Relations Association.

Tasini, J. 1988 "Why Labor Is at Odds with the NLRB,"*New York Times*, October 30, Section III, p. 4.

U.S. Department of Labor, Bureau of Labor Statistics. 2018. *News Release: Contingent and Alternative Employment Arrangements Summary*. 7 June. Washington, DC: U.S. Government Printing Office.

U.S. Department of Labor, Bureau of Labor Statistics. 2020a. *News Release: Union Members-2019.* 22 January. Washington, DC: US Government Printing Office.

U.S. Department of Labor, Bureau of Labor Statistics. 2020b. "Work Stoppages." https://www.bls.gov/web/wkstp/annual-listing.htm

Uberoi, N., K. Finegold, and E. R. Gee. 2016. *Health Insurance Coverage and the Affordable Care Act, 2010-2016.* ASPE Issue Brief. Washington, DC: U.S. Department of Health and Human Services.

University of Pennsylvania. 2020. "Openness at Constant Prices for the United States (OPENRPUSA156NUPN).", June 1. FRED, Federal Reserve Bank of St. Louis, https://fred.Stlouisfed.org/series/OPENRPUSA156NUPN

Von Wilpert, M. 2017. *City Governments are Raising Labor Standards for Working People-And State Legislatures are Lowering Them Back Down.* Washington, DC: Economic Policy Institute.

Wade, R. H. 2012. "Return of Industrial Policy?" *International Review of Applied Economics* 26 (2): 223–239. doi:10.1080/02692171.2011.640312.

Wenger, J. B. 2001. *Divided We Fall: Deserving Workers Slip Through America's Patchwork Unemployment Insurance System.* Washington, DC: Economic Policy Institute.

Zuckerman, M. 2019. *Finding Workers Where They Are: A New Business Model to Rebuild the Labor Movement.* New York: Century Foundation.

The U.S.–China trade imbalance and the theory of free trade: debunking the currency manipulation argument

Isabella Weber🆔 and Anwar Shaikh

ABSTRACT

The U.S.–China trade imbalance is commonly attributed to a Chinese policy of currency manipulation. However, empirical studies failed to reach consensus on the RMB misalignment. We argue that this is not a consequence of poor measurement but of theory. At the most abstract level the conventional principle of comparative cost advantage suggests real exchange rates will adjust so as to balance trade. Therefore, the persistence of trade imbalances tends to be interpreted as arising from currency manipulation facilitated by foreign exchange interventions. By way of contrast, the absolute cost theory explains trade imbalances as the outcome of free trade among nations that have unequal real costs. We argue that a disparity in real costs is the root cause of the U.S.–China trade imbalance.

1. Introduction

The history of globalizing capitalism shows the re-occurrence of lasting trade imbalances under different monetary regimes (Bordo 2005). Large trade imbalances accumulated again on the eve of the 2008 global economic crisis with the U.S.–China imbalance being the most drastic case (Figure 2). This reversed position of the U.S. and Chinese external imbalances is reflected in their bilateral trade balance. In the period leading up to the crisis, China increased its share in the U.S. trade deficit from around one-fifth in 2002, the first year after China's accession to the World Trade Organisation (WTO), to about one-third in 2008. Ten years after the crisis, global trade continues to be unbalanced even though China's current account surplus has sharply decreased to 0.4% of its GDP (International Monetary Fund (IMF) 2019a, 2), while the US deficit with China has reached another record in 2018 at USD 419 billion (U.S. Department of the Treasury, 2019, p. 1-2).

Reducing the U.S. trade deficit is at the top of the Trump Administration's foreign economic policy goals. In 2019 the U.S. President imposed substantial tariffs on Chinese goods and is threatening to raise them even further. U.S. authorities have long been concerned with trade imbalances, and administrations of different political orientations have consistently attributed the U.S. trade deficit to currency manipulation by their trade partners. The 1988 Omnibus Trade Act requires the U.S. Treasury to conduct semi-annual evaluations of unfair exchange rate devaluations by major trading partners. Trade surplus countries including Japan, Korea, Germany, Switzerland, China and India, are on the Treasury's 'Monitoring List' which has been expanded to include 21 countries (U.S. Department of the Treasury 2019a, 3). In 2019, the Treasury officially designated China as a currency manipulator in 'violation of China's G20 commitments to refrain from competitive devaluation' and pledges to take action (U.S. Department of the Treasury 2019b).

Many leading economists perceive global imbalances as a threat to the world's economic stability and as a root cause of the 2008 global economic crises (e.g. Obstfeld and Rogoff 2009, 2005; Lin, Dinh, and Im 2010). The most widespread explanation for the accumulation of large external imbalances on the eve of the crisis, specifically the large trade surplus of China with respect to the U.S., is that they were due to Chinese currency manipulation. According to this argument, the Chinese government reduced the value of the Renminbi (RMB) through exchange rate interventions. This lowered the costs of China's exports to the United States and raised the costs of U.S. imports to China, thereby *artificially* causing the tremendous trade imbalance between the two countries. This view deems a Chinese 'beggar-thy-neighbor devaluation' (Krugman 2009) to be the 'single largest cause of the U.S. trade deficit and of unemployment' (Scott, Jorgensen, and Hall 2013, 3). It follows from this logic, that in order to accelerate growth and restore full employment, the United States would have to reduce its large trade deficit. It is claimed that this could be done at no cost to the U.S. budget, if the United States prevented other countries, primarily China, from manipulating their currency and allow the Renminbi to return to a 'competitive' level (Bergsten and Gagnon 2012).

Proponents of the currency manipulation hypothesis point to China's massive accumulation of international exchange reserves as evidence for undervaluation (Bergsten and Gagnon 2012; IMF, 2010, p. 19; Bergsten 2006, 2007, 2010; Krugman, 20; Bergsten and Gagnon 2012). China is indeed accumulating international exchange reserves at a high rate (see Figure 3 in the Appendix). The total of Chinese international exchange reserves increased almost 17-fold from 2000 to 2010. By 2013, the Chinese reserves had reached a level of USD 3.6 trillion, 60% of which was estimated to be held in dollar-denominated assets, making China the biggest creditor of the U.S. at the time (Chinn 2013).

However, the link between China's supposed degree of undervaluation of the Renminbi (RMB) and its current account surplus is not established directly. Instead, the vast literature bases itself on the standard assumption of international trade theory that in unfettered exchange, trade would automatically balance. This assumption in turn underpins the various attempts to estimate the undervaluation of the Chinese RMB that presumably account for its persistent trade surplus. We show in this paper that such attempts have failed to achieve any consensus. Estimates of the degree of misalignment vary widely and some studies even find that the RMB is overvalued (Cheung and He 2019; Cheung 2012; Cheung, Chinn, and Fujii 2010a, 2010b; Dunaway and Leigh 2006; Cline and Williamson 2007).

This paper argues that the great unity across the political spectrum with regard to currency manipulation as the cause of trade imbalances is rooted in a shared Ricardian outlook on international trade. We analyze the Ricardian Comparative Cost Theory

(CCT) underpinnings of the currency manipulation argument and present an Absolute Cost Theory (ACT) of trade as an alternative view, initially at the *same* level of abstraction as the Ricardian one so as to highlight the fundamental differences. Ricardian trade theory predicts that trade naturally balances through automatic exchange rate adjustments. Trade imbalances are therefore often interpreted as the result of various interventions that impedes this adjustment (see Section 5). In contrast, the ACT from Smith and Harrod predicts that trade imbalances will be persistent and that trade surplus countries will end up as international creditors and trade deficit countries as international debtors – all through the workings of free trade.

The paper is structured in the following manner. The next section presents the two opposing theories of free trade beginning at the same level of abstraction as Ricardo's and then moving to more concrete considerations. The third section briefly reviews the central position that the currency manipulation argument has occupied in U.S. foreign economic policy over the last three decades. The fourth section evaluates the empirical studies that estimate the degree of exchange rate misalignment based on the *extended purchasing power parity* (PPP) and the *macroeconomic fundamental approach* and analyzes the CCT foundation of these models. Section 5 discusses a variety of alternate CCT-based explanations for trade imbalances, and contrasts their explanatory power with that of the ACT. The final section summarizes the opposed perspectives on the U.S.-China trade imbalance arising from the two basic theories of international trade.

2. Comparative and absolute cost theories of free trade

Since the times of the classical political economists, two alternative theories of free trade have competed: The Comparative Cost Theory (CCT) and the Absolute Cost Theory (ACT). The ACT was employed by Adam Smith (1904) and revived by Roy Harrod (1957).[1] The CCT, on the other hand, is rooted in David Ricardo's (1951) challenge[2] to Smith's theory and continues to be the foundation of the neoclassical standard theories of international trade as well as the currency manipulation argument.[3] The two competing theories lead to fundamentally opposed outlooks on the outcomes of free trade: CCT predicts trade to balance automatically, while ACT predicts that free trade under conditions of unequal cost competitiveness results in persistent trade imbalances. From the point of view of CCT, persistent large trade imbalances like those of the last two decades appear to be the result of policy intervention. From the point of view of ACT, these imbalances are the outcome of free trade itself.

The CCT and ACT differ in some of their basic assumptions. They share the assumption that labor is not mobile internationally. CCT also assumes that capital is not mobile, so the capital account of the balance of payments is zero. The overall balance of payments is the sum of the current account, the capital account and the financial account. A trade (current account) deficit means that a country is spending more abroad for imports than it receives for its exports (financial account surplus). The capital account being zero, balance of payments equilibrium requires that the trade deficit be offset by the net money outflow – i.e. that the current account deficit be offset by a financial account surplus. This money flow adjusts the relative nominal costs such as to balance trade and closes the system. In contrast, ACT assumes that capital is mobile internationally and that the trade balance is determined by real costs. The balance of payment is covered by an interest rate-differential induced capital flow and real costs remain unaltered. This capital flow from the surplus to the deficit country closes the system and trade continues to be unbalanced.

The key difference between CCT and ACT has to do with the *long-run* relation of the nominal exchange rate to the real exchange rate. In CCT, in the case of fixed exchange rates the money outflow induced by a trade deficit lowers the money supply. On the supposition of the Quantity Theory of Money, this will lower the price level. In the case of flexible exchange rates, the money outflow will depreciate the exchange rate without changing the price level. In either case, a trade deficit lowers the terms of trade (the real exchange rate), and assuming the elasticities meet the appropriate conditions, this improves the trade balance. The deficit country becomes more competitive as its prices fall (fixed exchange rate) or as the exchange rate depreciates (flexible exchange rate), while the surplus country becomes less competitive. In either case the real exchange rate adjusts. *In the Comparative Cost argument, the nominal exchange rate regulates the terms of trade.* These processes continue until the trade deficit disappears, i.e. until trade is balanced. As Ricardo (1951) points out, this adjustment mechanism would negate any initial differences in absolute costs of production (Shaikh 2016, pp. 491–510).

Various concrete mechanisms have been proposed that bring about such a terms of trade adjustment. Ricardo assumes a gold standard in which the state maintains a fixed relation between its currency and gold (fixed gold peg) and hence between its currency and those of other countries (fixed exchange rates). In this context, a trade deficit increases the demand for gold as a means of international settlement. By absorbing this demand at a fixed gold price, the state removes a certain amount of money from circulation, thereby decreasing the money supply and reducing the price level. In the cash-balance approach, a decrease in the money supply reduces the cash balances of individuals and firms. This leads them to curtail their spending, reduce aggregate demand and thereby cause prices to fall (Yeager 1966, p, 64). Alternately, the drop in aggregate demand leads to a rise in unemployment, a fall in money wages, hence a fall in prices corresponding to the fall in costs (Amin 1974, 47). In the monetarist argument, assumption of a given full employment output paired with a stable velocity of money implies that a fall in the quantity of money will lead to a fall in the price level (Shaikh 2016, pp. 189–191). All of these arguments pertain to the condition in which the state does not to counter these monetary effects, i.e. to sterilize the effects of the trade imbalance (see Section 5.1 of this paper).

The ACT differs in three key aspects from the CCT as regards the long-run relation between the nominal and the real exchange rate. First, the long-run nominal exchange rate is unaffected by the trade balance. Second, the terms of trade are not affected by a change in the nominal exchange rate. Third, long-run real wages are determined by the political balance between workers and capital not by the nominal exchange rate. Let us demonstrate these points.

First, a country with higher absolute costs will run a trade deficit,[4] and the resulting money outflow will initially depreciate the nominal exchange rate of the deficit country. But unlike in CCT this is only a transitory effect in ACT, since the same money outflow will reduce liquidity in the trade deficit country and raise the interest rate.[5] In the trade surplus country, the money inflow will raise liquidity and lower the interest rate. Short-term capital will then flow from the trade surplus (low-interest rate) country to the trade deficit (high-interest rate) country, reducing the interest rate differential and reappreciating the nominal exchange rate of the deficit country. These countervailing capital flows will continue until risk-adjusted interest rates are equalized at a point where the capital flow in each country offsets its trade imbalance, i.e. until the trade and capital flows offset each other, and the nominal exchange rate returns to its initial level – *all through the workings of capital markets under free trade*. Hence, from the perspective of ACT the trade deficit does *not* need to affect the nominal exchange rate on which Ricardo's balancing mechanism depends over the medium run.

The second thing that distinguishes ACT from CCT is the argument that even if there was a persistent depreciation of the nominal exchange rate, say through government policy (see Section 5 of this paper), *the terms of trade would still not fall*. Consider the relative prices of exports and imports (p_x, p_m) on the assumption that mobility of real capital equalizes international profit rates (r). Let $k_x \cdot e$ be the nominal unit costs of Chinese exports expressed in dollars through the exchange rate e (dollars/yuan), so that the unit price of Chinese exports expressed in US dollars is $p_x \cdot e = k_x \cdot e(1 + r)$.[6] Let real costs of Chinese exports be $k_{x_r} = \frac{k_x}{p_c}$, relative to (say) Chinese consumer prices p_c. Then the price of Chinese exports is $p_x \cdot e = k_{x_r}(1 + r)p_c \cdot e$. At the same time, the price of Chinese imports expressed in US dollars will be the dollar price of US exports $p_{x^*} = k_{x^*_r}(1 + r)p_{c^*}$. It follows that the terms of trade of China will be $\frac{p_x \cdot e}{p_{x^*}} = \frac{k_{x_r}(1+r)}{k_{x^*_r}(1+r)} \left(\frac{p_c \cdot e}{p_{c^*}} \right)$. If we assume that the prices of consumer goods are internationally equalized and that both countries have the same basket of goods (see Section 4 on PPP), as Ricardo does at this level of abstraction, then $\frac{p_c \cdot e}{p_{c^*}} \approx 1$, so $\frac{p_x \cdot e}{p_{x^*}} = \frac{k_{x_r}(1+r)}{k_{x^*_r}(1+r)}$, i.e. *the terms of trade will depend solely on relative real costs*. If some consumer goods are not internationally traded, then we may amend the consumer goods term in the terms of trade equation to distinguish between tradable and nontradable goods, so that

$$\frac{p_x \cdot e}{p_{x^*}} = \frac{k_{x_r}(1+r)}{k_{x^*_r}(1+r)} \left(\frac{p_c \cdot e}{p_{c^*}} \right) = \frac{k_{x_r}(1+r)}{k_{x^*_r}(1+r)} \left(\frac{p_{c_{NT}} \cdot e / p_{c_T} \cdot e}{p_{c^*_{NT}} / p_{c^*_T}} \right) \left(\frac{p_{c_T} \cdot e}{p_{c^*_T}} \right) \approx \frac{k_{x_r}(1+r)}{k_{x^*_r}(1+r)} \left(\frac{p_{c_{NT}} \cdot e / p_{c_T} \cdot e}{p_{c^*_{NT}} / p_{c^*_T}} \right)$$ on the assump-

tion that international trade equalizes the price of some common basket of goods. This incorporates the Balassa–Samuelson effect $\left(\frac{p_{c_{NT}} \cdot e / p_{c_T} \cdot e}{p_{c^*_{NT}} / p_{c^*_T}} \right)$ into the determination of the terms of trade via relative real costs (Shaikh 2016, pp. 509–519). With the long-run terms of trade pinned by relative real costs, trade imbalances will be persistent. These arguments are borne out by empirical evidence showing persistent trade imbalances and the long-run gravitation of real exchange rates around relative real costs (Shaikh 2016, pp. 508–516, 528–532 and Figures 1.4–11.7).

Finally, since real costs depend on real wages and productivity, it is certainly possible that in the short run workers may not be able to raise their nominal wages to compensate for a rise in import prices arising from a Chinese exchange rate depreciation. Then Chinese real wages may fall, which will improve its terms of trade. However, if workers have achieved a defensible standard of living, then real wages will rise back up and the terms of trade will revert to their previous levels. The real issue here is whether real wages can be pushed down by *any* means, not just by exchange rate depreciation. The long-run real wages depend not on the state's exchange rate policy but on the level of class struggle and the relative bargaining power of workers and capital. Note that the ACT argument is also perfectly consistent with the existence of value-chains since each national link in the chain is subject to the forces of real costs. In fact, these links are key in pinning the real exchange rate to competitive costs.

From the perspective of ACT, unit labor costs are key to explaining the US–China trade imbalance. Literature surveys have long suggested that relative unit labor costs are the best indicator of trade competitiveness (e.g. Turner and Van't Dack 1993; Turner and Golub 1997). Here we focus on the decade following China's WTO accession in 2001 when the debate over currency manipulation peaked. In this period China's unit labor cost has been a small fraction of that in the United States. Golub et al. (2018, 1517) define relative unit labor costs as the ratio of domestic and foreign unit labor requirements times the ratio of domestic and foreign average manufacturing wages expressed in domestic currency.[7] Any relative unit labor cost greater than one means that the domestic economy is more competitive than the foreign economy: it can produce the same value added at a lower wage cost. Using this Golub et al. (2018, 1522) find that for the years 2000 to 2010 the US labor cost was 7.8 times higher than that of China.[8] Chinese productivity benefits from economies of scale and scope: China has a large supply of engineers and skilled workers, an integrated supply chain for many parts of manufacturing goods, and an ability to change production methods very rapidly in comparison to US factories. China's cost competitiveness is increasingly strengthened by advanced and improving infrastructure and high and rising R&D and skill levels (Mohiuddin 2017). It should be added that these cost advantages need not be permanent: Chinese unit labor costs have risen faster than those of the US, so China's overall cost advantage has diminished despite increasing productivity, and in some industries has even reversed (Hou, Gelb, and Calabrese 2017, pp. 5, 7, 40).

The Absolute Cost approach also provides an alternate route to Thirlwall's Law. This law is an empirical proposition that the growth rate of national output (g_Y) is roughly equal to the growth rate of exports (g_X) divided by the elasticity of imports (ε_M): $g_Y = \frac{g_X}{\varepsilon_M}$. Thirlwall bases himself on a Keynesian foundation, in which exports are an autonomous element of demand, so that the law is taken to mean that export growth drives output growth. On the further assumption that international trade is balanced $\left(\frac{X}{M} = 1\right)$ and the real exchange rate is constant, this defines a balance of payments constraint. The empirical strength of the original formulation, despite its assumption of balanced trade and a constant real exchange rate, initially led the author to believe that capital flows were relatively unimportant and that the real exchange rate played a minor role in balance of trade adjustments (Thirlwall and Hussain 1982, 498). However, subsequent elaborations of the model extended the Thirlwall result to the case of persistent trade imbalances covered by capital flows and slowly changing real exchange rates (Thirlwall 2011, pp. 335, 339–340). The latter two assumptions are *exactly* the ones implied by ACT and applicable to the US–China case: absolute cost differences lead to persistent trade imbalances, and long-run real exchange rates are regulated by slowly changing relative real costs of exports and imports. This also implies that capital flows and nominal exchange rate changes do not affect the long-run real exchange rate, and hence do not affect the balance of trade in a persistent way.

Our presentation of the ACT has addressed the same basic issue as the CCT at comparable levels of abstraction, so as to bring out the theoretical and empirical differences. Orthodox economists themselves have long noted that the standard theory fails at an empirical level. This failure has given rise to a large number of models that often contradict each other and seldom test well. Yet the basic logic of the CCT continues to dominate economics and to have a major influence on economic policy.

3. Monitoring currency manipulation – an integral element of U.S. foreign economic policy

The accusation against China for manipulating its currency, rooted in the CCT perspective on international trade, is central to President Trump's belligerent foreign trade policy. At the Republican convention at which he was nominated as the candidate, he pledged he would stop China's 'devastating currency manipulation' and added 'they are the greatest currency manipulators ever!' (Rauhala 2016). President Trump's Director of Trade and Industrial Policy, Peter Navarro (2011), proclaims 'Death by Currency Manipulation': 'China's manipulation of its currency, the yuan, is the tap root of everything wrong with the U.S.-China trade relationship.' (p. 67). This culminated in the designation of China as a currency manipulator in 2019 by the U.S. Treasury as part of the escalation of the trade war (2019b), ironically at a time when many economists (e.g. Setser 2019; Summers 2019) found that China is no longer manipulating the RMB and the IMF reported that China's external position is in line with fundamentals (2019b, 28).

The rhetoric of Trump and his advisors represents a new level of aggression, yet the claim of currency manipulation has long been a central concern of U.S. governments. Since the 1988 Omnibus Trade Act the U.S. Treasury issues semi-annual reports to 'consider whether countries manipulate the rate of exchange between their currency and the U.S. dollar for purposes of preventing effective balance of payments adjustments or gaining unfair competitive advantage in international trade' (Section 3004). A general pattern emerges from the U.S. Treasury reports over the last three decades: The U.S. government exerts continues pressure by threatening to accuse the East Asian surplus economies of currency manipulation and demands that they enter into bilateral negotiations on a wide array of market liberalization policies.[9] Initially the reports targeted the Asian Newly Industrialized Economies (NIE) (i.e. Korea, Taiwan, Hong Kong and Singapore), with whom the U.S. was running increasing trade deficits (U.S. Department of the Treasury 1989, 12). China has been on the monitoring list since 1991 but after 1994, when China unified its dual-track exchange rate, the Treasury had not found the legal criteria for currency manipulation fulfilled until 2019 (U.S. Department of the Treasury 2019b).

In the aftermath of the 2008 crisis, the tensions between the U.S. and China over the question of currency manipulation had reached a new height under the presidency of Barack Obama. Obama followed the logic of currency manipulation when he warned in 2010: 'One of the challenges that we've got to address internationally is currency rates and how they match up to make sure that our goods are not *artificially* inflated in price and their goods are *artificially* deflated in price. That puts us at a huge competitive disadvantage.' (Weaver 2010, emphasis added). The Obama administration enacted the Trade Facilitation and Trade Enforcement Act of 2015 in addition to the 1988 Act to strengthen the U.S. ability to take action against currency manipulation. Yet, all subsequent Treasury reports including that in 2019 were unable to establish that China was manipulating its currency.

4. Measuring the RMB misalignment

Despite great attention attributed to the RMB misalignment by policymakers, the degree of adjustment needed to reduce the Chinese current account surplus to a certain target level remains highly contested. The logic of the currency misalignment argument assumes that if there were no market distortions, the real exchange rate would converge to a certain *equilibrium level* that would bring the current accounts in line with selected *macroeconomic fundamentals*. However, several literature reviews have demonstrated that there is no consensus on how to determine the equilibrium exchange rate that brings about this external balance nor on the level of adjustment needed (Cheung and He 2019; Cheung 2012; Cheung, Chinn, and Fujii 2010a, 2010b; Dunaway and Leigh 2006; Cline and Williamson 2007). Since currency misalignment is defined as the deviation of the real exchange rate from its equilibrium level, there is consequently also no common measure of the currency misalignment (Cheung 2012). While there was a proliferation of studies aiming to estimate the RMB misalignment in the decade following China's accession to the WTO in 2001 and the analysis in this section focuses on this period, 'the search for a consensus on whether the Renminbi is undervalued continues' (Almås et al. 2017, 19).

This lack of consensus results from the current state of exchange rate economics. There is no generally accepted exchange rate model (Cheung, Chinn, and Fujii 2010b). Most widely applied are various incarnations of the Purchasing Power Parity (PPP) and the macroeconomic fundamental approaches (Ahmed 2009; Cheung, Chinn, and Fujii 2010a, 2010b; Dunaway and Leigh 2006; Cline and Williamson 2007). Both leave considerable room for judgment with regard to the model specifications (Cheung 2012; Dunaway and Leigh 2006). Consequently, studies using these approaches in order to estimate the misalignment of the RMB yield widely varying results. The following two sections analyze the PPP and the macroeconomic fundamentals approach on a theoretical level and provide an overview of the estimation results by studies conducted since China's accession to the WTO in 2001.

4.1. The purchasing power parity (PPP) approach

The PPP approach derives the equilibrium exchange rate directly from a comparison of price levels and estimates the degree of misalignment as the deviation of the actual real exchange rate from this equilibrium level, which also indicates the adjustment needed to overcome the current account imbalance. The PPP hypothesis is based on the law of one price (LoP), which states that in the absence of trading barriers and transaction costs *competition equalizes the prices of similar bundles of tradable goods across economies*. If e is the nominal exchange rate, P^* the foreign price index and P the domestic price index, then with no misalignment, the same bundle of goods would have the same price across countries denominated in a common currency[10]:

$$eP^* = P \qquad (1)$$

Since price level data is generally in terms of index numbers, this implies that the real exchange rate (Q) converges to a *stationary* value[11]:

$$Q = \frac{eP^*}{P} = \text{ constant} \tag{2}$$

The nominal exchange rate is expected to move so as to adjust the price levels so that their ratio is constant in the long run. If currencies are 'misaligned' by PPP standards, the real exchange rate as defined above will be non-stationary. According to standard trade theory, this might be either due to market interventions that prevent the equalization of prices or to currency interventions that hinder the nominal exchange rate from adjusting. The crucial challenge for an empirical evaluation of PPP is to find price indexes that represent comparable basket of goods in the countries of comparison (Shaikh and Antonopoulos 2012; Shaikh 2016, pp. 528–535).[12] Another fundamental obstacle to the PPP approach in the context of the debate over the RMB-dollar misalignment is that the United States can never be blamed for over- or undervaluation, as the dollar is the numeraire with an exchange rate always equal to the market rate (Cline and Williamson 2007).

PPP hypothesis has been found to be very weak empirically because real exchange rates do not converge to any stationary level (Rogoff 1996; MacDonald and Ricci 2001).[13] Therefore, a number of extensions have been suggested in hope of a better specification of the trade-balancing equilibrium exchange rate. These are usually classified as extended PPP approaches (Dunaway and Leigh 2006) or enhanced PPP approaches (Cline and Williamson 2007). We refer to both as *extended PPP approaches*. Most common is the introduction of the Balassa–Samuelson effect to accommodate the deviation of the real exchange rate from purchasing power parity (Bosworth 2004; Dunaway and Li 2005).

The ACT argument is different from that of the PPP. The latter takes the Law of One Price (LOP) to imply that real exchange rates, adjusted for the Balassa–Samuelson effect, will be stationary over the long run – on the assumption that baskets of goods are roughly similar in both countries. As noted, the PPP hypothesis performs very poorly at an empirical level because real exchanges are is decidedly not stationary, even over very long periods. The ACT argument is that the real exchange rate will track relative real cost, and since the latter can change over time, the real exchange rate should not be stationary. From this point of view, differences in relative real costs are an expression of differences in production 'baskets', so in order to adjust for differences in baskets we should look to see if the ratio of the real exchange rate to the underlying real production costs is stationary – which is exactly what the empirical evidence shows (Shaikh 2016, pp. 519, 531 text and Figure 11.6).

4.2. The macroeconomic fundamentals approach

The *macroeconomic fundamentals* approach attempts to arrive at various alternate specifications of the degree of real exchange rate 'misalignment' (e.g. Ahmed 2009; Borowski and Couharde 2003; Chinn and Prasad 2003; Cline 2007; Cline and Williamson 2007, 2010; Dunaway and Leigh 2006; Dunaway and Li 2005). The so-determined exchange rates are often called behavioral effective exchange rates (BEER)

(Cline and Williamson 2007).[14] We base our analysis on the influential standardization of this approach by the IMF (2006). There are three branches here: the macroeconomic balance (*MB*) approach; the equilibrium real exchange rate (*ERER*) approach; and the external sustainability (*ES*) approach.

The *MB* approach relies on a smoothed pooled sample of many countries spanning decades to econometrically 'explain' the sample current account through variables such as the fiscal balance, population growth, output growth, relative GDP, economic and banking crises, etc. The estimated sample coefficients are then applied to individual countries in light of their own levels of the explanatory variables in order to estimate the current account which should obtain in the country, its 'CA norm'. Finally, previously estimated elasticities of imports and exports are used to estimate the 'equilibrium' real exchange rate which would be needed to make a country's existing CA equal to its imputed norm. The difference between the actual real exchange rate and its MB equilibrium rate is the degree of its misalignment.

The reduced-form equilibrium real exchange rate (*ERER*) follows a similar approach but focuses directly on estimates of model-justified exchange rates. Panel regressions are used to estimate the relationship between real exchange rates and a set of fundamentals such as net foreign assets, productivity differentials, the terms of trade, trade restrictions, etc. The resulting cointegrating relation is used along with country-specific levels of the fundamentals to define ERER equilibrium real exchange rates, and the gap between these and actual exchange rates is deemed to be the degree of misalignment (IMF, 2006, pp. 13–18).

Finally, the external sustainability (*ES*) approach focuses on the sustainability of a country's CA. For each country this approach estimates a CA/GDP ratio which would be stabilize its net foreign assets (NFA) at some 'benchmark' levels in light of its estimated growth rate, inflation rate, and rates of return on external assets and liabilities. This leads in turn to an estimate of the ES real exchange rate needed to bring the actual current account into line with the estimated sustainable one. Once again, the difference between the actual exchange rate and the estimated equilibrium rate is the former's putative degree of misalignment (International Monetary Fund (IMF) 2006, pp. 18–23).

4.3. The unreliability of estimates of exchange rate misalignment

Given the challenges in measuring exchange rate misalignment, it is hardly surprising that there is no agreement on the degree of misalignment. Indeed, there is not even a consensus on whether the RMB is overvalued or undervalued. The estimates of RMB misalignment in percent for the period of 2000 to 2011 as collected in the four literature reviews (Cline and Williamson 2007; Dunaway and Li 2005; Cheung, Chinn, and Fujii 2010a; Cheung 2012) are plotted in Figure 4 and Figure 5 for estimates based on the extended PPP and the macroeconomic fundamentals approach separately. All PPP approaches that go beyond a simple PPP-based analysis are classified as extended PPP (including Penn effect, and various versions of BEER approaches). All approaches that follow the procedure sketched above as macroeconomic fundamental approach are labeled as such (including FEER). As demonstrated in earlier literature reviews, similar approaches yield a wide range of results even for the same year of estimation and even if conducted by the same authors. For example, MacDonald and Dias (2007) using the same BEER approach estimate a required appreciation of the RMB ranging from 27.3 to 46.6% depending on the definition of the trade-balancing scenario.

All literature reviews point to the fact that the estimation of currency misalignment is highly sensitive to the choice of variables, the equation specification, the sample period and country sets in the panel econometric estimations. For example, Cheung, Chinn, and Fujii (2010a) show that for 2009 an undervaluation of RMB varying from 1.6 to 38% can be generated using the same extended PPP approach while altering the sample period from 1990–2009 to 1980–2009, respectively. Further, sensitivity tests for the extended PPP approach suggest that China's real equilibrium exchange rate can vary widely when dropping one country from the panel or changing the proxies, because the estimated coefficients for explanatory variables change (Dunaway and Leigh 2006). Authors who have used the same methodology and similar specification when applying the macro-economic fundamental approach also deliver estimates of substantial variation (Dunaway and Leigh 2006; Bénassy-Quéré et al. 2004; Cline and Williamson 2007; Cheung, Chinn, and Fujii 2010b). For example Wang (2004) finds only a slight under-valuation whereas Goldstein (2004) estimates the RMB to be undervalued by between 15% and 30%. We share the conclusion of Schnatz (2011) that the methodologies used for assessing the 'fair value' of a currency vary significantly with the specific assumptions chosen by the modeler.

One crucial challenge is to define the trade elasticities, which determine the claimed link between the exchange rate appreciation and the current account. For example, Ahmed (2009) finds that a 20% appreciation of the RMB induces a USD 400 billion decrease in the Chinese exports after four years. Cheung, Chinn, and Fujii (2010a), on the other hand, only find an impact of USD 50 billion for the same appreciation. Consequently, the robustness of estimations of the equilibrium RMB exchange rate is very weak (Dunaway and Leigh 2006, p. 3; Cheung, Chinn, and Fujii 2010b). This causes serious problems of statistical significance. Cheung, Chinn, and Fujii (2007) concludes: 'One general observation is that, when one implements the standard operating procedure of accounting for sampling uncertainty in making inferences, there is no evidence supporting the claim that the RMB is substantially undervalued, using conventional significance levels.' (p. 20). The International Monetary Fund (IMF) (2006), for example, states: 'While the econometric model captures the broad trends in real exchange rate behavior, estimates of equilibrium exchange rates are unavoidably subject to significant uncertainty.' (p. 18)

We argue that the failure of the various PPP and macroeconomic fundamentals approaches to capture the alleged RMB misalignment (see Section 4) is not simply a matter of measurement, but of substantive theory. The presumption of CCT underlies the ways in which PPP and the macroeconomic fundamental approach are utilized to estimate exchange rate misalignments.

For the case of PPP, the Ricardian assumption of automatically balancing trade is imputed even though this is not required by the LoP. The LoP only says that prices of internationally traded goods will be roughly the same in different countries. This is perfectly consistent with unbalanced trade, because one country's exports can be different from its imports, even if every commodity in both sets has the same international price. The LoP in turn implies PPP only if countries have the same basket of internationally traded goods, which they assuredly do not. Since the LoP does not imply balanced trade, neither does PPP.

It follows that one cannot use PPP as a proxy for a trade-balancing equilibrium exchange rate, as is done in the currency manipulation argument. Extended PPP also assumes a stationary equilibrium of balanced trade in the long run, but accounts for existing imbalances via various supposed macroeconomic fundamentals – on which basis it esti-mates the currency re-alignment needed for trade balance. Neither approach considers structural differences in cost competitiveness as the cause of persistent trade imbalances.

All three branches of the macroeconomic fundamentals approach discussed in the previous section share the conclusion that actual exchange rates are misaligned or even manipulated because the observed current account adjusted for full capacity utilization does not match some theoretically expected one. And all three conclude that the putative current account gap can be closed through an appropriate adjustment of the real exchange rate which would occur if governments would abstain from exchange rate interventions. These are CCT assumptions to the core.

CCT not only underlies the PPP and macroeconomic fundamental approaches, it is also the foundation for several other explanations of the US–China trade imbalance.

5. Other comparative cost explanations of the U.S.–China trade imbalance

ACT implies that real cost differences give rise to trade imbalances covered by endogen-ous capital flows. This is consistent with the US–China trade relation. On the other hand, CCT implies balanced trade, in which case various exogenous factors must account for the observed trade imbalance: foreign exchange intervention, the role of the dollar as international reserve currency, neo-mercantilist development policy, or a global savings glut. This section shows how all these strands of literatures on persistent trade imbalances are ultimately rooted in the logic of CCT.

5.1. Foreign exchange intervention theory

One explanation of the current account deficit in the US is that it arises from various state interventions in China and elsewhere (e.g. Gagnon 2017). The foreign exchange inter-vention theory often assumes that capital inflows can be sterilized because they represent temporary 'shocks' to the system. For example, Bayoumi, Gagnon, and Saborowski (2015, pp. 147–150, 172) begin with the standard CCT assumption that in the absence of private financial capital flows, a net official outflow will cause a current account surplus in China and hence a deficit in the US. However, the Chinese official capital outflow will decrease its capital stock and raise the marginal product of capital, which will in turn stimulate a private capital inflow into China until it just offsets the net official outflow. It follows that in 'a world with efficient markets and perfect capital mobility flows', China's net official outflows designed to sterilize a current account surplus will have no effect on the current account because they are fully offset by private flows[15]: sterilization will not affect the current account surplus. On the other hand, 'when capital mobility is imperfect', a net official outflow from a country will not be fully offset and will therefore give rise to a current account surplus. On this CCT reasoning, China's trade surplus, and hence the US trade deficit is *caused* by China's net official outflow to the US. This is their explanation for a variety of empirical findings, including their own, of an association between net official flows and the current account balances. It follows that 'a policy change to cease net official flows and allow the exchange rate to appreciate would lead to a significant decline' (p. 172) in China's current account surplus.

In our argument, the absolute cost disadvantage of the US gives rise to a trade deficit offset by covering endogenous capital inflows. From this starting point, a net official flow from China into the US will raise the nominal exchange rate of the USD. The inflow will also increase liquidity in the US and reduce its interest rate below the bilateral arbitrage equality level. The appreciation of the nominal exchange rate will only have short-term effects on the US cost disadvantage, so the US trade deficit will revert. On the other hand, the reduction in the US interest rate will inhibit some of the endogenous capital inflow from China. The exogenous capital inflow will therefore *crowd out* the endogenous one until the net capital inflow covers the trade deficit, with the US interest rate now below that of China – as has been observed over the period in which the bilateral trade imbalance rose significantly (IMF 2019c). This explains the previously cited empirical associations between net official flows and current account imbalances. Notice that this ACT result follows from the US cost disadvantage, not from imperfections in capital markets as in the CCT explanation.

The foregoing argument has been predicated on the free movement of Chinese short-term capital. However, China's outbound capital account had been liberalized only to a very limited extent in the first decade after the WTO accession in 2001 (He et al. 2012, 1). Consequently, the short-term capital outflows that we would expect from the ACT perspective to come forth to counterbalance the trade surplus had the capital account been fully liberalized is limited by capital controls. The increase in official reserves we have observed is necessary to close the gap. In this sense, Chinese foreign exchange intervention could be viewed as a *surrogate* for a market mechanism, rather than a market distortion (Harrod 1957, 85–6).

5.2. Dollar as international reserve currency

An alternate claim is that the US trade deficit is due to the role of the USD as an international reserve currency (e.g. McKinnon 2001; Bergsten 2009). This 'Triffin Paradox' says that the reserve currency role of the USD generates exogenous capital inflows into the US as foreigners seek to acquire dollar-denominated foreign exchange reserves mostly consisting of US Treasury securities (McKinnon 2001, 231). This appreciates the US currency above its trade-balanced value and gives rise to a US trade deficit. At the same time, it depreciates the RMB below its own corresponding value and gives rise to a Chinese trade surplus – all through the workings of international financial markets (Bordo and McClauley 2018). Once again, the underlying assumption is the CCT notion that exogenous capital inflows – as opposed to endogenous capital inflows that ACT would expect to be brought about by a trade imbalance – cause the US trade deficit (Bergsten 2009, 20). Our own argument is that an exogenous capital inflow will crowd out the endogenous one without eliminating the current account deficit, as laid out in the preceding discussion of foreign exchange inter-vention theory. Here, we only add the observation that the US ran a trade *surplus* from 1960 to 1970 despite the fact that the dollar was a reserve currency (Shaikh 2016, p. 534, Figure 11.7; McKinnon 2001), which is further evidence against the reserve currency hypothesis.

5.3. Neo-mercantilist development policy

While the currency manipulation argument blames China for distorting international trade flows and undermining the stability of the global economy, a contemporary version of the mercantilist theory sees the same RMB misalignment as part of an export-oriented development strategy inherent in the initial and revived Bretton Woods system (see e.g. Dooley, Folkerts-Landau, and Garber 2003, 2007). This only differs from the currency manipulation theory on the question of motivations: the currency manipulation argument claims that China misaligns its currency in order to create a trade surplus, the neo-mercantilist perspective says China does it in order to foster export-led growth. Both arguments are rooted in the CCT claim that China's trade surplus is due to its interference in free trade, as opposed to the ACT claim that it is rooted in an absolute cost advantage. We might also ask, if currency manipulation has supposedly enabled some East Asian countries to achieve export-led growth on the unprecedented scale, why is it that large parts of the developing world – in particular in Africa, South- and Southeast Asia, and Central America – continue to run substantial trade deficits (United Nations Commission on Trade and Development (UNCTAD) 2019, 17) instead of simply depreciating their currencies in order to gain export competitiveness? From our perspective, this is because building up high productivity export-oriented industries is a slow and expensive process of industrialization and infrastructure building. Low wages are not sufficient: Despite low wages in Sub-Saharan Africa, unit labor costs there continue to be higher than in China, and their trade deficits have been persistent (Golup et al., 2018).

5.4. Global saving glut hypotheses

The savings glut thesis of Ben Bernanke (2005, 2007) claims that it is an increase in net savings abroad – especially in China – and a shortfall of savings in the U.S. which led to net capital exports to the U.S. and thereby caused its current account deficit. As an accounting identity, an excess of domestic investment over the sum of domestic savings (the difference between private disposable income and private consumption and government savings) must be equal to the current account deficit.[16] In Bernanke's reading the causation flows from domestic excess investment to the current account deficit: countries whose domestic savings fall short of their domestic investment needs will import more than they export. Therefore, Bernanke suggests that the 'proximate cause of the increase in the U.S. external deficit was a decline in U.S. saving; between 1996 and 2004, the investment rate in the United States remained almost unchanged at about 19% of GDP, whereas the saving rate declined from 16% to 1/2% to slightly less than 14% of GDP'.

In Bernanke's framework 'domestic investment not funded by domestic saving must be financed by capital flows from abroad, and, indeed, the large increase in the U.S. current account deficit was matched by a similar expansion of net capital inflows'. However, Bernanke points out that on a global level 'national current account deficits and surpluses must balance out, as deficit countries can raise funds in international capital markets only to the extent that other (surplus) countries provide those funds.' This leads him to his view on the cause of the Chinese trade surplus: 'it is not surprising that the widening of the U.S. current account deficit has been associated with increased current account surpluses in the rest of the world ... (and) much of the increase in current account surpluses during this period took place in developing countries rather than in the industrial countries'. In these countries, desired savings exceeded desired investment. In the particular case of China, savings rates rose faster than investment rates, 'leading to an increase in that country's current account surplus'. This is what Bernanke calls 'the global savings glut' (Bernanke 2007, 1–2).

In our case, the causation is exactly the reverse. The ACT hypothesis implies that the cost competitiveness of Chinese producers gives rise to China's trade surplus, which is why its national savings exceeds its domestic investment – i.e. why it becomes an exporter of financial capital. Ferguson and Schularick (2011) point to studies by Kuijs (2005) and Wolf (2009) that the increase in Chinese savings came from retained corporate earnings of private and state enterprises, not from households as Bernanke would have it. This finding is perfectly consistent with the ACT: If Chinese cost competitiveness results in a relative trade surplus we would expect an increase in corporate savings.

6. Conclusion

The currency manipulation argument remains a main point of reference for policy-makers in the U.S. despite the fact that the link between foreign exchange market interventions, exchange rate devaluation and current account imbalances has not been established empirically. Most studies focus on the estimation of exchange rate under-valuation approximated with respect to a statistically derived equilibrium exchange rate. However, there is no consensus on the degree nor the direction of misalignment.

Standard trade models trying to estimate the RMB misalignment are rooted in the Ricardian notion of comparative advantage. This holds true for the extended PPP and the macroeconomic fundamental approaches, the most widely used ones the estimation of currency misalignment. The currency manipulation argument and the misalignment measurements rely on the CCT assumption that in the absence of market interventions real exchange rates would change such as to make the trade balance adjust to the capital account. The same assumption underpins other explanations of the US–China trade imbalance such as the savings or money glut thesis, as well as those that see the US deficit as a result of the USD's role as international reserve currency.

Some economists who had previously argued vigorously that it would be vital for the U.S. economy to prevent China from manipulating its currency have recently come to take a more cautious stance. For example, Paul Krugman who in 2010 laid out the currency manipulation argument and called for 'Taking on China and its Currency' has come to downplay the importance of the trade deficit with China (Krugman 2018). The fact that actual patterns of trade are so different from those derived from orthodox trade theory has become a pressing concern and we share with many economists the goal of showing why this is the case. But, we think that the problems cannot be solved when staying within the framework of the theory of comparative cost.

Absolute Cost Theory (ACT) based on real cost differences expects persistent trade imbalances even in the context of free international competition. Trade imbalances tend to cause interest rate differentials that induce short-term capital flows in the opposite direction to the net trade flow, thus clearing the balance of payments. Persistent trade imbalances themselves arise from differences in real costs of production: countries that have higher absolute real costs will tend to run a trade deficit and become international debtors. The opposite happens to countries with superior price competitiveness. Exogenous capital flows and foreign exchange interventions can temporarily change the nominal exchange rate, but not the long-run real exchange rate and the trade balance, both of which remain tied to real cost competitiveness.

Notes

1. Keynes also rejects the theory of comparative advantage (Milberg 1994, 2002).
2. Essay 2 shows how Ricardo's arguments were developed in debate with Thornton. Thornton's theory presents some kind of a middle ground between ACT and CCT but cannot be clearly located due to the lack of a value theory in his work.
3. Marx (1961) in Chapter 2, Section C, of *A Contribution to the Critique of Political Economy* discusses James Steuart and Thomas Tooke as presenting arguments similar to what is identified here as ACT and supports the validity of their arguments against those presented by Montesquieu, Hume, Ricardo and Mill all broadly in line with what we call CCT. Marx (1975) continues his discussion of currency circulation and trade in Chapter 34 of Capital Volume III with an abundance of historical detail on the English 1844 Bank Law. In this discussion Marx articulates his support for ACT but does not present a theory of his own.
4. This is precisely the proposition that orthodox economists such as Krugman and Obstfeld (1994) have long dismissed as a 'myth' because they adhere to the theory of comparative cost.
5. Marx makes this point in Volume III of Capital. 'If gold is exported, then, according to this Currency Theory, commodity-prices must rise in the country importing this gold, and thereby the value of exports from the gold-exporting country on the gold-importing country's market; on the other hand, the value of the gold-importing country's exports would fall on the gold-exporting country's market while it would rise on the domestic market, *i.e.*, the country receiving the gold. But, in fact, a decrease in the quantity of gold raises only the interest rate, whereas an increase in the quantity of gold lowers the interest rate; and if not for the fact that the fluctuations in the interest rate enter into the determination of cost-prices, or in the determination of demand and supply, commodity prices would be wholly unaffected by them.' (Marx 1975, 547)
6. For simplicity in exposition, this abstracts from fixed capital.
7. In formal terms: Let a be the inverse of productivity, i.e. the unit labor requirement in manufacturing: $a = \frac{L}{Q}$, where Q is value added and L is the labor required. Note that these are absolute levels, not index numbers. The relative unit labor cost (RULC) is defined as: RULC $= \frac{aw}{a*w*e} = \left(\frac{a}{a*}\right)\left(\frac{w}{w*e}\right)$, where we is the average labor compensation per worker in manufacturing and e is the exchange rate expressing domestic currency per unit of foreign currency.
8. This finding is consistent with Ceglowski and Golup (2012) and Ferguson and Schularick (2011). Ceglowski and Golup (2012, 15) also show that China's relative unit labor costs are lower than that in South-Korea, Japan, Mexico, and the EU. The breakdown of unit labor costs in terms of real wages and real output used by Ceglowski and Golub (2012) is consistent with our foregone analysis. They use producer prices but suggest that consumer prices would be preferable. The only difference is that they assume that PPP holds in general whereas we suggest to limit this to tradable goods which gives rise to our incorporation of the Samuelson-Balassa effect.
9. For example, the 2019 report the Treasury urges China to pursue 'meaningful structural reforms to reduce the role of the state in the economy' and to 'permit a greater role for market forces' (U.S. Department of the Treasury 2019a, 23).
10. Note that the composition of goods as well as the ratio of tradables to nontradables must be the same in both price indexes P^* and P. This is further discussed in Section 2.

11. See Shaikh (2016, pp. 517–527) for a detailed discussion of this point.

12. The simplest and methodologically most questionable PPP approach is the Big Mac index, compiled by *The Economist*. It tries to get around the problem by using a basket containing just one good, the Big Mac.

13. The real exchange rate is not stationary over the short run (Isard 1995, pp. 63–65)(Isard 1995). It does revert to a 'target level' over runs of 10–20 years, but this is not the PPP level (Engel 2000, 21). Even if there was a reversion to a non-stationary mean, the 'speed of convergence is extremely slow' (Rogoff 1996). For a more detailed discussion see Shaikh and Antonopoulos (2012) and Shaikh (2016, pp. 522–527).

14. The IMF calls a similar approach 'reduced form equilibrium real exchange rate' in its 'Methodology for CGER Exchange Rate Assessments' (2006). Ahmed (2009) subsumes what Cline and Williamson (2007) call BEER under the label extended PPP approaches.

15. Various studies have found that sterilization is 'difficult to execute and sometimes even self-defeating', and that 'when capital is highly mobile, attempts at sterilization ... cannot work for long if the capital inflows persist' (Jang-Yung Lee 1997).

16. In national accounts, aggregate domestic and foreign demand (C + I + G + X) is equal to aggregate domestic and international supply (Y + M), where C = consumption, I = investment, G = government spending, X = exports, Y = domestic supply (national income) and M = imports. Further, let T = aggregate taxes. Then, relative to national income (Y), we can say that the excess of investment I over the sum of domestic savings (the difference between private disposable income (Y-T) and private consumption (C) and government savings (G-T) is equal to the current account deficit (M-X). Expressed formally: $\frac{I-[(Y-T-C)+(T-G)]}{Y} = \left(\frac{M-X}{Y}\right)$.

17. Figure 4 and Figure 5 are compiled based on the selection of studies contained in the four literature reviews (Cline and Williamson 2007; Dunaway and Li 2005; Cheung, Chinn, and Fujii 2010a; Cheung 2012). If a source reported more than one estimate of the RMB misalignment, each estimate was treated as a separate data point in the scatter plots. If the estimate of the RMB misalignment is reported as a range, the maximum and minimum value are reported in the scatter plots as two separate estimates.

18. See footnote 12.

Disclosure statement

No potential conflict of interest was reported by the author(s).

ORCID

Isabella Weber (iD) http://orcid.org/0000-0003-0694-8823

References

Ahmed, S. 2009. "Are Chinese Exports Sensitive to Changes in the Exchange Rate?" Board of Governors of the Federal Reserve System International Finance Discussion Papers, No. 987.

Almås, I., M. Grewal, M. Hvide, and S. Ugurlu. 2017. "The PPP Approach Revisited: A Study of RMB Valuation against the USD." *Journal of International Money and Finance* 77: pp. 18–38.

Amin, S. 1974. *Accumulation on A World Scale: A Critique of the Theory of Underdevelopment.* Vol. 2. New York: Monthly Review Press.

Arndt, S., and J. Richardson, eds. 1987. *Real Financial Linkages among Open Economies.* Cambridge, MA: MIT.

Bayoumi, T., J. Gagnon, and C. Saborowski. 2015. "Official Financial Flows, Capital Mobility, and Global Imbalances'." *Journal of International Money and Finance* 52: 146–174.

Bénassy-Quéré, A., P. Duran-Vigneron, A. Lahrèche-Révil, and V. Mignon. 2004. "Burden Sharing and Exchange Rate Misalignments,'." In *Dollar Adjustment: How Far? Against What?*, edited by C. F. Bergsten and J. Williamson, 69–94. Washington: Institute for International Economics.

Bergsten, C. 2006. "The US Trade Deficit and China", Testimony before the Hearing on US-China Economic Relations Revisited Committee on Finance, United States Senate, Peterson Institute for International Economics, Washington.

Bergsten, C. 2007. "Currency Misalignments and the US Economy", Testimony before the Subcommittees on Trade, Ways and Means Committee, Commerce, Trade and Consumer Protection, Energy, and Commerce Committee, and Domestic and International Monetary Policy, Trade and Technology, Financial Services Committee, House of Representatives, Washington.

Bergsten, C. 2009. "The Dollar and the Deficits: How Washington Can Prevent the Next Crisis", Foreign Affairs, 88, 20–38.

Bergsten, C. 2010. "Correcting the Chinese Exchange Rate: An Action Plan", Testimony before the Committee on Ways and Means, US House of Representatives, Peterson Institute for International Economics, Washington.

Bergsten, C., and J. Gagnon 2012. "Currency Manipulation, the US Economy and the Global Order", Peterson Institute for International Economics Policy Brief, No. PB1225.

Bernanke, B. 2005. "The Global Saving Glut and the U.S. Current Account Deficit", Speech 77, Board of Governors of the Federal Reserve System (U.S.).

Bernanke, B. 2007. "Global Imbalances: Recent Developments and Prospects", Bundesbank Lecture, September 11.

Bordo, M. 2005. "Historical Perspective on Global Imbalances", National Bureau of Economic Research (NBER), Working Paper Series, No. 11383.

Bordo, M., and R. McClauley 2018. "Triffin: Dilemma or Myth?", National Bureau of Economic Research (NBER), Working Paper Series, No. 14195.

Borowski, D., and C. Couharde. 2003. "The Exchange Rate Macroeconomic Balance Approach: New Methodology and Results for the Euro, the Dollar, the Yen and the Pound Sterling." *Open Economies Review* 14 (2): 169–190.

Bosworth, B. 2004. "Valuing the Renminbi", Tokyo Club Rewearch Meeting, February.

Ceglowski, J., and E. Golub. 2012. "Does China Still Have a Labor Cost Advantage?" *Global Economy Journal* 12 (3): 1–30.

Cheung, Y. (2012. "Exchange Rate Misalignment: The Case of the Chinese Renminbi", CESifo Working Paper, Monetary Policy and International Finance, No. 3797.

Cheung, Y., M. Chinn, and E. Fujii 2007. "The Overvaluation of Renminbi Undervaluation", National Bureau of Economic Research, Working Paper Series, No. 12850.

Cheung, Y., M. Chinn, and E. Fujii 2010a. "Measuring Misalignment: Latest Estimates for the Chinese Yuan", La Follette School Working Papers, No. 2010–010.

Cheung, Y., M. Chinn, and E. Fujii 2010b. "Measuring RMB Misalignment: Where Do We Stand?", Hong Kong Institute for Monetary Research Working Paper, No. 24/2010.

Cheung, Y., and S. He. 2019. "Truths and Myths about RMB Misalignment: A Meta-analysis." *Comparative Economic Studies* 61 (3): 464–492.

Chinn, D. (2013. "American Debt, Chinese Anxiety", The New York Times, October 20.

Chinn, D., and E. Prasad. 2003. "Medium-Term Determinants of Current Accounts in Industrial and Developing Countries: An Empirical Exploration." *Journal of International Economics* 59: 47–76.

Cline, W. R. 2007. "Estimating Reference Exchange Rates". presented at a workshop at the Peterson Institute for International Economics, Washington.

Cline, W. R., and J. Williamson 2007. "Estimates of the Equilibrium Exchange Rate of the Renminbi: Is There a Consensus And, if Not, Why Not?", Paper presented at the Conference on China's Exchange Rate Policy, Peterson Institute for International Economics, Washington DC, October 12.

Cline, W. R., and J. Williamson 2010. "Notes on Equilibrium Exchange Rate", Peterson Institute for International Economics Policy Brief, PB 10–2, January.

Dooley, M., D. Folkerts-Landau, and P. Garber 2003. "An Essay on the Revived Bretton Woods System", National Bureau of Economic Research (NBER) Working Paper 10331.

Dooley, M., D. Folkerts-Landau, and P. Garber 2007. "The Two Crises of International Economics", National Bureau of Economic Research (NBER) Working Paper 13197.

Dunaway, S., and L. L. Leigh 2006. "How Robust are Estimates of Equilibrium Real Exchange Rates: The Case of China", International Monetary Fund Working Paper, WP/06/220.

Dunaway, S., and X. Li 2005. "Estimating China's Equilibrium" Real Exchange Rate", International Monetary Fund Working Paper, WP/05/202.

Engel, C. 2000. "Long-Run PPP May Not Hold After All." *Journal of International Economics* 57: 243–273.

Ferguson, N., and M. Schularick. 2011. "The End of Chimerica." *International Finance* 14 (1): 1–26.

Gagnon, J. 2017. "Do Governments Drive Global Trade Imbalances?", Peterson Institute for International Economics, Working Paper, December.

Goldstein, M. 2004. "Adjusting China's Exchange Rate Policies", Paper presented to the International Monetary Fund Seminar on China's Foreign Exchange System at Dalian, China.

Golub, S., J. Ceglowski, A. Mbaye, and V. Prasad. 2018. "Can Africa Compete with China in Manufacturing? the Role of Relative Unit Labor Costs." *The World Economy* 41: 1508–1528.

Harrod, R. 1957. *International Economics*. 3 ed ed. Chicago: University of Chicago Press.

Harvey, J. T. 1996. "Orthodox Approaches to Exchange Rate Determination: A Survey." *Journal of Post Keynesian Economics* 18 (4): 567–583.

He, D., L. Cheung, W. Zhang, and T. Wu. 2012. "How Would Capital Account Liberalisation Affect China's Capital Flows and the Renminbi Real Exchange Rates?", Hong Kong Institute for Monetary Research Working Paper, No. 09/2012.

Hou, J., S. Gelb, and L. Calabrese 2017. "The Shift In Manufacturing Employment In China', Supporting Economic Transformation (SET) Background Paper", online: https://set.odi.org/wp-content/uploads/2017/08/SET-China_Shift-of-Manufacturing-Employment-1.pdf

International Monetary Fund (IMF). 2006. "Methodology for CGER Exchange Rate Assessments", International Monetary Fund Research Department.

International Monetary Fund (IMF). 2010. "People's Republic of China: 2010 Article IV Consultation -staff Report". IMF Country Report, No. 10/238.

International Monetary Fund (IMF). 2018. "World Economic Database. Reports for Selected Country and Subject", online: https://www.imf.org/

International Monetary Fund (IMF). 2019a. "External Sector Report: The Dynamics of External Adjustment", Washington D.C., July.

International Monetary Fund (IMF). 2019b. "People's Republic of China: Article IV Consultation – Press Release", IMF Country Report No. 19/266, Washington D.C.

International Monetary Fund (IMF). 2019c. "International Financial Statistics: Interest Rates Selected Indicators – United States and China", online: https://data.imf.org/regular.aspx?key=61545855

Isard, P. 1995. *Exchange Rate Economics*. Cambridge: Cambridge University Press.

Krugman, P. 2009. "World Out of Balance", The New York Times, November 15.

Krugman, P. 2018. "The Art of the Flail", The New York Times, April 5.

Krugman, P., and M. Obstfeld. 1994. *International Economics*. New York: Harper Collins.

Kuijs, L. 2005. "Investment and Saving in China", The World Bank Office Beijing, Working Paper Series 3633.

Lee, J.-Y., 1997. "Sterilizing Capital Inflows". *Economic Issues* 7 (March). onlinehttps://www.imf.org/external/pubs/ft/issues7/index.htm

Lin, J., H. Dinh, and F. Im. 2010. "US–China external imbalance and the global financial crisis." *China Economic Journal* 3 (1), 1-24.

MacDonald, R., and L. Ricci 2001. "PPP and the Balassa Samuelson Effect: The Role of the Distribution Sector", Area Conference on Macro, Money and International Finance, CESifo Conference Centre, Munich.

MacDonald, R., and P. Dias 2007. "Behavioural Equilibrium Exchange Rate Estimates and Implied Exchange Rate Adjustments for Ten Countries", presented to a workshop on Global Imbalances at the Peterson Institute for International Economics, February, Washington.

Marx, K. 1961. *Marx Engels Gesamtausgabe, Band 13*. Berlin: Dietz Verlag.

Marx, K. 1975. *Capital: Volume III*. International Publishers Co: Inc, New York.

McKinnon, R. 2001. "The International Dollar Standard and the Sustainability of the U.S. Current Account Deficit." *Brookings Papers on Economic Activity* 2001 (1): 227–239.

Milberg, W. 1994. "'Is Absolute Advantage Passé? Towards a Post-Keynesian/Marxian Theory of International Trade'." In *Competition, Technology and Money: Classical and Post-Keynesian Perspectives*, edited by M. A. Glick, 219–236. Aldershot: Edward Elgar Publishing.

Milberg, W. 2002. "'Say's Law in the Open Economy: Keynes's Rejection of the Theory of Comparative Advantage'." In *Post Keynesian Econometrics, Microeconomics and the Theory of the Firm: Beyond Keynes, Volume One*, edited by S. C. Dow and J. Hillard, 239–253. Aldershot: Edward Elgar Publishing.

Mohiuddin, O. 2017. "China Still Lucrative for Businesses despite the Rising Wage Rates", Euromonitor International, online: https://blog.euromonitor.com/china-still-lucrative-businesses-despite-rising-wage-rates/

Navarro, P. 2011. *Death by China: Confronting the Dragon – A Global Call to Action*. New Jersey: Pearson Education.

Obstfeld, M., and K. Rogoff. 2005. *'Global Current Account Imbalances and Exchange Rate Adjustments'*. Washington: Brookings Institute.

Obstfeld, M., and K. Rogoff 2009. "Global Imbalances and the Financial Crisis: Products of Common Causes", CEPR Discussion Paper, No. 7606.

Rauhala, E. 2016. "Trump Blasts China, China, China, in Republican Convention Speech", The Washington Post, July 22.

Ricardo, D. 1951. *On the Principles of Political Economy and Taxation, First Published 1817*. Cambridge: Cambridge University Press.

Rogoff, K. 1996. "The Purchasing Power Parity Puzzle." *Journal of Economic Literature* 34: 647–668.

Schnatz, B. 2011. "'Global Imbalances and the Pretence of Knowing Fundamental Equilibrium Exchange Rates'." *Pacific Economic Review* 16 (5): 604–615.

Scott, R., H. Jorgensen, and D. Hall 2013. "Reducing U.S. Deficit Will Generate a Manufacturing-Based Recovery for the United States and Ohio. Ending Currency Manipulation by China and Others Is the Place to Start", Economic Policy Istitute Report, Ohio.

Setser, B. 2019. *Is China Manipulating Its Currency? Council on Foreign Relations*. Accessed 10 September 2019. Retrieved from https://www.cfr.org/in-brief/china-manipulating-its-currency

Shaikh, A., and R. Antonopoulos. 2012. "'Explaining Long-Term Exchange Rate Behavior in the United States and Japan'." In *Alternative Theories of Competition: Challenges to the Orthodoxy*, edited by J. Moudud, C. Bina, and P. Mason, 201-228. Abington: Routledge.

Shaikh, A. 2016. *Capitalism: Competition, Conflict, Crises*. Oxford University Press: New York, NY.

Smith, A. 1904. *An Inquiry into the Nature and Causes of the Wealth of Nations*. 5 ed. Methuen & Co: London. first published 1776.

Stein, J. 1995. *Fundamental Determinants of Exchange Rates*. Oxford: Clarendon Press.

Summers, L. 2019. "By Naming China a Currency Manipulator, Mnuchin Has Damaged His Credibility", Washington Post, 6 August, online: https://www.washingtonpost.com/opinions/2019/08/06/by-naming-china-currency-manipulator-mnuchin-has-damaged-his-credibility/?noredirect=on

Thirlwall, A. P. 2011. "Balance of Payments Constrained Growth Models: History and Overview." *PSL Quarterly Review* 64 (259): 307–351.

Thirlwall, A. P., and N. M. Hussain. 1982. "The Balance of Payments Constraint, Capital Flows and Growth Rate Differences betweenDeveloping Countries." *Oxford Economic Papers, New Series* 34 (3): 498–510.

Turner, A. G., and S. S. Golub. 1997. "Towards a System of Unit Labor Cost-based Competitiveness Indicators for Advanced, Developing and Transition Countries." In

International Monetary Fund, Staff Studies for the World Economic Outlook, pp. 47–60. Washington, DC: IMF.

Turner, P., and J. Van't Dack. 1993. "Measuring International Price and Cost Competitiveness (Bank for International Settlements Economic Paper No. 39, November)".

U.S. Department of the Treasury. 1989, April. "Report to the Congress on International Economic and Exchange Rate Policy". From U.S. Department of the Treasury Office of International Affairs: https://www.treasury.gov/resource-center/international/exchange-rate-policies /Documents/Report%20to%20the%20Congress%20on%20International%20Economic%20and %20Exchange%20Rate%20Policies%20-%20April,%201989.pdf

U.S. Department of the Treasury. 2019. "Report to Congress. Macroeconomic and Foreign Exchange Policies of Major Trading Partners of the United States". May, 2019 Accessed 30 December 2019. Retrieved from https://home.treasury.gov/system/files/206/2019-05-28-May-2019-FX-Report.pdf.

U.S. Department of the Treasury. 2019a, April. "Report to Congress. Macroeconomic and Foreign Exchange Policies of Major Trading Partners of the United States". From U.S. Department of the Treasury Office of International Affairs: https://home.treasury.gov/system/files/206/2019-05-28-May-2019-FX-Report.pdf

U.S. Department of the Treasury. 2019b. "Press Release: Treasury Designates China as Currency Manipulator', from U.S". Department of the Treasury: https://home.treasury.gov/news/press-releases/sm751

United Nations Commission on Trade and Development (UNCTAD). 2019. "Key Statistics and Trends in International Trade 2018", online: https://unctad.org/en/PublicationsLibrary/ditc tab2019d2_en.pdf

Wang, T. 2004. "'Exchange Rate Dynamics'." In *China's Growth and Integration into the World Economy: Prospects and Challenges*, edited by E. Prasad, 21-28. Washington: International Money Fund.

Weaver, M. 2010. "China Hits Back at Obama's Claim that Yuan Is Undervalued", The Guardian, 4. February.

Wolf, M. 2009. *Fixing Global Finance*. New Haven: Yale Univeristy Press.

Yeager, L. 1966. *International Monetary Relations: Theory, History and Policy*. New York: Harper & Row.

Appendix

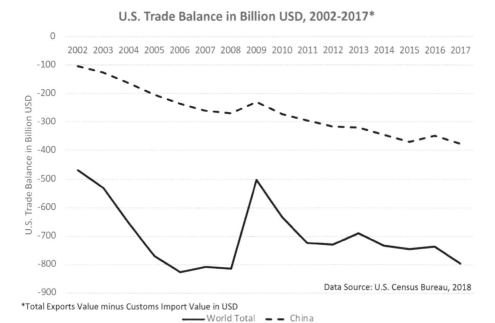

Figure 1.

Figure 2.

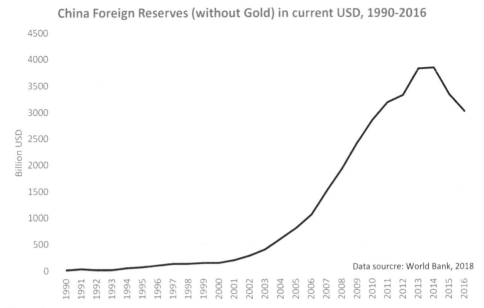

Figure 3.

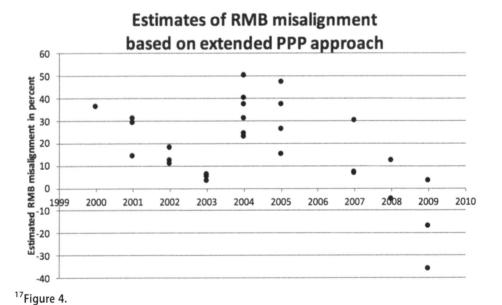

[17]Figure 4.

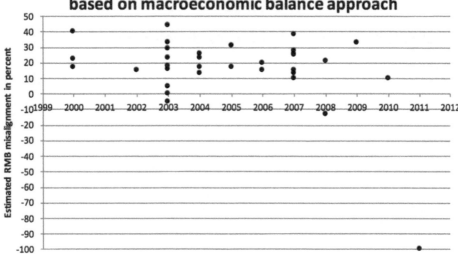

[18]Figure 5.

Global value chains – a ladder for development?

Petra Dünhaupt and Hansjörg Herr

ABSTRACT

For countries of the Global South, global value chains offer an opportunity to integrate into international trade and to industrialise relatively easily. However, in this contribution, we argue that this is not sufficient for a catching-up development, i.e. to reach the GDP per capita levels of the countries of the Global North. On the contrary, there is a risk that countries will remain trapped in low value-added activities. The theoretical argument is supported by case studies of four industries in six countries. For catching-up, countries need comprehensive horizontal and vertical industrial policy.

1. Introduction

The early movers in capitalist development still belong to the top group of industrialised countries today, known as the Global North in comparison to the Global South. From World War II until today, only a small group of countries have managed to initiate a catching-up process. Examples are Japan in the 1950s and 1960s and later, South Korea and other Asian Tigers. A very rough indicator of catching-up is the development of a country's real GDP per capita in relation to real GDP per capita of the US. Looking at this indicator, the overall result is that most countries in the Global South have the potential to both increase real GDP per capita and reduce extreme poverty, but there has not been much convergence.[1] In fact, the absolute differences in real GDP per capita between most countries of the Global South compared to developed countries have *increased*. Instructive is also the development of real earnings of unskilled workers. In an OECD project, the real wages of unskilled construction workers, taken as representative for all unskilled workers, were analysed between the 1820s and 2000s. The increases in real wages in many countries of the Global South compared to the Global North were alarmingly low.[2]

There are great hopes that the spread of global value chains (GVCs) and foreign direct investment (FDI) will help countries to catch up to GDP per capita levels of the Global North. Today, more than two-thirds of world trade occurs through GVCs (Dollar 2019, 1). This contribution discusses whether or not these hopes are justified.

In the second section, traditional and new trade theory are discussed with the aim of arriving at a better understanding of GVCs. Section three focuses on the analysis of GVCs. More specifically, it is analysed which traditional models can be used to analyse GVCs and which new concepts must be added. Section four presents case studies of GVCs. Section five concludes.

2. Traditional and new trade theory

According to *traditional trade theory*, based on comparative advantage, developing countries specialise on the export of low-tech, low-skill, labour-intensive goods. In contrast, developed countries concentrate on high-tech, high-skill, capital-intensive production. According to Ricardo (1817), the comparative advantage results from different levels of productivity, whereas Heckscher (1919) and Ohlin (1933) stress different endowments of capital, land and labour. They demonstrate that, under very strong assumptions, such as the full use of all factors of production and constant returns to scale, a switch to free trade increases world output and the welfare of all nations. However, these assumptions are almost never realised. The full use of all factors of production is historically in capitalist economies an exceptional situation and in developing countries a fantasy. The assumption of constant returns to scale contradicts all empirical evidence in most economic sectors.

According to these models, a switch to free trade leads to lower wages of unskilled workers in developed countries, as jobs requiring low skills are moved to developing countries, and to lower wages of skilled workers in developing countries, as jobs requiring high skills are moved to developed countries. 'Thus, the issue of whether globalisation is welfare enhancing comes back to the question ... is it possible to ensure, either through redistributive taxes or changes in institutions/rules, that workers are not made worse off' (Korinek and Stiglitz 2017, 17). We can transfer the logic of winners and losers to the global level. The world as a whole may benefit from trade, but not all countries: '*The popular school has assumed as being actually in existence a state of things which has yet to come into existence*. It assumes the existence of a universal union ..., and deduces therefrom the great benefits of free trade' (List 1909, 102). If there were a world state, losing regions would be supported by the state centre or more prosperous regions via transfers, education, infrastructure investment, etc. Also, labour would move from the underdeveloped to the more developed regions without any restrictions, creating a tendency to adjust wages to the same level in the whole nation.[3] Between independent nations, such mechanisms do not exist. This implies that an underdeveloped nation must, above all, use its own power to develop. Consequently, development is, following List, primarily a national task – until today.

After World War II, Singer (1949) and Prebisch (1950) argued that integration into global markets has a number of negative effects on developing countries, mainly as a result of the market pushing them towards the production and export of non-reproducible natural resources (e.g. crude oil) or reproducible natural resources (e.g. coffee). This has a negative effect because, firstly, income elasticity for primary and therefore inferior products is low compared to manufacturing products; secondly, exporters are confronted with demand oligopolies; thirdly, productivity increases in the production of reproducible natural resources lead to falling prices and mainly benefit industrial countries. In contrast, industrial production in the Global North involves the permanent development of new products, bringing with it the potential to earn high income. Singer and Prebisch assumed that, under free trade, it would be difficult for the Global South to industrialise and to catch up, since their dependence on producing primary products would cause their terms of trade to consistently deteriorate.

From the 1970s on oil prices and prices for rare minerals became very volatile. A good example is the development of the inflation adjusted price for one barrel crude oil. From the 1950s until the early 1970s it was below 30 and dropped to almost 20 US dollars. Then the two oil price shocks increased the price to 120 US dollars in 1980 before in 1986 the price dropped back to around 45 US dollars. With some volatility it remained at that level until it jumped to over 160 US dollar US dollars in 2008, before collapsing back to around 50 US dollar in 2014 and 40 US-dollar in 2020, the actual nominal price (Macrotrends 2020). Nevertheless, many countries with such exports, including Venezuela, Russia, Nigeria and the Islamic Republic of Iran, were not able to trigger a successful process of industrial development. These countries suffer from a persistent overvaluation of their exchange rates, which negatively affects the industrial sector, even if they realise current account surpluses and a shrinking industrial base. In addition, the relatively low price elasticities of natural resources lead to high price volatility, which is reinforced by speculation in futures markets. Finally, natural resource wealth stimulates rent-seeking of domestic elites and foreign companies and can lead to the so-called 'resource curse' (Humphreys, Sachs, and Stiglitz 2007).

Let us now turn to the *new trade theory*. Krugman (1981) developed a model to explain international trade with external economies of scale. External economies of scale are based on the interrelation of specialised firms, a specialised and qualified workforce, joint research of firms and networks with research institutes and universities, good infrastructure, cooperation and trust among firms, and personal contacts between researchers and managers. In short, there are economic clusters with positive external effects, synergy and network effects, and possibilities to reduce information and transaction costs. In such clusters, production costs are lower and, more importantly, innovative power is higher than outside such clusters. As a result, economic dynamics are concentrated within these clusters, which become stronger and stronger, and eventually the market mechanism causes regions and whole countries without such clusters to be left behind.

Krugman (1979) assumed external economies of scale and, at the same time, constant returns to scale for individual firms.[4] Under these assumptions, and the fact that, due to historical circumstances, clusters in one group of countries (now the developed countries) developed first, international trade leads to uneven development. Growth of GDP per capita in developed countries is systematically higher than in developing countries because of different productivity growth paths in the two groups of countries. In general, the disaster for developing countries or latecomers is that clusters are superior in creating new innovations, leading to new technologies and new products. It then becomes apparent that industrial policy in developing countries must create and foster the development of clusters in order to increase the productive power and capacity of the country.

In almost all major industries, in addition to external economies of scale, internal economies of scale exist. Therefore, many successful clusters consist of a combination of large and small firms. Internal economies of scale exist, for example, in the case of indivisibilities which play a role in the use of certain technologies (an assembly belt cannot be used for a small volume of production), research (research institutes need to have a certain size to be efficient) and branding (TV marketing is only viable for larger companies). In the cases of Facebook and Twitter, strong network effects lead to large economies of scale. In the case of internal economies of scale, low average costs and thus low selling prices can only be achieved if high volumes are produced. Incumbent firms in such markets are protected by high barriers to entry for new firms. Markets with internal economies of scale endogenously develop oligopolistic or even monopolistic structures. Firms in oligopolistic and monopolistic markets can and will use their market power to set prices at levels which maximise profits. Other rent-seeking strategies may be followed, including the creation of cartels, following the price leadership of one firm, competing with non-price measures etc. Thus, especially for new products introduced in the market, high prices can be achieved (Besanko et al. 2017).

Based on what Crouch (2016, 3) calls a 'corporate neoliberalism' rent-seeking possibilities have never been so widespread as today. This is based on increasing financialisation and power of shareholders, concentration in markets and bigger role of especially multinational companies and last not least a significant rise of branding and product diversification (Stiglitz 2012; Herr 2019b). They contributed substantially to the increasing inequality of income distribution and wealth concentration in developed countries (Stiglitz 2012; for empirical development in the tradition of Thomas Piketty see also Sawyer 2015). The increasing role of multinational enterprises (MNE) is an intrinsic part of the type of capitalism which has developed from the 1980s onwards, when the more regulated type of capitalism in the tradition of the New Deal was transformed into a more neoliberal one (Dullien, Herr, and Kellermann 2011). Crouch (2016, 3) named the new type of liberalism 'corporate liberalism', as it is very much distinct from the classical liberalism which stressed competition and saw rent-seeking as violating the rule of the market.

Rent-seeking based on economies of scale and technological leadership not only harms consumers and smaller firms in the Global North, but it is also harmful for the Global South, which is forced to pay higher prices and accept a further erosion of its terms of trade. External and internal economies of scale and technological leadership support each other and create high barriers for the catching-up of developing countries.

3. The analysis of GVCs

3.1. Basics of GVCs

GVCs started to dominate international trade in the 1990s. The background for it was the deregulation and liberalisation of international trade (Sawyer and Arestis 2008) and the financial system together with innovations in telecommunication and transportation.

In GVCs, the production process is divided into different tasks, which are allocated all over the world. For example, a product crosses several times boarders until it is finished or a finished good is assembled in one country using a large number of intermediate products produced in different countries. The governance of production processes is complex and usually not market based. As a rule, GVCs are governed by lead firms which in many cases are MNEs. The geographical dispersion of tasks depends, as does trade in general, to a large extent on comparative advantages and economies of scale. In the case of horizontal specialisation, firms outsource to specialised firms, which can perform tasks better because of technological leadership or economies of scale. For developing countries, vertical specialisation in GVCs is more typical. Here, the motivation is to cut costs, especially wage costs, but also other costs resulting from ecological, labour market and other regulations. Lead firms also exploit lax regulations in many countries of the Global South and the traditional societal structures that keep wage dispersion high and employment relations precarious. Developing countries have a comparative advantage in simple and low-skilled tasks, while developed countries have a comparative advantage in complicated and high-skilled tasks (Feenstra 2010). The distribution of tasks as a result of these factors is graphically represented in the so called 'smile curve', developed by Shih (1996), which is depicted in Figure 1. Typically, high value-added pre-fabrication tasks (e.g. research, design, logistics, and finance) and high value-added post-fabrication tasks (e.g. selling, marketing and high quality after sales services) in GVCs are taken over by developed countries. Developing countries have a comparative advantage in simple fabrication. More complicated fabrication, like in the precision metal industry or machine building industry, will stay in developed countries. Relatively simple services in all fabrication stages can be shifted to developing countries, such as part of book-keeping or call centres.

To save costs, lead firms can establish wholly owned firms or joint ventures in developing countries to complete simple tasks in the production process. While there are of course other motivations for FDI, especially the desire to enter large markets or to secure resources not available in the home country (Dunning 1993), in vertical GVCs, cost savings are the driving force. A second option for lead firms is subcontracting to legally independent firms. This gives rise to different governance forms. Governance in GVCs is defined as the 'authority and power relationships that determine how financial, material, and human resources are allocated and flow within a chain' (Gereffi 1994, 97).

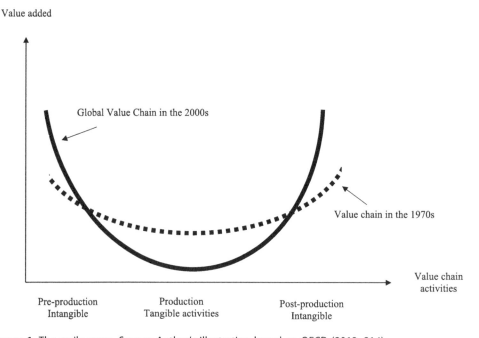

Figure 1. The smile curve. Source: Author's illustration based on OECD (2013, 214).

Market relationships are most likely to exist when tasks can be easily codified, product specifications are relatively simple and suppliers can take over the task with little input from buyers and asset specificity. Modular governance exists when a supplier is able to take full responsibility for the production of the task and can guarantee sufficient and stable quality. In relational governance structures, product specifications cannot be codified, transactions are complex and tacit knowledge must be exchanged. Captive governance structures dominate when capabilities of suppliers are relatively low and intervention in and control over the tasks of producers by lead firms are high. Captive suppliers usually take over a narrow range of tasks. Hierarchical relationships exist in the case of FDI. If there are no suitable suppliers in developing countries because of technological standards, the complexity of production and/or the size of firms, a lead firm or large intermediate firm will choose FDI (Gereffi, Humphrey, and Sturgeon 2005). Depending on the industry and the specific lead firms' strategies, one can usually find several governance types even in the same GVC.

For example, Apple and other electronics firms use the Taiwanese contract manufacturer Foxconn as a supplier of electronic devices. Foxconn, in turn, produces in many FDI firms in developing countries, as a rule subsidiaries fully owned. As another example, captive governance exists widely in the garment sector because tasks are relatively simple and can be standardised, production technologies are widespread, and economies of scale are not so high. In such cases lead firms may choose subcontracting, as this shifts risks such as demand volatility to subcontractors. A good example of how lead firms use this 'flexibility' can be found in the cancellation of orders and even refusal to pay for already ordered garment goods during the COVID-19 crisis 2020, leaving suppliers in Bangladesh and other countries to recuperate the costs of production (Anner 2020).

Integration into GVCs can help countries in the Global South to industrialise relatively easy. They do not need to build a complete industry to start production; they can specialise on individual components and assembly activities (Baldwin 2011). According to World Bank (2020), the share of industrial employment in per cent of total employment decreased in high-income countries from 31.0% in 1991 to 22.7% in 2019. In low- and middle-income countries in the same period, the share increased from 19.5% to 23.1%. However, the extent of industrialisation in the Global South is very uneven. In low-income countries, industrial employment increased only from 8.9% to 10.7% of total employment. Nevertheless, the industrialisation of the Global South is now much higher than expected by Prebisch and Singer. Empirical studies show that export-oriented sectors in the Global South often have wage levels above the national average and better working conditions (World Bank 2016). However, this is not always the case, as activities at the lower tiers of GVCs, from subcontracting of subcontractors to home work, can involve extremely precarious conditions (Anner 2015; von Broembsen and Harvey 2019). Overall, GVCs increase wage dispersion in the Global South, mostly benefitting skilled workers.

The relationship between GVC-participation and employment is a further important topic. Using average values between 2008 and 2013, Farole (2016) showed that almost no relationship exists between the GVC-participation index[5] and the employment share of adult population – in fact, a higher GVC-participation is related to a slightly lower employment share. Interpreting this result, Rodrik (2018, 5) writes:

It appears that exports are creating fewer and fewer jobs, and GVCs are certainly not helping. This is disappointing from a number of perspectives. It puts a damper on the idea of trade as an engine of growth. It suggests that the technological and organizational benefits associated with exports are not being disseminated throughout the economy. And since exports tend to be associated with better-paying jobs, it raises concerns about wage levels and inclusion.

From a Keynesian perspective this result is not a surprise. Exports related to GVCs may increase, but other sectors may require higher imports and shrink. Deeper integration in world markets is not an engine of growth and employment *per se*. For example, in the 1950s and 1960s, growth dynamics in most Western countries were higher than in the past decades, characterised by deep integration in world markets. It is the combination of increasing foreign trade and simultaneously increasing foreign trade *surpluses* that usually stimulates macroeconomic growth and employment. However, such mercantilist strategies are not possible by all countries. Overall, integration in global trade does not, in the typical developing country, create sufficient jobs to lead to inclusive growth and the absorption of traditional sectors.

3.2. Limited catching-up in GVCs

Integration into GVCs does not automatically lead to economic catching-up (World Bank 2016; see the introduction). Friedrich List's (1909) argument, that less developed countries specialise in low-tech, low-skill labour-intensive productions and that this reduces the chances of developing their national productive powers, also counts here. In GVCs, they no longer produce the low-tech products; they produce the low-tech *tasks* of all products. As a result, especially external economies of scale are difficult to create in the Global South.

However, a certain *economic upgrading* can be expected in GVCs. Humphrey and Schmitz (2002) distinguish between four types of economic upgrading in GVCs: product upgrading (producing a task with higher quality), process upgrading (using better technology), functional upgrading (taking over higher value-creating functions) and inter-sectoral upgrading (moving production to related or new industries).

The literature states that the possibilities for economic upgrading depend on the type of governance in GVCs (Gereffi, Humphrey, and Sturgeon 2005). In subcontracting relations, product upgrading can be considered as frequent and, in some cases, even process upgrading can be expected. The explanation for this is that even in captive governance constellations, lead firms most likely share knowledge and transfer skills to their suppliers because they have a high interest in a high quality of the product and, in some cases, consumers demand the fulfilment of certain ecological or social standards. In relational governance, there is intensive information flow between lead firms and sub-contractors and the likelihood of technology spill-overs in the field of product and process upgrading is high. In modular governance constellations, suppliers already possess a certain skill-level and technological knowledge. In many cases modular suppliers are themselves MNEs that have subsidiaries in developing countries. The OECD (2013: 20 f.) stresses that new and better intermediate and capital goods which are better suited to produce tasks as well as more competition added to productivity increases in

GVCs. However, lead firms have no interest in functional and inter-sectoral upgrading because this would create competitors in their core competencies.

For catching up, especially functional and inter-sectoral upgrading are of paramount importance. As was shown graphically in Figure 1, the difference in value-added between simple tasks and more sophisticated tasks has increased. In captive governance, prospects for functional and inter-sectoral upgrading are very limited, especially when a subcontractor mainly produces for one lead firm. Relational governance implies the transfer of knowledge. In modular governance, the supplier already possesses a certain level of technological knowledge. But even in these two governance types it may be extremely difficult to enter high value-creating activities. Even for a relatively high-quality garment producer in Vietnam, it is difficult to establish a well-known brand; the same problem exists for a high-quality component supplier in the electronic industry. Important is also the independence of the supplier. Independent firms have an incentive to upgrade, whereas completely dependent firms are trapped in the production of a given task for a dominant demander (Humphrey and Schmitz 2002).

The biggest hopes for technology and skill transfers are in the area of FDI. It is very likely that a lead firm will transfer technology and skills in the field of product and process upgrading to its subsidiaries. But who will benefit? Jumps in productivity in developing countries via high-tech production involving simple tasks can lead to unexpected effects. If most of the products of the tasks in a country are exported, a new technology will cause the price of the tasks decrease. Let us assume a balanced trade balance. In the extreme case where the task dominates exports, the terms of trade can deteriorate to such an extent that, despite higher real exports, the import volume of the country shrinks, the country has a smaller amount of goods in the domestic economy at its disposal, and the country suffers from immiserising growth (Bhagwati 1958). This implies that all the positive effects of productivity increases are experienced abroad.

A key problem is that FDI firms in many cases remain isolated without creating backward and/or forward linkages with the domestic economy. In this case the integration of the country in GVCs produces only very small advantages for the whole economy (Hirschman 1958). MNEs, which are not anchored in domestic economies, can very quickly leave the country if wages increase too much or other negative developments for the company occur. The OECD (2013, 35) describes this as the 'increasingly footloose character of MNE activities. (...) The risk is particularly acute for small emerging and developing economies where access to the domestic market or local knowledge is of limited importance to MNEs' location decisions.'

According to Amsden (2001, 207), MNEs invest virtually nothing in local research and development in developing countries – unless they are forced to do so. Lead firms defend their technological superiority. Such a strategy is rational from the firm perspective, as it supports current and future rent-seeking. In addition, conditions for research, developing new products, design, branding, etc. are usually much better in the home country of the MNE, in which it is integrated in highly efficient economic clusters. The successful catching-up of Asian countries and the lack of substantial catching-up in Latin America can, to a large extent, be explained by differences in ownership. In Latin America, leading firms in important industries are usually owned by FDI firms, whereas in Asia, states supported domestically owned firms and tried to create national champions (Shapiro 2007). Raj-Reichert (2019) argued that for the Malayan electronics industry and particularly contract manufacturers, the excessive reliance on FDI can be seen as the reason for the industry's inability to upgrade and for the middle-income trap that Malaysia seems to be caught in since the early 2000s.

3.3. Value grabbing in GVCs

Inherent in GVCs is a further characteristic, which in traditional trade theory is neglected and also not sufficiently reflected in the governance classification of Gereffi, Humphrey, and Sturgeon (2005): typical vertical GVCs are characterised by monopsony or oligopsony structures (Milberg and Winkler 2013). Lead firms or big first-tier suppliers (or even second- or third-tier) can dictate the price of the task to their suppliers and set it to a level which minimises profits of suppliers. The resultant low profit margins in countries of the Global South reduce the scope for investment and dampen domestic demand. Hence, it is relatively clear that in a typical developing country with weak institutions, GVCs can easily lead to business practices which hinder or even prevent social upgrading. For example, in the garment industry, the prices of tasks permanently decrease as a result of a race to the bottom which further embeds poor working conditions and violations of workers' rights in these countries (Anner 2015).

In case of FDI, large parts of firms' profit will be transferred to the lead firms' headquarters abroad. The OECD (2013, 36) summarises this succinctly: 'When countries participate in global value chains mainly through affiliates of foreign MNEs, these firms continue to capture much of the value. They often own and control the knowledge-based assets that create value in the GVC: brands, designs and patents, but also organisational and distribution networks'. MNEs can have a monopolistic or oligopolistic position in their selling market. In this case they operate in a 'rent-seeking' paradise, with rent-seeking on both their buying and selling markets. However, it is also possible for MNEs to be exposed to severe competition in their selling markets.

4. Case studies

The following case studies cover the automobile, garment, electronics and IT sectors in various countries during the decades before 2019.

4.1. Automobile sector

Global lead firms in the automobile sector come almost completely from the Global North. By means of FDI, they produce in all regions of the world. Governance between lead firms and their first-tier suppliers (so-called original equipment manufacturer (OEMs), like the German company Robert Bosch) is relational, largely due to the complexity of tasks. Power relationships between lead firms and first-tier suppliers are relatively balanced. First-tier suppliers, which are often MNEs themselves, have a number of own suppliers with partly domestic ownership. Governance between first- and second-tier suppliers is relational or captive. Suppliers in the second or third tiers are confronted with oligopsony of monopsony structures. South Africa is a good example for this (Mashilo 2019). After the end of apartheid in 1994, the world's leading car manufacturers and OEMs established production sites in South Africa. Automobile clusters were also established in Brazil, with foreign lead firms in the centre and foreign OEMs and domestic firms as second- and third-tier suppliers. Both countries were not able to establish own, relevant car brands. The consequence is that – very much in line with the smile curve – most of the design, research or branding is done in foreign headquarters. Attempts to take over higher value-creating tasks by domestic firms failed.

The exemplar for the more recent development of a domestic car industry is China (Lo and Wu 2014). China started to use FDI in the early 1980s and more intensively in the early 1990s to modernise its automotive sector. However, until the end of the 1990s, only joint ventures with no more than 50% foreign ownership were allowed. Before China joined the WTO in 2001, local content requirements forced FDI firms to source local products. This supported the development of Chinese first- and second-tier suppliers. China also persisted in the development of its own, large domestic brands, which benefitted from the know-how and skills spill-overs that could not be prevented by foreign firms. An advantage was that large Chinese car manufacturers were state-owned and could be integrated together with FDI in a long-term oriented industrial policy strategy. The Chinese financial system, which is until today dominated by state-owned banks, was an integral part of the industrial policy strategy. Even though Chinese automobile brands have so far been unable to establish a position in the global car market, this cannot be excluded in the future. Ultimately, Chinese car manufacturers managed to compete on the domestic market with global brands. They followed a kind of leapfrogging strategy to manage functional and even inter-sectoral upgrading, for example in the field of electro mobility (Dünhaupt et al. 2019).

With the support of industrial policy and protection from foreign competition, India has a long tradition of developing its own brands in the car industry. This was successful in the sense that Indian brands dominate the domestic market. However, except for some niche markets in developing countries, Indian cars are not widely exported and not competitive globally. Governance and power relationships between lead firms and first-tier suppliers can be characterised as relational governance, and suppliers in lower tiers suffer from captive governance (Herr et al. 2020).

In Brazil and South Africa, strong trade unions and government enforcement of labour laws led to relatively good working conditions in the sector and, compared with the national standard, relatively high wages. These are highest in the assembly plants of the global brands, lower in the first-tier suppliers, and again lower in further tiers. Freedom of association is realised. In China the situation is very much differentiated (Lüthje 2014). Overall and compared with other sectors, development of real wages in the automotive sector is relatively good. In state-owned companies, wages and working conditions are relatively good, state-controlled trade unions play a social role and a kind of paternalistic corporate governance exists. In private firms (which are also partly FDI-firms) and first- and second-tier suppliers, leasing work plays an increasing role, which divides the workforce into regular workers and workers in precarious conditions. Freedom of association is not realised in China. Wildcat strikes occur frequently as a kind of substitute for the lack of collective wage bargaining. In India leasing work in the car industry has exploded in the last decades. A large part of the workforce must accept precarious working conditions, real wage increases for most of the workers in the sector have been low, and working conditions are bad. Trade unions are politically divided and weak (Herr et al. 2020).

4.2. Garment sector

In the garment sector, big brands and big retailers have almost completely shifted production to the Global South. Due to the relatively simple technological standard required in the sector, subcontracting plays a relatively big role, which shifts the pressure for quick delivery and volatile production volumes to firms in the Global South. Big intermediate firms are common. For example, Li & Fung, a company from Hong Kong, manages 15000 suppliers from all over the world. Lead and large intermediate firms are in a powerful position, as the competition between suppliers worldwide is very high. This has led to a race to the bottom regarding working conditions and social standards, with massively falling prices for garment products (Anner 2015). A specific characteristic to the sector is that lead firms, for example big retailers, in many cases sell their products in very competitive markets.

For a long time only captive governance structures dominated the garment sector. However, in the last decades more modular governance developed, with bigger suppliers being able to deliver finished goods, including design (Gereffi and Frederick 2010; Elms 2013). However, this shift toward more modular governance did not change the power asymmetries in GVCs and the enormous pressure on suppliers to cut costs. Tiers in the garment sector are differentiated across many levels, including home work. Compared to the national standard, wages are low, jobs are precarious and working conditions are poor.

The Vietnamese garment industry is at the bottom of the GVC and mostly takes over the tasks of cutting, sewing and trimming. In the sector there are a number of big domestic companies, former state-owned firms, and a number of FDI firms, many from Asia. But most of the 2,500 exporting companies are relatively small. Functional upgrading is very limited; own important brands do not exist. Frequent wildcat strikes aim to improve wages and the bad working conditions in the sector (Do 2020).

With a share of total world garment exports of over 30% in 2018, China is the biggest garment exporter in the world, followed by Bangladesh (6,8%), Vietnam (6.4%), India (3.3%) and Turkey (3.1%) (Lu 2020). The sector in China is dominated by medium-sized and small producers in around 150 clusters. Most companies are domestically owned; some of the bigger ones are Chinese state-owned companies or have foreign owners (in many cases Chinese owners, not living in mainland China). Especially after the crisis 2008/09 the sector underwent a structural change. A general higher wage level in China, higher minimum wages and a shortage of workers who had other options led to increasing wage costs. Many domestic Chinese producers actively started to build up brands, particularly for the domestic market, as well as mechanising production and increasing productivity substantially. The Chinese government supported this process via industrial policy measures. Some Chinese firms shifted productions to neighbouring countries via FDI. However, it remains that around 60% of producers in China only take over simple tasks. Compared with other sectors, working conditions in the garment sector are poor. For example, the role of temporary workers increased substantially and excessive overtime in periods of high demand is widespread. A further shrinking of employment in the Chinese garment sector has to be expected as a result of economic upgrading (Butollo 2014; Liu 2020).

In Bangladesh around 80% of exports and 25% of GDP come from the garment sector. While FDI played an important role in the past, domestic producers have since come to dominate. Captive governance structures dominate the sector. Productivity increased substantially, but, similar to Vietnam, functional upgrading could only be achieved by a very limited number of bigger companies (Moazzem and Sehrin 2016; Curran and Nadvi 2015).

Workers in the garment sector in countries of the Global South are mostly young women and many are domestic migrant workers. In the countries analysed, real wages increased along with the national trend. In China, based on the general increase of macroeconomic productivity, there were substantial real wage increases due to increasing minimum wages and a shortage of workers. This is held as the main explanation for why Chinese firms started outsourcing garment production. As a rule, working conditions in the garment industry in the Global South are extremely poor. Symbolic of this is the Rana Plaza accident in 2013, which left more than 1100 dead. Pressure from consumer organisations led to improvements of working conditions in some areas – for example the multi-stakeholder Bangladesh-Accord to improve safety standards – but overall, working conditions remained poor. One explanation of this is the lack of independent trade unions in China and Vietnam and weak and partly anti-union policy in Bangladesh. Enforcement of labour laws is also generally poor (Herr et al. 2020).

4.3. Electronic hardware sector

In the electronic hardware industry, for some time a production model has been established in which product innovation is mostly decoupled from manufacturing. Since the end of the 1980s, most electronics companies had abandoned manufacturing and started to focus on their core competencies, i.e. technology leadership, design, branding and sales. Multinational contract manufacturers have taken over not only the actual production, but also procurement, assembly and logistics, as well as parts of the design and production-related engineering. Components in the GVC are purchased from a large number of component manufacturers. On the one hand, these are other large MNEs, such as Microsoft or Intel, and on the other hand, these are small manufacturers who produce components with little added value (ILO 2014). However, there are still lead firms which produce in-house. For example, the two large Korean electronics companies, Samsung and LG, have electronic goods manufactured in their production facilities worldwide. Overall, the electronic hardware industry is mainly characterised by modular governance between lead firms and contract manufacturers, and hierarchical governance in the case of FDI.

Vietnam as a low-cost production location attracted substantial FDI in the electronic hardware sector. The FDI firms in Vietnam assemble imported intermediate products and export their output. Samsung, for example, assembles in two locations in Vietnam one third of its mobile telephones sold worldwide. In 2015, these two locations employed over 325,000 persons, 70% of them women. No relevant forward or backward linkages were developed. Value creation in Vietnam remained low. Even after over twenty years of production in the sector there has been no economic upgrading. Samsung has no incentive to upgrade production in its FDI firms in Vietnam; in its profit maximising strategy it has chosen Vietnam to take over simple tasks. Domestic firms in the sector in Vietnam are unimportant (Do 2017).

MNEs also established production arms in the electronic hardware sector in Brazil. However, the motivation was not to exploit cheap labour, as wages are high in comparison to other locations such as Vietnam. Production was relocated to Brazil because of its high import tariffs for final products and low tariffs for components in the sector. Similar to Vietnam, almost all intermediate products are imported. Assembled products are sold in Brazil or exported to other Latin American countries. To a limited extent domestic firms are integrated in a captive governance relation. There has been no substantial upgrading for decades (van Wetering, Gomes, and Schipper 2015).

Working conditions in the electronic hardware sector in Vietnam are poor and even worse than in the garment sector. Employees are mainly young women from the countryside. Typically, they work for some years in the factories. Wages for the majority of workers are also low compared to national standards. Excessive overtime is the most prevalent means of increasing monthly wages for workers. Strike actions are less than in the garment sector (Do 2017). The situation is completely different in Brazil. Strong trade unions and enforced labour laws have led to relatively good working conditions and, compared with other sectors in Brazil, relatively high wages (van Wetering, Gomes, and Schipper 2015). As a result, MNEs in Brazil have not succeeded in fostering the poor labour standards and anti-union policies practiced, for example, in Vietnam.

4.4. IT service sector

The IT service sector is very much fragmented, with some MNEs but also many medium-sized and small companies. This is a result of the diverse products the sector sells and its highly innovative dynamic. The governance structure is also diverse and depends on the tasks taken over in GVCs. From the 1990s on, IT services have been increasingly outsourced to the Global South (Snowdon and O'Donoghue 2018).

India is by far the most important country for service outsourcing in GVCs. The qualification level of employees in India in the sector is relatively high and wages compared to the Global North are relatively low. India was able, supported by industrial policy, to develop own global champions in the sector and develop IT clusters. In 2017, of the 25 revenue strongest IT companies in the world, five came from India. There has been substantial economic upgrading in the sector, with some Indian companies managing to establish relational governance models with lead firms in the Global North. However, until today India mainly takes over simple IT tasks in GVCs. In many cases captive and monopsony structures exist, for example in the widely common call centres in Indian IT clusters (Noronha and D´Cruz 2020).

China has positioned itself more broadly than India. Besides substantial upgrading in electronic hardware, China also shows substantial upgrading in the IT service sector. Huawei, for example, has become a world leading IT company in the field of hardware and software. In comparison to the automotive sector, China managed substantial economic upgrading in the IT sector without any FDI. Comprehensive industrial policy and high demand for IT products stand behind the success of the Chinese IT industry. Mainly because of political reasons, but also because of language barriers (for example, in call centres), China is not integrated in GVCs in the IT service sector. However, it is very likely that China will become a global player in the hardware and IT service sector, at least in developing countries, with its potential for inter-sectoral expansion (Lo and Wu 2014; Zhu and Morgan 2018).

In India, social upgrading in the IT service sector is ambivalent. Wages are high compared to the national standard and jobs belong to the formal sector. However, burdens for employees include excessive overtime, night work, and jobs below the qualification level. In China, payment is high in national comparison as well. In both countries, due to the diversification of IT services, wage dispersion is high and working conditions vary substantially. Trade unions play no substantial role in the IT service industry in either country.

4.5. Summary of case studies

Substantial functional and inter-sectoral upgrading – without comprehensive horizontal and vertical industrial policy in form of government support and protection – did not occur in the country cases. In the cases looked at, China achieved substantial economic upgrading triggered by different methods: in the automobile sector using FDI, in the IT sector developing top players without FDI, in the garment sector as a result of pressure from higher wages. India, to a lesser extent, is also an illustrative example of upgrading and industrial policy, especially in the IT sector. Vietnam, Bangladesh, South Africa and Brazil are cases of integration in GVCs without tendencies of relevant functional or inter-sectoral upgrading. In the cases presented, working conditions and wages compared with national standards depend mainly on trade union strength and government enforcement of labour laws.

5. Conclusion

The main message of this paper is that market mechanisms in vertical GVCs support industrialisation and also product and process upgrading under certain conditions. But the market mechanism does not lead to a catching-up of countries in the sense that they approach productivity, real GDP per capita levels, and the innovative power comparable with developed countries. We can agree with Baldwin (2013, 39) when he writes: 'Smile curve economics suggests that the fabrication stages in manufacturing may not be the development panacea as they once were'. This is supported by the theoretical analysis and case studies presented above. The main arguments are:

Comparative advantages push countries form the Global South to low-tech, labour-intensive productions with low economic dynamism. This is also the case in GVCs. Absolute advantages of some developing countries in the export of natural resources adds to the lack of innovative and high-tech industries and the low demand for skills in developing countries.

External economies of scale give clusters and firms with massive government support in the Global North a huge advantage, as it is difficult for the Global South to create such clusters. Internal economies of scale lead endogenously to oligopolistic and monopolistic structures, as incumbent firms are protected from newcomers simply because of their size. Power concentration in markets leads to rent-seeking. Lead firms and big intermediate firms in GVCs are almost always in a position to exploit their power *vis-à-vis* their suppliers. In many cases lead firms can also pursue rent-seeking in their seller markets. In addition, lead-firms are technology leaders and own intellectual property which both contribute to increasing rent-seeking power.

GVCs and especially FDI can help in the areas of product and process upgrading. FDI, in spite of some positive productivity and skill effects, follows comparative advantages. It is not helpful in functional and inter-sectoral upgrading, as lead firms do not transfer core competences. In addition, FDI firms do not automatically create forward- and backward linkages with the domestic economy. Furthermore, there is the danger of immiserising growth, with the positive effects of technology transfer to the Global South regarding simple tasks mostly being realised in the Global North.

GVCs lead to higher inequality in both the Global North and South. For the latter, monopsony and oligopsony structures, along with profit transfers in case of FDI, lead to value grabbing and a lack of profit or wage income. In addition, wage dispersion increases if large economic sectors are excluded from prosperous development. All of these dampen consumption and investment demand, investment in training and education, and GDP growth (Ostry 2015). Beneficiaries of value grabbing are the MNEs and their shareholders and the consumers in the Global North, depending on the market constellation in which MNEs are positioned.

Better working conditions and higher real wages do not automatically follow economic upgrading. The case studies above show that strong trade unions and good, enforced labour laws are essential for social upgrading.

One conclusion of this is that for catching-up, countries need comprehensive horizontal and vertical industrial policy. Showcases of catching-up after World War II were Japan in the 1950s, later South Korea and Taiwan, and still later, China. All used extensive vertical and horizontal industrial policy in conjunction with a highly regulated financial system to develop their economies (Stiglitz 1996). Following comparative advantages by starting simple productions in GVCs is a recommendable strategy – but one which should not privilege foreign companies at the expense of domestic ones. However, this is not enough. Active policies to take over higher value-creating functions in GVCs are essential, and industrial policy must violate the market logic of comparative advantages by creating and building comparative advantages, clusters and large national firms which are able to compete internationally. An important element of an industrial policy package is generating sufficient demand for the industries which are to be developed.

Adjustment of the level of the exchange rate can play a key role as a means of avoiding current account deficits. It should function as a general protection in a world of low tariffs, while vertical industrial policy is used to support selected sectors in GVCs and beyond. Under WTO rules – and even free trade and investment treaties – comprehensive industrial policy is possible.[6]

Notes

1. From the 1960s until today India, Pakistan or Bangladesh have remained at levels of around or below 10% of US per capita GDP. Latin American countries stagnated at levels of 40% or lower. The majority of African countries stagnated below 10% of US GDP per capita. Some Asian countries, as mentioned, were able to reach real GDP per capita levels comparable to the US. China is a prime example for reducing the GDP per capita gap with the US (Herr 2018; Penn World Table 2019).
2. As a measure for real wages a subsistence basket of goods was used. Between the 1820s and 2000s the number of such baskets increased in Western Europe for an unskilled construction

worker from 12.6 to 163.3, in Eastern Europe from 7.2 to 38.7, in Latin America and Caribbean from 5.7 to 26.5, in East Asia from 3.3 to 26.6, in South and South-East Asia from 4.1 to 10.0, in Middle East and Northern Africa from 6.5 to 71.6, in Sub Sahara from 2.9 to 18.3 and in USA, Canada and Australia from 43.9 in the 1860s to 169.8 in the 2000s (de Zwart, van Leeuwen, and van Leeuwen-li 2014).

3. Stolper and Samuelson (1941) showed in the framework of the Heckscher-Ohlin model that free trade, even without migration, can lead to the same remuneration of the factors of production. The Stolper-Samuelson theorem suffers especially from the unrealistic assumption that in all countries the same technology exists and that international trade creates a shortage of unskilled workers in developing countries. Given the typically massive underemployment in developing countries, increasing demand for unskilled labour does not increase wages (Lewis 1954).

4. Such a model is attractive, as it can only lead to normal profits for firms and avoids analysis of monopolies and oligopolies.

5. The participation index is calculated as the sum of the share of imported inputs in the overall exports of a country (looking backward in the GVC) plus the share of exported goods and services used as imported inputs to produce other economies' exports (looking forward in GVC) (OECD 2013, 30).

6. For modern industrial policy strategies see Herr (2019a) and Dünhaupt and Herr (2020).

Acknowledgements

We thank the Hans-Böckler Foundation for financial support for the research project "Global Value Chains – Economic and Social Upgrading". We especially thank Christina Teipen and Fabian Mehl who together with us carried out the project at the Berlin School of Economics and Law. We also thank our project partners for their collaboration, and an anonymous referee for helpful comments

Disclosure statement

No potential conflict of interest was reported by the author(s).

Funding

This work was supported by the Hans Böckler Stiftung.

References

Amsden, A. H. 2001. *The Rise of the Rest: Challenges to the West from Late-Industrializing Economies.* Oxford: Oxford University Press.

Anner, M. 2015. "Stopping the Race to the Bottom: Challenges for Workers' Rights in Supply Chains in Asia." *FES Briefing Paper.* Hanoi: FES Vietnam.

Anner, M. 2020. "Abandoned? The Impact of Covid-19 on Workers and Businesses at the Bottom of Global Garment Supply Chains." *Research Report, PennState Center for Global Workers' Rights.*

Baldwin, R. 2011. "Trade and Industrialisation after Globalisation's 2nd Unbundling: How Building and Joining a Supply Chain are Different and Why It Matters." *NBER Working Paper 17716.* Accessed 21 April 2020. https://www.nber.org/papers/w17716

Baldwin, R. 2013. "Global Supply Chains: Why They Emerged, Why They Matter, and Where They are Going." In *Global Value Chains in a Changing World*, edited by D. K. Elms and P. Low, 13–60. Geneva: World Trade Organisation.

Besanko, D., D. Dranove, M. Shanley, and S. Schaefer. 2017. *Economics of Strategy*. 7th ed. New York: John Wiley and Sons.

Bhagwati, J. 1958. "Immiserizing Growth: A Geometrical Note." *Review of Economic Studies* 25: 201–205. doi:10.2307/2295990.

Butollo, F. 2014. *The End of Cheap Labour? Industrial Transformation and "Social Upgrading" in China*. New York: Campus.

Crouch, C. 2016. *The Knowledge Corrupters. Hidden Consequences of the Financial Takeover of Public Life*. Malden: Polity Press.

Curran, L., and K. Nadvi. 2015. "Shifting Trade Preferences and Value Chain Impacts in the Bangladesh Textiles and Garment Industry." *Cambridge Journal of Regions, Economy and Society* 8: 459–474. doi:10.1093/cjres/rsv019.

de Zwart, P., B. van Leeuwen, and J. van Leeuwen-li. 2014. "Real Wages since 1820." In *How Was Life? Global Well-being since 1820*, edited by J. L. van Zanden, J. Baten, M. d`Ecrole, A. Rijpma, C. Smith, and M. Timmer, 73–86. Paris: OECD Publising.

Do, C. 2017. "The Missing Link in the Chain? Trade Regimes and Labour Standards in the Garment, Footwear and Electronics Supply Chains in Vietnam." *Friedrich-Ebert Foundation Singapore*.

Do, C. 2020. "Social and Economic Upgrading in the Garment Supply Chain in Vietnam, IPE (Institute for International Political Economy Berlin)." *Working Paper, no. 137*.

Dollar, D. 2019. "Executive Summary, In: Global Value Chain Development Report 2019." 1–8. Geneva: World Trade Organisation.

Dullien, S., H. Herr, and C. Kellermann. 2011. *Decent Capitalism. A Blueprint for Reforming Our Economies*. London: Pluto Press.

Dünhaupt, P., and H. Herr. 2020. "Catching up in a Time of Constraints - Industrial Policy under World Trade Organization Rules, Free Trade Agreements and Bilateral Investment Agreements." *Friedrich-Ebert Foundation Singapore*.

Dünhaupt, P., H. Herr, F. Mehl, and C. Teipen. 2019. "Entwicklungschancen durch Integration in globale Wertschöpfungsketten: Ein Länder- und Branchenvergleich." *WSI Mitteilungen* 72: 403–411. doi:10.5771/0342-300X-2019-6-403.

Dunning, J. H. 1993. *Multinational Enterprises and the Global Economy*. Harlow: Addison-Wesley.

Elms, D. K. 2013. "Views of GVCs Operators." In *Global Value Chains in a Changing World*, edited by D. K. Elms and P. Low, 161–169. Geneva: World Trade Organisation.

Farole, T. 2016. "Do Global Value Chains Create Jobs?" *IZA World of Labour, no. 291*.

Feenstra, R. C. 2010. *Offshoring in the Global Economy. Microeconomic Structure and Macroeconomic Implications*. Cambridge: MIT Press.

Gereffi, G., J. Humphrey, and T. Sturgeon. 2005. "The Governance of Global Value Chains." *Review of International Political Economy* 12: 78–104. doi:10.1080/09692290500049805.

Gereffi, G. 1994. "The Organization of Buyer-Driven Global Commmodity Chains: How U.S. Retailers Shape Overseas Production Networks." In *Commodity Chains and Global Capitalism*, edited by G. Gereffi and M. Korzeniewicz, 95–122. Westport, CT: Greenwood Press.

Gereffi, G., and S. Frederick. 2010. "The Global Apparel Value Chain, Trade and the Crisis." In *Global Value Chains in A Postcrisis World - A Development Perspective*, edited by O. Cattaneo, G. Gereffi, and C. Staritz, 157–208. Washington, DC: World Bank.

Heckscher, E. 1919. "The Effect of Foreign Trade on the Distribution of Income." *Ekonomisk Tidskrift*, 497–512.

Herr, H. 2018. "Underdevelopment and Unregulated Markets: Why Free Markets Do Not Lead to Catching-up, In: European Journal of Economics and Economic Policy." *Intervention* 15: 208–219.

Herr, H. 2019a. "Industrial Policy of Economic and Social Upgrading in Developing Countries." *Friedrich-Ebert Foundation Singapore*.

Herr, H. 2019b. "Karl Marx's Thought on Functional Income Distribution: A Critical Analyses Form a Keynesian and Kaleckian Perspective, In: European Journal of Economics and Economic Policies." *Intervention* 16: 272–285.

Herr, H., C. Teipen, P. Dünhaupt, and F. Mehl. 2020. "Wirtschaftliche Entwicklung und Arbeitsbedingungen in globalen Wertschöpfungsketten." *Working Paper Forschungsförderung, Nr. 175, Hans-Böckler-Stiftung*.

Hirschman, A. O. 1958. *The Strategy of Economic Development*. New Haven: Yale University Press.

Humphrey, J., and H. Schmitz. 2002. "How Does Insertion in Global Value Chains Affects Upgrading in Industrial Clusters." *Regional Studies* 36: 1017–1027. doi:10.1080/0034340022000022198.

Humphreys, M., J. Sachs, and J. E. Stiglitz, eds.. 2007. *Escaping the Resource Curse*. New York: Columbia University Press.

ILO. 2014. "Ups and Downs in the Electronics Industry: Fluctuating Production and the Use of Temporary and Other Forms of Employment." *International Labour Office, Sectoral Policies Department*. Geneva: ILO.

Korinek, A., and J. Stiglitz. 2017. "Artificial Intelligence and Its Implications for Income Distribution and Unemployment." *NBER Working Paper Series, No. 24174*.

Krugman, P. 1979. "Increasing Returns, Monopolistic Competition, and International Trade." *Journal of International Economics* 9: 469–479. doi:10.1016/0022-1996(79)90017-5.

Krugman, P. 1981. "Trade, Accumulation, and Uneven Development." *Journal of Development Economics* 8: 149–161. doi:10.1016/0304-3878(81)90026-2.

Lewis, W. A. 1954. "Economic Development with Unlimited Supplies of Labour." *Manchester School or Economics and Social Studies* 10: 139–191. doi:10.1111/j.1467-9957.1954.tb00021.x.

List, F. 1909. *The National System of Political Economy*. Translated by Sampson S. Lloyd. London: Longmans, Green and Co. (German edition 1841).

Liu, X. 2020. "Chinese Apparel Industry: Economic and Social Upgrading." *IPE (Institute for International Political Economy Berlin) Working Paper*.

Lo, D., and M. Wu. 2014. "The State and Industrial Policy in the Chinese Economic Development." In *Transforming Economies: Making Industrial Policy Work for Growth, Jobs and Development*, edited by M. Xirinachs-J., I. Nübler, and R. Kozul-Wright, 307-326. Geneva: ILO.

Lu, S. 2020. "FASH455 Global Apparel & Textile Trade and Sourcing." Accessed 19 April 2020. https://shenglufashion.com/2019/08/16/wto-reports-world-textile-and-apparel-trade-in-2018/

Lüthje, B. 2014. "Labour Relations, Production Regimes and Labour Conflicts in the Chinese Automotive Industry." *International Labour Review* 153: 535–560. doi:10.1111/j.1564-913X.2014.00215.x.

Macrotrends. 2020. "Crude Oil Prices – 70 Year Historical Chart." Accessed 07 June 2020. https://www.macrotrends.net/1369/crude-oil-price-history-chart

Mashilo, A. 2019. "Auto Production in South Africa and Components Manufacturing in Gauteng Province." *Global Labour University Working Paper, No. 58*.

Milberg, W., and D. Winkler. 2013. *Outsourcing Economics. Global Value Chains in Capitalist Development*. Cambridge: University Press.

Moazzem, K. G., and F. Sehrin. 2016. "Economic Upgrading in Bangladesh's Apparel Value Chain during the Post-MFA Period: An Exploratory Analysis." *South Asia Economic Journal* 17: 73–93. doi:10.1177/1391561415621824.

Noronha, E., and P. D´Cruz. 2020. "The Indian IT Industry: A Global Production Network Perspective, IPE (Institute for International Political Economy Berlin)." *Working Paper, No. 134*.

OECD. 2013. *Interconnected Economies: Benefiting from Global Value Chains*. Paris: OECD Publishing.

Ohlin, B. 1933. *Interregional and International Trade*. Cambridge: Harvard University Press.

Ostry, J. D. 2015. "Inequality and the Duration of Growth." *European Journal of Economics and Economic Policies* 12: 147–157.

Penn World Table. 2019. https://www.rug.nl/ggdc/productivity/pwt/pwt-database

Prebisch, R. (1950): The Economic Development of Latin America and Its Principal Problems. Lake Success, NY: United Nations Department of Economic Affairs

Raj-Reichert, G. 2019. "Global Value Chains, Contract Manufacturers, and the Middle-Income Trap: The Electronics Industry in Malaysia'." *Journal of Development Studies*. doi:10.1080/00220388.2019.1595599.

Ricardo, D. 1817. *On the Principles of Political Economy and Taxation*. London: John Murray.

Rodrik, D. 2018. "New Technologies, Global Value Chains, and the Developing Economies." *Pathways for Prosperity Commission Background Paper Series, no. 1*. Oxford, UK.

Sawyer, M., and P. Arestis. 2008. "Financial Liberalization: The Current State of Play." In *Issues in Finance and Industry: Essays in Honour of Ajit Singh*, edited by P. Arestis and J. Eatwell, 74-93. New York: Palgrave Macmillan..

Sawyer, M. 2015. "Confronting Inequality: Review Article on Thomas Piketty on 'Capital in the 2st Century." *International Review of Applied Economics* 29: 878–889. doi:10.1080/02692171.2015. 1065227.

Shapiro, H. 2007. "Industrial Policy and Growth, United Nations Department of Economic and Social Affairs." *DESA Working Paper No. 53.*

Shih, S. 1996. *Me-Too Is Not My Style: Challenge Difficulties, Breakthrough Bottlenecks, Create Values*. Taipei: Acer Foundation.

Singer, H. 1949. "Economic Progress in Under-developed Countries." *Social Research* 16: 236–266.

Snowdon, J., and O. O'Donoghue. 2018. "What Does the IT Services Market Look like in 2018? HFS Research, 30. 08.2018." Accessed 15 March 2019. https://www.hfsresearch.com/market-analyses/what-does-the-it-services-market-look-like-in-2018

Stiglitz, J. E. 1996. "Some Lessons from the East Asian Miracle." *World Bank Research Observer* 11: 151–177. doi:10.1093/wbro/11.2.151.

Stiglitz, J. E. 2012. *The Price of Inequality*. New York: Norton.

Stolper, W. F., and P. Samuelson. 1941. "Protection and Real Wages." *Review of Economic Studies* 9: 58–73. doi:10.2307/2967638.

van Wetering, H., M. Gomes, and I. Schipper. 2015. "Brazil, the New Manufacturing Hotspot for Electronics? Good Electronics Network." Accessed 24 April 2019. https://reporterbrasil.org.br/wp-content/uploads/2016/09/Brazil-the-new-manufacturing-hotspot-for-electronics.pdf

von Broembsen, M., and J. Harvey. 2019. "Decent Work for Homeworkers Is Global Supply Chains: Existing and Potential Mechanisms for Workercentred Governance." *Global Labour University Working Paper, No. 54.*

World Bank. 2016. "Jobs in Global Value Chains." *Jobs Notes Issue No. 1.*

World Bank. 2020. "World Bank Data." Accessed 17 April 2020. https://data.worldbank.org/indicator/SL.IND.EMPL.ZS?locations=XT

Zhu, J., and G. Morgan. 2018. "Global Supply Chains, Institutional Constraints and Firm Level Adaptations: A Comparative Study of Chinese Service Outsourcing Firms." *Human Relations* 71: 510–535. doi:10.1177/0018726717713830.

The impact of capital flow reversal shocks in South Africa: a stock- and-flow-consistent analysis

Konstantin Makrelov, Rob Davies and Laurence Harris ⓘD

ABSTRACT

South Africa has a very well-developed financial sector and high reliance on capital flows. The country saw large capital outflows as the Covid-19 crisis developed, accompanied by a large depreciation of the rand and spikes in bold yields. We employ a stock- and flow-consistent model to study the impact of capital flow reversal shocks on the South African economy. The model includes a richer representation of institutional balance sheets than existing models. The financial sectors behaviour in the model draws on the theoretical frameworks, which highlight the relationship between bank capital, the risk-taking behaviour of the financial sector, lending spreads and economic activity. We specify a dynamic adjustment model of household expectations with properties that differ from the way in which expectations are formed in either stock- and flow-consistent or (DSGE) models. Household expectations resemble bounded rationality. The financial accelerator mechanism operates through the balance sheets of all institutions in the economy. We find that a reversal in capital flows can affect the domestic economy through its impact on domestic liquidity, on the risk-taking behaviour of the financial sector, and on the demand for assets.

1. Introduction

The Covid-19 crisis highlighted again the importance of capital flows for many emerging markets and the impacts of capital flows reversal shocks. South Africa as many other countries saw large capital outflows at the beginning of the Covid-19 crisis, accompanied by a strong depreciation of the rand and a rise in bond yields. Non-resident ownership of government bonds has declined by 12 percentage points to 30% of total debt.

In this paper, we examine how financial sector dynamics affect the economic impact of a capital flow reversal shock in South Africa. The country has high reliance on capital inflows and foreign debt is mostly rand denominated. The emphasis is on how a capital flow reversal shock transmits through the financial sector and how the presented model can incorporate discontinuity in economic behaviour. The reasons for capital flow reversals include both domestic and global factors, with perceptions of risk being an increasingly important driver (Ahmed and Zlate 2014; Eller, Huber, and Schuberth 2020;

Forbes and Warnock 2012; Rey 2015).[1] Recent research indicates that the risk-taking channel of monetary policy identified by Borio and Zhu (2012) is a major driver of capital flows.[2] More expansionary monetary policy in the US reduces the risk premiums and borrowing costs for global banks, leading to higher capital flows into emerging markets and more favourable financial conditions (Anaya, Hachula, and Offermanns 2017; Baskaya et al. 2017; Bräuning and Ivashina 2017; Bruno and Shin 2015; Byrne and Fiess 2016; Chari, Stedman, and Lundblad 2017). This effect is stronger for emerging markets with deep and liquid financial markets and is affected by countrys macroeconomic fundamentals(Cerutti, Claessens, and Puy 2019; Dahlhaus and Vasishtha 2020; McQuade and Schmitz 2017). Domestic banks also experience cheaper global funding, which allows them to expand domestic credit (Baskaya et al. 2017).

Eichengreen and Gupta (2016) and Ghosh, Ostry, and Qureshi (2016) find that an increase in risk aversion significantly raises the probability of a sudden stop.[3] The impact of increased risk aversion and uncertainty, especially in the case of sudden stops, can be severe, including banking and sovereign debt crises with large output and employment losses (Cavallo et al. 2015; Eichengreen and Gupta 2016; Magud and Vesperoni 2015; Reinhart and Reinhart 2008).[4] The effect can be long-lasting as the impact on investment is strong and negative, largely due to a deterioration in financial conditions and uncertainty regarding the availability of funding (Converse 2018; Joyce and Nabar 2009).

Calvo (1998) presents the theoretical mechanism of how capital flow reversal shocks generate large economic losses. Such effects of a capital flow reversal shock may be exacerbated by a transmission mechanism that includes the risk-taking channel and a financial accelerator mechanism (Bernanke, Gertler, and Gilchrist 1999). In this case, the foreign sector is the lender and the domestic economy is the borrower. In expansionary phases, an increase in capital inflows reduces domestic credit constraints and increases lending. This, in turn, increases asset prices and capital gains, and improves the net worth of domestic institutions, encouraging further inflows.[5] In a second-round effect, this translates into further improvements in net worth. Unlike the standard (closed economy) financial accelerator mechanism, movements in the currency can reinforce the financial accelerator mechanism. Higher capital inflows lead to an appreciation of the currency, improving the balance sheets of those firms whose debt is denominated in foreign currency. This, along with higher asset prices, encourages higher inflows (Brunnermeier et al. 2012).[6] A reversal in capital flows starts a process which works the opposite way, with the exchange rate and asset prices exacerbating credit constraints. Asymmetric information and moral hazard problems in the banking sector increase, and foreign funders become more likely to pull their funds out (Goldstein and Turner 1996; Mishkin 1996). Companies engage in distress sales, driving down asset prices, encouraging further capital outflows and exchange rate depreciation, and setting off a downward spiral (Joyce and Nabar 2009). In this case, the depreciation of the currency, which is expected to stabilise the economy through its impact on competitiveness, may increase the output losses through its impact on the net worth of firms with short-term foreign currency-denominated liabilities (Blanchard et al. 2010). These dynamics can be triggered by the continuous deterioration of some economic indicators and can represent a threshold point or discontinuity as the capital reversal shock hits the economy (Akıncı and Chahrour 2018; Calvo 2003). This makes the assessment of capital flow reversal shocks more difficult in standard modelling frameworks.

Balance sheet dynamics are central to the negative effects associated with a capital flow reversal or an extreme sudden-stop situation (Calvo, Izquierdo, and Loo-Kung. 2006; Calvo, Izquierdo, and Mejia 2004). Mishkin (1999) argues that a reversal in capital flows has large economic impacts only if it affects the balance sheets of economic agents, and in particular the balance sheets of banks.

This importance of the financial sector reflects its rapid growth over the last 50 years. Increased liberalisation and deregulation of the financial sector have led to a very large increase in financial flows across countries and the emergence of financial instruments, supposedly developed to reduce risk in the financial system. The outcome of these developments has often been to increase the probability and severity of sudden stops and the transmission of financial crisis across country borders (Arestis and Sawyer 2001).

The strength of these impacts is linked to the level of financial development (or financialization in one of the senses defined by (Sawyer 2013)). The level of financial development can have both positive and negative effects on the probability and the size of the impact associated with capital flow reversals.[7] High levels of financial development have made it easier for economies to diversify across small shocks, but they have also exposed them to large systemic shocks, which generate significant movements in asset prices (Rajan 2005). It is for these reasons that Sawyer (2016) argues that the levels of financialization have become excessive, generating risk, uncertainty and inequality rather than supporting growth.

Despite the importance of the financial sector in driving capital flows and in the transmission mechanism of capital flow reversal shocks, the existing literature on reversal episodes generally lacks a model of such financial dynamics. Mendoza (2006) reviews several small dynamic stochastic general equilibrium (DSGE) models which rely on a debt-deflation mechanism and credit constraints to generate large negative economic impacts in response to sudden-stop episodes. Similarly, Fornaro (2015) and Ottonello (2013) develop a small general equilibrium models with rational expectations, nominal wage rigidities, and a financial accelerator mechanism. A different but related strand of DSGE models aims to study portfolio allocation between foreign and domestic assets, and how this generates capital flows (Devereux and Sutherland 2009; Tille and Van Wincoop. 2010). While all of these models tend to generate a financial accelerator effect, the financial sector is not modelled explicitly and there is no consistent stock-flow accounting. While there are various criticisms of DSGE models,[8] the absence of consistent balance sheet dynamics makes them less suitable for the analysis of capital flow dynamics. Calvo, Izquierdo, and Mejia (2004), Eggertsson and Krugman (2012), and Borio and Zhu (2012) argue, disaggregated balance sheet dynamics are important for studying the impacts of capital flow reversal shocks, fiscal policy, and the general risk behaviour of agents in the economy.

We study the impacts of capital flow reversal shocks in a model which explicitly models the financial flows and balance sheets in an economy. Specifically, we develop a small general equilibrium model that builds on Devarajan and Go (1998) and is stock- and flow-consistent in the tradition of Backus et al. (1980) and Godley and Lavoie (2012).[9] The model has several financial instruments and institutions. Consumption and production behaviour is micro-founded in agents inter-temporal optimisation, allowing us to capture how changes in preferences, technology and resource constraints affect outcomes. Prices exhibit a degree of stickiness, and there is a monetary policy reaction function based on a Taylor rule. These features are similar to the New Keynesian DSGE models but, unlike the DSGE models, ours is not stochastic.

Two important features of our model make it different from the traditional stock- and flow-consistent models and DSGE models. . First, our analysis of financial sector behaviour is based on modern theories of financial transmission mechanisms developed in the wake of the 2008 global financial crash, with modifications appropriate for application to South Africa (Borio and Zhu 2012; Woodford 2010).[10] Second, we specify a dynamic adjustment model of household expectations, which resembles bounded rationality. Our specification of bounded rationality reflects limited information on which expectations are formed, and which are revised over time.

The effects of a capital flow reversal shock in our model are larger compared to other studies evaluating the impact of capital flow reversal on the South African economy. Smit, Grobler, and Nel (2014) find small impacts. Frankel, Smit, and Sturzenegger (2008) find a similar impact, which is sensitive to the monetary policy response. A higher repurchase rate (repo rate) exacerbates the impact. They justify the muted impacts by South Africas small holding of foreign currency-denominated debt with short maturities, which is not modelled explicitly. The econometric models used in the analysis have no financial dynamics.,[11,12]

We generate a larger impact as the outflow of foreign savings still reduces the availability of savings in the domestic economy, lowers the level of intermediation, and pushes loan spreads up and asset prices down. Even in the absence of large stocks of foreign currency-denominated debt, a capital flow reversal shock does have an impact on the domestic economy through its impact on liquidity and demand for domestic financial assets.

We also illustrate how our framework handles discontinuity in expectations by shortening the household optimisation period. We move from bounded rationality to more myopic expectations. The simulation illustrates the role of expectation formation in amplifying capital flow reversal shocks. During the Covid-19 crisis, high levels of uncertainty have changed the expectation formation process of households as households ability to smooth consumption is reduced. In our framework, household expectations change from model-consistent (over 10 periods) to more myopic in the presence of a large negative shock.

The structure of the paper is as follows. The next section provides a short overview of South Africa's recent experience with external capital flows. This is followed by section 3, which presents the model. Section 4 presents the data employed in the analysis, while section 5 discusses the calibration strategy. The results are presented in section 6, and section 7 concludes .

2. South African external capital flows

Our analysis is based on the South African economy with its well-developed financial sector. The Global Competitiveness Report 2019 ranks South Africa 19th in terms of its level of financial sector development.[13] The South African rand is the 20th most-traded currency globally, and the country has one of the highest market-capitalisation-to-GDP ratios.[14]

South Africas deep and liquid financial markets support economic development and facilitate funding for both private and public institutions. This indicates that an analysis of the macroeconomic shocks in the South African context needs to consider financial sector behaviour.

Figure 1 shows portfolio flows into Emerging Markets and South Africa. Total portfolio flows fell sharply in March as global risk aversion increased due to the developing Covid-19 crisis, indicating a sudden stop episode. On an aggregate level portfolio flows into Emerging Markets have recovered somewhat. South Africa also saw a large retreat in March but without a recovery, suggesting that domestic factors such as a deteriorating fiscal situation, are also contributing to the exit of non-resident investors. Non-residents own about 30% of rand-denominated debt, down by 12 percentage points from its peak.

Taking a longer view of capital movements suggest that the structurally low levels of domestic savings in South Africa have led to the country placing high reliance on foreign savings. This has increased South Africa's vulnerability to capital flow volatility.[15] While South Africa has a relatively low level of foreign currency-denominated debt (just over 20% of GDP), foreign capital flows are an important contributor to asset prices and borrowing cost movements, and thus influence domestic institutional decisions to accumulate assets and liabilities. Another important characteristics is that South African investors are limited in terms of their foreign asset exposure, which also limits the size of repatriation flows in the presence of a sudden stop in inflows. This increases the country's vulnerability to capital flow reversal shocks (Agosin, Díaz, and Karnani 2019).

The flow-of-funds data produced by the South African Reserve Bank (SARB) provide information on foreign saving inflows and their allocation to various financial instruments. Net foreign saving inflows slowed down significantly between 2008 and 2009 as the global financial crisis hit South Africa (Figure 2). As central banks in advanced economies embarked on unconventional monetary policy, net saving inflows accelerated into South Africa. The inflows slowed down again over 2016 and 2017 as policy

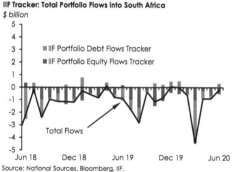

Figure 1. Portfolio flows.

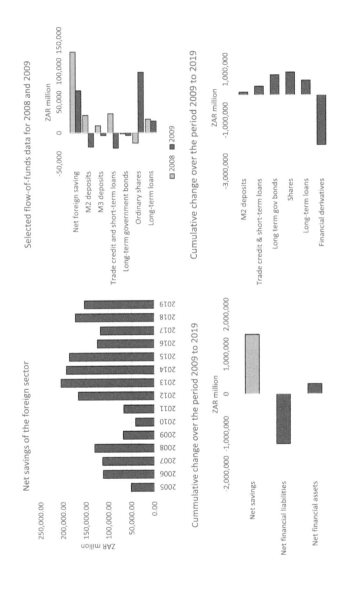

Figure 2. Adjustment to lower net foreign savings. Source: South African Reserve Bank

uncertainty increased in South Africa and central banks in advanced economies started to reverse some of the unconventional monetary policy interventions.

The reductions in capital inflows from 2008 to 2009 and from 2016 to 2017 and due to the Covid-19 crisis in 2020 are significant, constituting capital flow reversal shocks to the economy. The top-right panel of Figure 2 indicates that the adjustment in net foreign savings in 2009 took place mainly through deposits, short-term loans and shares. The increase in shareholding may reflect carry-trade opportunities (see Hassan (2015)). The average repo rate in 2008 was 11.6% compared to 8.4% in 2009, while the JSE All-Share Index declined by 13.0% in 2009. The combination of lower equity prices, a weaker rand and good dividend payouts may have created expectations of higher future returns among foreign investors.

The bottom two panels of Figure 2 present cumulative gross flows. The bottom-left panel indicates that there has been a significant decline in the net financial liabilities of the foreign sector over the period. These are the foreign assets of South African residents. This likely reflects capital value losses but also a return of South African assets as the currency depreciated. Foreign saving inflows recorded a large increase over the period. There were also large compositional changes in the foreign sector holding of South African financial assets. Over the period from 2009 to 2019, the foreign holding of long-term government bonds, shares and long-term loans recorded a significant increase. Financial derivatives include instruments such as options and swaps. The negative value reflects that, at maturity, the value of the derivative instrument (calculated as the net value of all transactions) falls as the instrument is exercised.

The outcomes suggest that capital flow reversal has an important impact through the financial sector and asset prices. The foreign sector affects bond and equity markets as well as the money multiplier through its impact on loan extension and direct deposits into the financial system. It can cause credit constraints to tighten. Assessing the impact of capital flow reversal shocks requires tracing the impacts through the financial sector.

Our framework tries to capture these relationships. We have several institutions. Households interest in retirement and life funds is a liability for the financial sector. In turn, the financial sector purchases bonds and equities. Apart from the financial sector, the foreign sector is a large purchaser of local bonds in our framework. The demand for specific assets is a function of the economic environment, relative returns and price changes affecting the balance sheets of institutions. The framework also captures the specific flows of interest and dividend income associated with the stock of bonds, bank deposits and equities.

3. Model

The model dynamics build on the simple computable general equilibrium (CGE) model developed by Devarajan and Go (1998) and incorporate elements of DSGE models as well as stock and flow models in the tradition of Backus et al. (1980) and Godley and Lavoie (2012). The model also incorporates elements of the theoretical models developed by Borio and Zhu (2012) and Woodford (2010). Makrelov et al. (2020) provide a detailed model description and discussion of all the equations as well as model calibration.

- Six types of institutions (agents) make real and financial decisions:
- the representative household;
- the representative firm (non-financial corporation);

- the representative financial corporation;
- government;
- the reserve bank (central bank); and
- the rest of the world.

The different agents meet in the financial, product and factor markets.

In the product market, the supply of goods and services is driven by producers maximising profits subject to a CES production function. We have one domestically produced good. Demands arise from the household, government, investment and net exports. Prices of imports, exports and the domestically produced good adjust to ensure flow equilibrium in the product market. Similar to Devarajan and Go (1998), the model includes three macroeconomic balances: the government balance, the external balance and the savings-investment balance. These are in addition to accounting rules that ensure stock and flow consistency on the financial side.

In the factor markets, the demands for capital and labour are driven by the real borrowing costs in the economy as well as the deviation of aggregate demand from its steady state. The real borrowing costs reflect the prevailing credit conditions and, along with aggregate demand, proxy the current economic conditions. Higher real rates reduce the demand for factors of production directly and indirectly through their impact on aggregate demand. Labour demand tends to be more sensitive to changes in real borrowing costs and aggregate demand than capital as capital is generally activity specific and thus immobile. The factor returns adjust reflecting the imperfect substitutability of capital and labour.

Households maximise (optimise) consumption over a period of 10 quarters, subject to a future wealth target and all equations in the model. They receive factor income, dividends, interest income, social contributions and other income. They make decisions about consumption (savings), investment and asset and liability accumulation

The external sector interacts with the domestic economy in both the financial and product markets. Exports and imports are modelled as imperfect substitutes to the domestically produced good and are driven by changes in relative prices. We uses Constant Elasticity of Transformation function to define export behaviour and an Armington specification for imports. Some of the foreign liabilities of the domestic economy are fixed in foreign currency units, while others vary with the level of domestic economic activity. The exchange rate ensures the closure of the external balance. It also affects the liability side of the foreign sector (expressed in local currency units) and along with exogenous changes to foreign savings leads to changes in the financial wealth of the external sector.

Financial instruments are grouped in five categories: equities, loans, cash and deposits, bonds, and other. Financial behaviour in our framework is based on flow-of-funds dynamics. In every period, agents experience a change in their financial wealth resulting from their decisions to save and invest, accumulate net liabilities (sources of funds), and changes to the equity price.[16] Financial wealth and any changes in it are held in portfolios comprising agents preferred combinations of the financial instruments from the categories that are available to them. The asset demand specification for the financial and foreign sectors is based on a Tobin asset demand function (Backus et al. 1980; Godley and Lavoie 2012; Tobin 1982). Market equilibrium is achieved either through changes in interest rates (loan or bond markets) or through residual supply. In the case of equities for example, the non-financial sector provides equities on demand while the equity price is determined by macroeconomic factors as in Chen, Roll, and Ross (1986). Money is endogenous in our framework.[17]

Critical in our model is the link between the supply of loans and the willingness of banks to hold higher capital. This relationship is linked to economic activity and it tries to operationalise the channels identified by Borio and Zhu (2012).

Our assumptions about expectations are different from those of mainstream DSGE models. The household has model-consistent expectations (similarly to DSGE models) within each period. However, the ability of the representative household to foresee the future is limited to 10 periods (2.5 years)[18] and the formation of expectations can vary between periods. This renders the model suitable for analysing non-linearities such as sudden stop shocks. Newly formed expectations can be introduced, for example by shortening or increasing the optimisation period, or by changing the value of coefficients or the structure of equations between periods. As the household solves for each period, new information about the economy becomes available, which is incorporated into the next periods optimisation. Our expectation formation resembles bounded rationality (Simon 1955, 1982). The choice of bounded rationality, over behavioural norms as in the traditional stock and flow models or rational expectations as in DSGE models, is based on experimental research, which suggests that agents exhibit bounded rationality (Assenza et al. 2014; Hommes 2011; Roos and Luhan 2013).

The data used in the calibration of the model are a series of financial macro SAMs and institutional balance sheets over the period 2002 to 2016 and National Accounts data for South Africa. The choice of period reflects the introduction of inflation targeting and the immediate period after the 2008 global financial crisis. Our approach of real sector parameters closely follows the approach outlined by Devarajan and Go (1998). The substitution elasticities are based on the recent estimates for South Africa produced by Kreuser, Burger, and Rankin (2015) and Saikkonen (2015).

The coefficients for the asset demand function are based on those used by Godley (1996) and Godley and Lavoie (2012). The coefficients reflect a stronger response of equity and bonds to changes in relative prices. Our strategy here is to utilise the coefficients generated by other studies, bearing in mind the limitations of this approach, or to get some sense of the relationship through simple econometric estimates, which are further calibrated in the model to generate a consistent baseline. These coefficients are the same across the two simulations and are listed in Annexure A.

4. Results

We simulate the impact of a decline in net foreign savings by 2% of GDP over four quarters and explain the transmission mechanism in our stock- and flow-consistent model. We compare the result from the main simulation with a scenario where the capital flow reversal shock changes the household expectations from model-consistent to more myopic in line with the high levels of uncertainty around the Covid crisis. This aims to show how discontinuity in the behaviour of households affects the results, and illustrates the ability of our framework to capture discontinuous behaviour, which characterises sudden-stop behaviour.

The reversal in foreign saving flows reduces liquidity in the domestic market and requires the rebalancing of investment and domestic savings in order to maintain the equilibrium. In addition, following the mechanisms identified by Borio and Zhu (2012) and Woodford (2010), there is an increase in the probabilities of default and the perception of risk, and deterioration in valuations and the net worth of the financial sector, leading to lower levels of intermediation and higher lending spreads. The banks cash reserve ratio in our model jumps, reducing the money multiplier and the supply of loans by the financial sector (panel 1 in Figure 3). The reduction in the supply of loans increases the spread over the repo rate, and the loan rate rises.[19] This is depicted in panel 2 of Figure 2. The trend reflects the initial fall in foreign saving inflows and the subsequent recovery.

The increase in the reserve ratio is also driven by the fall in the value of financial sector assets, which is explained below.

An increase in the nominal loan rate translates into a higher real rate. This effect is strengthened by a fall in inflation and inflation expectations relative to the baseline (panel 1 in Figure 3). The output response dominates the exchange rate impact on inflation. This is also what we have observed in the first 7 months of 2020. The higher real rates affect the economy in several different ways:

(1) Firstly, investment falls across all institutions, which helps to rebalance savings and investment.
(2) The increase in the real rates negatively affects aggregate demand and the demand for factors of production, which decreases utilisation in the economy and, thus, production.
(3) The demand for loans decreases, thus decreasing the sources of funding available for investment in real and financial assets.
(4) Interest income increases.

Table 1 shows the impact on investment at 1 period (t + 1) and 10 periods (t + 10) after the shock. Investment by non-financial firms is initially 3.8% lower, and it is 3.9% lower in the outer years despite the recovery in net foreign savings. This result is in line with the long-term impacts on investment found by Joyce and Nabar (2009). This decline reflects the permanent decline in the equity price relative to the baseline (panel 2 Figure 4 below).

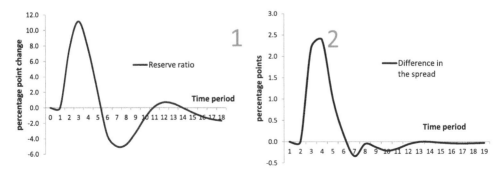

Figure 3. Interest rate spread and reserve ratio. Source: Model simulations (quarterly data)

Table 1. Impact on real expenditure.

Per cent deviation from baseline	Impact	
	t + 1	t + 10
Household expenditure	−2.43	−0.67
Investment		
Non-financial firms	−3.81	−3.88
Other institutions	−4.11	−4.14
Exports	0.25	−0.97
Imports	−4.74	−0.96
GDP	−0.48	−0.97

Source: Model simulations (quarterly data)

The fall in the equity price relative to the baseline reflects an expectation of lower inflation initially and lower growth in money supply, but, more importantly, the medium-term effect is driven by a permanently lower stock of capital and lower levels of capacity utilisation compared to the baseline.

Exports and imports follow the expected trends as the reduction in foreign savings translates into a depreciation of the exchange rate (see panel 6, Figure 4)[20]. The decline in imports is significantly larger than the increase in exports. It reflects not only the depreciation in the currency, but also the significant decline in aggregate demand.

The response is dependent on the assumed elasticities in the Constant Elasticity of Substitution and Armington functions. As the flow of foreign savings normalises and the exchange rate depreciation is reversed, the level of exports declines compared to the baseline while the level of imports recovers but remains below the baseline level. The normalisation of imports also reflects a recovery in household consumption in the outer years. As in Smit, Grobler, and Nel (2014), the main adjustment is through imports, which reduces the overall negative effect on GDP.

The lower inflation and lower utilisation of resources reduce the repo rate through the Taylor rule specification. This, in turn, reduces the real policy rate, and helps to alleviate some of the pressures from the increase in the real lending rates.

The savings of the financial and non-financial sectors increase by reducing dividend payments (panel 7 in Figure 4). Total dividend payments are close to 40% lower compared to the baseline, which negatively affects the income of all institutions, particularly the income of households. This decline in dividend income is offset somewhat for some institutions by an increase in interest income (panel 8, Figure 4). For the representative household, however, the combination of lower dividend income and higher interest expenditure reduces its ability to save and consume.

Factor income also declines as capacity utilisation declines. Figure 3 shows divergent trends in capital and labour wages. We assume full employment.[21] The labour force increases but economic activity is lower than in the baseline, and labour wages must fall in order for labour to be absorbed. In the case of capital, the effect is more positive over the entire horizon. This reflects the fall in investment and the slower pace of capital accumulation in the scenario.

Table 1 indicates that the immediate impact on household consumption is large and negative. However, as foreign savings normalise, household consumption recovers marginally. There are several forces that affect household behaviour. On the one side,

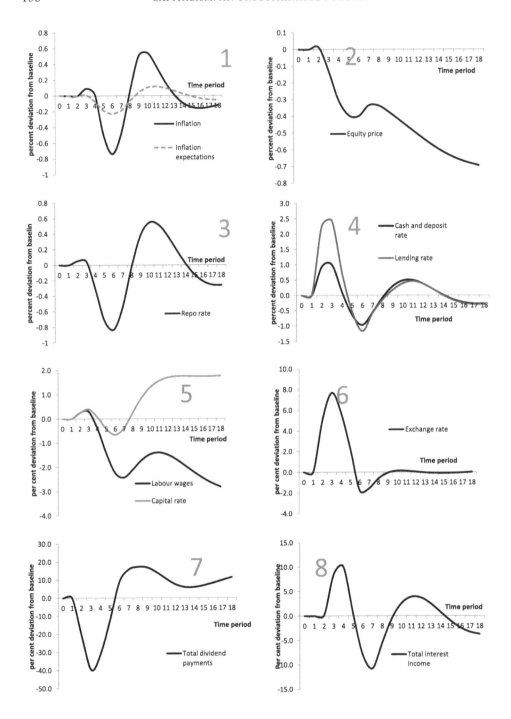

Figure 4. Impacts on rates and prices. Source: Model simulations (quarterly data)

household income falls as explained above, which translates into lower consumption. On the other side, the fall in the equity price and the lower provision of loans make it more difficult for the household to achieve its wealth target. The household needs to save more and consume less in order to compensate for the fall in the sources of funding and to achieve its desired level of future wealth and consumption. The lower level of expected inflation mitigates this impact somewhat in the short run as the household is targeting real wealth.

Figure 5 presents household optimisation[22] behaviour at three points in time: the time the shock takes place (*t*), 10 periods after the shock (*t* + 10), and 15 periods after the shock (*t* + 15). The largest differences in growth rates are in period *t*. At time *t*, the economy is faced with a sudden shock, and household wealth falls. The household must consume less and save more in order to achieve its target level of wealth. Household consumption declines due to lower income and fewer sources of funding, but there is also a greater need to save in order to achieve its target level of wealth. The growth rates in the simulation are significantly lower initially. The household, however, expects that the simulation growth rates will rise and eventually exceed those in the baseline given the cyclical structure of the economy and the likely response of monetary authorities to the shock. The stronger growth rate of financial wealth in the outer years of the optimisation period allows households to consume more. At this point, households cannot see that the capital flow reversal shock continues for four more quarters. This expectation of future improvements lowers the impact of capital flow reversal on household consumption in period *t*. It allows households to smooth their consumption.

By period *t + 10*, households face lower equity prices and the expected recovery in period *t* has not taken place. Household consumption growth is lower in order to maintain the growth in real financial wealth close to the baseline. By period *t + 15*, the growth rates in the baseline and the simulation are very similar for household consumption and wealth compared to the baseline. However, in level terms, household wealth and household consumption are permanently lower as the temporary credit constraint on the economy has reduced the stock of financial wealth compared to the baseline.

In the Tables 2 and 3, we present the impact on the stocks of assets and liabilities, and we explain the impacts at period *t* + 3.

The financial wealth held by the foreign sector in South Africa is affected mainly by a fall in savings and a depreciation in the currency. A decrease in foreign saving inflows

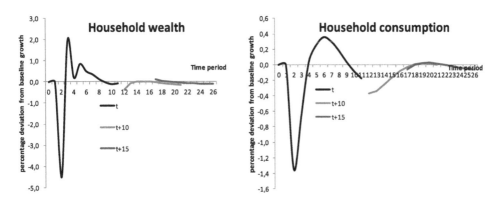

Figure 5. Household optimisation behaviour. Source: Model simulations (quarterly data)

reduces the purchases of South African financial assets by the foreign sector. The exchange rate increases the local currency value of foreign currency-denominated assets held by domestic institutions (these are liabilities for the foreign sector). The depreciation in the currency increases the value of bonds as well as cash and deposit liabilities relative to the baseline. We assume that their value is fixed in foreign currency units. The stocks of foreign loans and equity assets held by domestic residents are linked to domestic output. Lower domestic output and a weaker domestic currency discourage domestic institutions from increasing their holding of foreign equities and loans.

The foreign sector holding of South African financial assets declines across the board as the sources of funding and, in particular, the inflows of foreign savings decline. The decrease in the holding of bonds is smaller compared to the other asset classes, which reflects their higher relative return.[23]

It is this higher relative return on bonds that also encourages the financial sector to increase its holding of bonds (Table 2). It can also reflect some form of a flight to safety as bonds are associated with a lower risk of default. This impact works through the Tobin asset-demand function, which drives the demand for assets for the financial and foreign sectors.

The lower levels of cash and deposits received by the financial sector affect the financial accelerator mechanism in our framework. The extension of loans is lower because the financial sector chooses to hold more reserves, but also because the level of cash and deposits declines relative to the baseline. This mechanism creates a direct link between money creation and economic activity and it reflects our approach to modelling money as endogenous in our framework. In addition, this negative effect on financial wealth is compounded by a decline in equity liabilities, driven by the slower creation of equity assets by households as well as the lower equity price. The loans extended to the financial sector experience no change. There are two effects that determine the impact on the demand for loans by the financial sector. Higher loan rates discourage borrowing, but higher interest income encourages borrowing. If we change the assumption of no change in the share of non-performing loans, then both effects operate in the same direction and demand falls. Overall, the pool of funds generated from the sources of funding and available for investment is lower relative to the baseline, despite the higher levels of net savings. All asset holdings for the sector, except bonds, decline.

The non-financial sector also funds its financial wealth through net savings, loans and equity sales. While savings increase initially, the lower demand for equities (due to their lower return) and lower levels of economic activity lead to a significant reduction in the

Table 2. Changes to the holding of financial assets.

Deviation from baseline	Assets			
	Equities	Bonds	Cash and deposits	Loans
t + 3				
Reserve Bank	−0.4	2.0	0.0	0.0
Financial sector	−0.7	5.4	−1.7	−2.9
Non-financial sector	−0.4	0.0	−1.6	−0.3
Households	−2.0	0.0	−2.2	0.0
Government	−0.4	0.0	−1.8	−0.1
Rest of the World	−19.3	−12.6	−22.5	−14.6

Source: Model simulations (quarterly data)

equity liability for the sector (Table 3). At the same time, the higher loan rate and lower income decrease the demand for loans. The equity and loan effects offset the positive impacts from higher savings. The financial wealth available for investing declines, which leads to a decline in the holding of assets across the board.

The decline in the cash and deposits holding of the household reflects lower income, which offsets the impact of higher cash and deposit rates. The fall in the transaction demand for money is higher than the increase in the demand for money as a store of value. The decline in financial wealth translates into a lower demand for equities on the asset side. The decline in the value of assets also reflects the fact that the representative household has achieved lower levels of wealth in the previous periods. The households anticipation of a recovery in the economy, based on its model-consistent expectations, has led to smoother consumption and lower savings in the initial periods of the optimisation horizon.

Government maintains its levels of spending, which translates into higher issuance of bonds given its falling income. The increase in bond issuance is also driven by the fall in the other sources of funds, such as loans. In terms of our specification, the decline in loan liabilities relative to the baseline is driven by the higher borrowing costs and the lower income of government. The marginal decrease in government equities, both on the asset and on the liability sides, reflects the lower equity price as the quantity of equities is modelled exogenously.[24] Similarly to the other institutions, the decrease in the sources of funding (liabilities) is matched by a decline in the uses of funding (assets).

The central bank sees a large increase in interest income as loan rates and cash and deposit rates rise. This increases the demand for loans as a source of funding, raising the financial wealth of the central bank and translating into higher purchases of bonds. Our assumption is that any increase in the financial wealth of the central bank translates into a greater holding of bonds. The stocks of all the other assets are assumed to be exogenous.[25]

In table 4, we outline the changes in net financial wealth, measured as the difference between the stock of financial wealth and the stock of financial liabilities divided by nominal GDP. The results indicate that the net financial position of the country improves as a result of a reversal in capital flows. This is not surprising, as the assets of the foreign sector are denominated in rand terms in our framework, whereas the liabilities (the foreign assets of the domestic sector) are denominated in foreign currency units. The domestic economy benefits from the depreciation in the currency. At the same time, the

Table 3. Changes to the holding of financial liabilities.

Deviation from baseline	Liabilities			
	Equities	Bonds	Cash and deposits	Loans
t + 3				
Reserve Bank	−0.4	0.0	−1.9	6.4
Financial sector	−1.4	0.0	−2.9	0.0
Non-financial sector	−10.3	0.0	0.0	−2.5
Households	0.0	0.0	0.0	−2.9
Government	−0.4	1.9	0.0	−2.2
Rest of the World	−4.8	2.2	2.2	−4.5

Source: Model simulations (quarterly data)

Table 4. Changes to net financial wealth.

Net financial wealth	Impact	
change as percent of GDP	t + 1	t + 10
Reserve Bank	0.0	0.0
Financial sector	1.1	0.3
Non-financial sector	3.4	4.2
Households	0.2	0.7
Government	−0.4	0.0
Rest of the World	−4.3	−5.2

Source: Model simulations (quarterly data)

decline in foreign savings reduces the sources of funding and the stock of assets held by the foreign sector relative to the baseline.

In our framework, which is stock- and flow-consistent, a deterioration in the net wealth of one sector must be matched by improvements in the net wealth of other sectors. In this case, the improvement takes place mainly through the balance sheets of the financial and non-financial sectors, which increase savings to offset the fall in foreign savings. For the financial sector, this effect is temporary. For the non-financial sector, the impact carries throughout the simulation period. The more permanent effect for the non-financial sector is explained by the higher net savings in the outer years, which are driven by permanently lower levels of investment.

The impact on domestic demand is similar to the mild scenario of Smit, Grobler, and Nel (2014). However, our shock is significantly smaller, almost half as large. Our transmission mechanism is different and assumes the amplifying effects of the financial sector, which in our framework work through the balance sheets of all the institutions in the economy. The results are relatively small compared to international experiences, as South Africa has a relatively low stock of foreign currency-denominated debt. This also minimises the probability of a capital flow reversal shock causing a banking crisis. However, our analysis indicates that, even in the absence of foreign currency-denominated debt, the financial sector has an important role in promulgating shocks through the economy.

In our analysis so far, we have assumed that there is no structural change in the behaviour of institutions in response to the capital flow reversal shock. A structural change reflects a discontinuity linked, for example, to bubble bursting or a very large economic shock as the Covid crisis. General equilibrium models are continuous models and cannot handle catastrophe theory-type dynamics. The nature of our framework, however, allows us to introduce such dynamics. While we cannot change the behaviour *within* periods as the household optimises, we can change behaviour *between* periods.

In response to a negative shock, the household shortens its optimisation period. Household expectations become more myopic. The simulation aims to capture the break in expectations described by Harris (1979). This is also in line with economic literature, which shows that agents switch between different forecasting rules and excess volatility can be explained by these changes (Grandmont 1998; Hommes 2011; Roos and Luhan 2013).

In Figure 6, we present the baseline consumption path indexed to 100 in period zero. Household optimisation, which starts at period zero, is labelled as *t-2*. Figure 6 shows that the consumption paths overlap over the solution period.

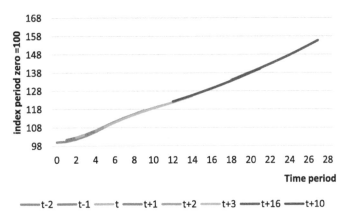

Figure 6. Consumption path in the base. Source: Model simulations (quarterly data)

The shock takes place in the second period. The optimisation path labelled t in Figure 7 has shifted to the right, reflecting that households need to save more and consume less. This path reflects the growth rates depicted in Figure 4.

The optimisation in the next period (*t + 1*) follows on *t*. As the outer periods of the optimisation horizon are reached, the path of household consumption gets closer to the path set in period *t-1*. The shift in the optimisation path is also affected by the household anticipating a recovery in the economy, as was highlighted earlier. This leads to a lower adjustment in household consumption expenditure in the initial periods of the optimisation horizon. The outer-year optimisation paths, labelled *t + 10* and *t + 16*, are close to the baseline path.

In the second simulation, we introduce discontinuity by changing household expectations from model-consistent to more myopic. The shock leads to households shortening

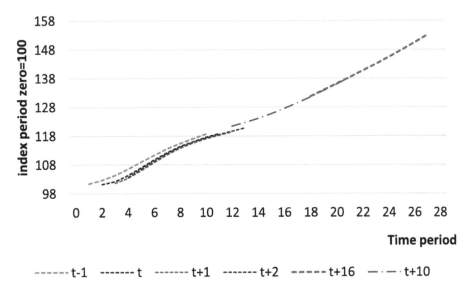

Figure 7. Consumption path following a reversal in capital flows without a change in the household optimisation horizon. Source: Model simulations (quarterly data)

their optimisation horizon, thus trying to achieve their target wealth over a shorter period of time. The results are presented in Figure 8. The shock is introduced by assuming that, in the period when the shock takes place, households reduce their horizon to three periods while trying to achieve the same level of financial wealth. The optimisation horizon reverses gradually as the shock is reversed.

The results show a larger shift to the right. The impact on consumption is larger, reflecting that the household saves more, which is a function of its expectations. Now, it cannot foresee a recovery in the economy and it has a shorter period of time to achieve its wealth target. The sudden change in expectations and the consequent behaviour of the household exacerbates the negative effects associated with the capital flow reversal shock. This is shown in Figure 9, which plots the ratio of consumption under simulation two to simulation one, indexed to 100 in the base year. The ratio depicted declines initially as consumption under this simulation is lower compared to simulation one (with less myopic expectations). The impact shows a larger decline in household consumption over the period associated with the capital flow reversal shock. The trend shows recovery as capital flows normalise and the optimisation period moves gradually from 3 periods back to 10 periods. In the outer years, household consumption in simulation two is slightly higher relative to simulation one as income is relatively higher. The higher levels of savings in the initial periods of simulation two provide for a higher stock of assets, which generates relatively higher interest and dividend income in the outer periods.

While in the first simulation household consumption was 2.4% lower relative to the baseline in period $t + 1$, now it is almost 4% lower. The higher savings by households, however, reduce some of the negative effects on investment associated with the reversal in foreign savings. Liquidity, proxied by growth in cash and deposits, is higher compared to simulation one, which reduces the negative impact on the financial sectors reserve ratio. This leads to a lower loan rate and higher investment compared to simulation one in the outer years.[26]

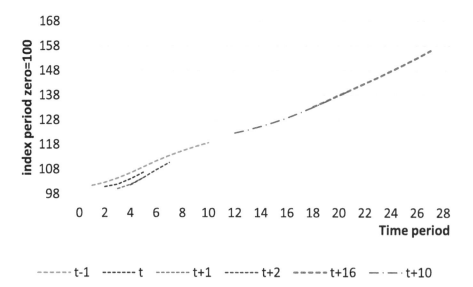

Figure 8. Consumption path following a reversal in capital flows with a change in the household optimisation horizon. Source: Model simulations (quarterly data)

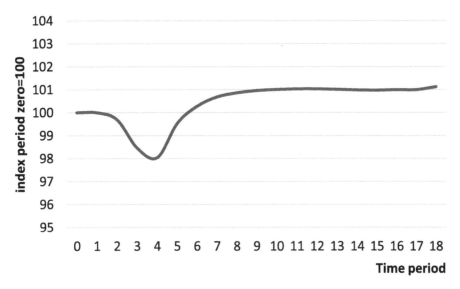

Figure 9. Ratio of consumption under simulation two to simulation one. Source: Model simulations (quarterly data)

5. Conclusion

The Covid crisis has generated very large swings in capital flows accompanied by large changes in exchange rates and asset prices in emerging market. Our analysis aimed to assess the impacts via the financial sector. Our main conclusion is that, even in the absence of a large stock of foreign currency-denominated debt, a capital flow reversal shock can still generate a sizable impact. A reduction in capital flows decreases liquidity in the domestic market and increases the need to raise the level of domestic net savings. Financial sector perceptions of risk increase, which encourages the sector to hold more reserves and reduces the supply of loans. This pushes up the lending spread and reduces the equity price. Economic activity declines. The real economy effects feed back to the financial sector through the balance sheets of all institutions, creating a financial accelerator effect.

The results can be significantly larger if there is a change in the expectation formation process of households. We introduce a discontinuity in our framework, which worsens the real economy impacts and increases volatility. As indicated before, such discontinuity can be linked, for example, to a large economic shock such as the Covid crisis.[27]

Our results also highlight the strong growth in the financial sector post the 1994 period in South Africa, its increased sophistication and integration into the global financial economy, which have increased its ability to amplify economic shocks.

Notes

1. Sarno, Tsiakas, and Ulloa (2016) find that global factors such as US interest rates and global risk aversion explain close to 80% of the variation in global bond and equity flows.
2. For South Africa, earlier studies find the interest rate, inflation differentials, exchange rate movements, political risk, the real GDP growth rate, the depth of capital markets, the effectiveness of law and order, and the government deficit as important drivers of capital flows (Ahmed, Funke, and Arezki 2005; Fedderke and Liu 2002; Wesso 2001). Aron, Leape,

and Tomas (2010) find that risk aversion is a significant driver of both equity and overall flows, and that good performance by the US stock market leads to higher flows into South Africa. The latter results, according to these authors, represent a global liquidity effect driven by higher profitability. Hassan (2015) links capital flows, in particular bond flows, to carry-trade opportunities (conditions characterised by a relatively high interest rate differential and low volatility).

3. At the same time, a period of higher-than-average flows, the so-called capital bonanza, also increases the probability of a capital reversal and a financial crisis as it tends to create excessive lending (Caballero 2016; Reinhart and Reinhart 2008).

4. Reinhart and Reinhart (2008) provide a short review of the different definitions of sudden stops used in the literature. Episodes of a sudden stop generally refer to a sudden and large decline in capital flows, which is accompanied by a significant increase in some measure of the external cost of funding (Calvo, Izquierdo, and Mejia 2004; Cavallo and Frankel 2008).

5. A number of studies provide evidence for a link between capital flow and asset prices. See, for examp IMF (2010, 2013a) and Sá, Towbin, and Wieladek (2011).

6. While the exchange rate plays a smaller role in countries that have a fixed exchange rate, the financial accelerator mechanism remains important. Magud and Vesperoni (2015) find that countries with fixed exchange rates have stronger credit extension associated with large capital inflows and create more foreign liabilities. In addition, the fixed exchange rate prevents the adjustment between the tradable and non-tradable sectors, which generates larger output losses than in economies characterised by a flexible exchange rate.

7. See for example Mendoza, Quadrini, and Ríos-Rull (2009), Aghion, Bacchetta, and Banerjee (2004) and Caballero (2016).

8. See for example Borio and Zhu (2012) and Duca and Muellbauer (2014).

9. Sawyer and Passarella (2017) employ a stock-flow consistent model, in the tradition of Godley and Lavoie, in their analysis of the monetary circuit.

10. In the model developed by Woodford (2010), the lending spread is a function of financial sector institutions capital. Raising the level of capital is costly, and leverage is limited by regulatory requirements. Shocks that impair the capital of an institution (a bank or, more generally, a financial intermediary) or that create leverage ratio regulatory requirements translate into higher lending spreads and lower volumes of lending and economic activity. Borio and Zhu (2012) also link the capital of the financial sector to bank behaviour. Breaching the minimum capital threshold is costly for a bank. In the face of a possible breach, banks will take defensive action to avoid the high costs, which will affect the availability and pricing of funding extended to customers. The economic cycle changes the strength of this effect, as probability of default, valuations, and the perception of risk change. In turn, this shifts the relative position of banks capital to the regulatory threshold and affects bank behaviour. The accelerator effects in both models are driven by the relationship between capital and economic activity. Higher economic activity reduces the probability of default and the perception of risk, and improves valuations. This reduces lending spreads, which encourages further improvements in economic activity.

11. The IMF (2013b) also emphasises the likely impact of capital flow reversal on South Africa as a result of US monetary policy normalisation. However, the IMF does not provide quantitative estimates of the economic impact. It argues that South Africa is likely to see a large outflow due to its twin deficits, but there are also mitigating factors such as government debt that is almost entirely denominated in rand terms and is mostly long-term.

12. A few studies look at the impacts of a sudden stop in a sample of countries, including South Africa. However, they do not provide specific impacts for South Africa (Calvo, Izquierdo, and Mejia 2004; Cavallo et al. 2015; Joyce and Nabar 2009; Reinhart and Reinhart 2008). These studies tend to argue that the impacts tend to be larger than those identified by Frankel, Smit, and Sturzenegger (2008) and Smit, Grobler, and Nel (2014), especially if the sudden stop is accompanied by a banking crisis.

13. The report is available at http://www3.weforum.org/docs/WEF_TheGlobalCompetitiveness Report2019.pdf.

14. See the Bank for International Settlements Triennial Central Bank Survey, available at https://www.bis.org/statistics/d11_3.pdf.
15. This section builds on the historical analysis of South African capital flows presented in Smit, Grobler, and Nel (2014).
16. These changes in financial wealth are used to build financial Social Accounting Matrices, which represent real and financial transactions and ensure flow consistency. The approach is described in section 4.
17. Arestis and Howells (1996) and Arestis and Sawyer (2001) provide theoretical support for this approach.
18. The assumption of 10 periods reflects the period that monetary shocks take to dissipate in an economy and the inflation expectations period generally targeted by central banks. We have assumed that this period also reflects the household expectation horizon.
19. This mechanism is also in line with the empirical findings in Rey (2015).
20. The exchange rate is modelled following the approach in Devarajan and Go (1998). In their model, if the real exchange rate is fixed, foreign savings need to adjust, and if the real exchange rate is flexible, then foreign savings are exogenous. In our model, we have exogenous foreign savings and a flexible real exchange rate. An exogenous reduction in capital flows generates an exchange rate response to ensure that the current account is in line with the new level of foreign savings.
21. This is a reasonable assumption in the market for high skilled workers. Various economic reports highlight South Africas skills constraint and its impact on the demand for low skilled and semi-skilled workers. See for example NPC (2013) and Faulkner, Loewald, and Makrelov (2013). The markets for low skilled and semi-skilled workers are not characterized by full employment. Given that we do not distinguish between different types of labour, our labour market assumptions are somewhat restrictive in order to improve the model tractability.
22. The optimization path is based on the households limited foresight of 10 quarters and it shows the model solutions for each quarter.
23. Stocks where a change is not recorded reflect the fact that the asset or liability instrument is modelled exogenously or that the institution does not hold the particular asset or liability.
24. Government equity liabilities reflect our aggregation of financial instruments. Other loan stock and preference shares were classified as equities.
25. The change in equities reflects the fall in the equity prices and not in the stock of equities.
26. The results are sensitive to the size of coefficients, particularly the ones driving financial behaviour. This is a similar finding to that of Adam and Bevan (1998). Smaller response of the reserve ratio to changes in the repo rate or the balance sheet of the financial sector, translate into the money multiplier in our framework becoming less sensitive to the economic cycle. This effect, however, depends on the elasticity of loan demand to lending rates. A more inelastic relationship requires larger lending spread changes to equilibrate the model supply and demand of loans, which amplifies the impact of economic shocks. Similarly, a more inelastic response of the demand for bonds to the bond yield in the Tobin asset demand function leads to larger shifts in bond yields in order to incentivise the foreign and financial sectors to purchase bonds.
27. While we have chosen to introduce discontinuities in the solution process by changing the household expectation formation, there are also other ways to introduce them. These include changes to the functional specification of the model as well as to some of the parameters.

Acknowledgements

The authors would like to thank participants at seminars at the South African National Treasury, South African Reserve Bank and SOAS, University of London as well as acknowledge financial support from the South African National Treasury and the United Nations World Institute for

Development Economic Research.

Disclosure statement

No potential conflict of interest was reported by the author(s).

Funding

This work was supported by UNU-WIDER and the South African National Treasury.

ORCID

Laurence Harris (iD) http://orcid.org/0000-0002-1921-7328

References

Adam, C. S., and D. L. Bevan. 1998. *Costs and Benefits of Incorporating Asset Markets into CGE Models: Evidence and Design Issues*. Institute of Economics and Statistics: University of Oxford.

Aghion, P., P. Bacchetta, and A. Banerjee. 2004. "Financial Development and the Instability of Open Economies." *Journal of Monetary Economics* 51: 1077–1106.

Agosin, M. R., J. D. Díaz, and M. Karnani. 2019. "Sudden Stops of Capital Flows: Do Foreign Assets Behave Differently from Foreign Liabilities?" *Journal of International Money and Finance* 96: 28–36.

Ahmed, F., N. Funke, and R. Arezki. 2005. "The Composition of Capital Flows: Is South Africa Different?" *IMF Working Paper Series*. No WP/05/40.

Ahmed, S., and A. Zlate. 2014. "Capital Flows to Emerging Market Economies: A Brave New World?" *Journal of International Money and Finance* 48 (Part B): 221–248.

Akıncı, Ö., and R. Chahrour. 2018. "Good News Is Bad News: Leverage Cycles and Sudden Stops." *Journal of International Economics* 114: 362–375.

Anaya, P., M. Hachula, and C. J. Offermanns. 2017. "Spillovers of US Unconventional Monetary Policy to Emerging Markets: The Role of Capital Flows." *Journal of International Money and Finance* 73: 275–295.

Arestis, P., and M. C. Sawyer. 2001. *Money, Finance and Capitalist Development*. Cheltenham, United Kingdom: Edward Elgar Publishing.

Arestis, P., and P. Howells. 1996. "Theoretical Reflections on Endogenous Money: The Problem with Convenience Lending." *Cambridge Journal of Economics* 20: 539–551.

Aron, J., J. Leape, and L. Tomas. 2010. "Portfolio Flows; Exchange Rate Policy and Capital Markets in South Africa." In *CSAE Conference: Economic Development in Africa*. Oxford, United Kingdom.

Assenza, T., T. Bao, C. Hommes, and D. Massaro. 2014. "Experiments on Expectations in Macroeconomics and Finance." In *Experiments in Macroeconomics,* edited by John Duffy, 11–70. Vol. 17. Emerald Group Publishing Limited. United Kingdom.

Backus, D., W. Brainard, G. Smith, and J. Tobin. 1980. "A Model of U.S. Financial and Nonfinancial Economic Behavior." *Journal of Money, Credit, and Banking* 12: 259–293.

Baskaya, Y. S., J. Di Giovanni, Ş. Kalemli-Özcan, J.-L. Peydro, and M. F. Ulu. 2017. "Capital Flows and the International Credit Channel." *Journal of International Economics* 108: S15–S22.

Bernanke, B. S., M. Gertler, and S. Gilchrist. 1999. "The Financial Accelerator in a Quantitative Business Cycle Framework." In *Handbook of Macroeconomics. Volume 1C,* edited by J. B. Taylor and M. Woodford, 1341–1393. Amsterdam; New York and Oxford: Elsevier Science, North-Holland. Handbooks in Economics, vol. 15.

Blanchard, O. J., H. Faruqee, M. Das, K. J. Forbes, and L. L. Tesar. 2010. "The Initial Impact of the Crisis on Emerging Market Countries [With Comments and Discussion]." *Brookings Papers on Economic Activity* 41(1): 263–323.

Borio, C., and H. Zhu. 2012. "Capital Regulation, Risk-Taking and Monetary Policy: A Missing Link in the Transmission Mechanism?" *Journal of Financial Stability* 8: 236–251.

Bräuning, F., and V. Ivashina. 2017. "US Monetary Policy and Emerging Market Credit Cycles" *Journal of Monetray Economics* 112: 57–76.

Brunnermeier, M., J. De Gregorio, B. Eichengreen, M. El-Erian, A. Fraga, T. Ito, P. R. Lane, J. Pisani-Ferry, E. Prasad, and R. Rajan. 2012. "Banks and Cross-border Capital Flows: Policy Challenges and Regulatory Responses." *Committee on International Economic Policy and Reform*, Washington DC.

Bruno, V., and H. S. Shin. 2015. "Capital Flows and the Risk-taking Channel of Monetary Policy." *Journal of Monetary Economics* 71: 119–132.

Byrne, J. P., and N. Fiess. 2016. "International Capital Flows to Emerging Markets: National and Global Determinants." *Journal of International Money and Finance* 61: 82–100.

Caballero, J. A. 2016. "Do Surges in International Capital Inflows Influence the Likelihood of Banking Crises?" *The Economic Journal* 126: 281–316.

Calvo, G. A. 1998. "Capital Flows and Capital-Market Crisis: The Simple Economics of Sudden Stops." *Journal of Applied Economics* 1: 35–54.

Calvo, G. A. 2003. "Explaining Sudden Stop, Growth Collapse and BOP Crisis: The Case of Distortionary Output Tax." *IMF Staff papers*.

Calvo, G. A., A. Izquierdo, and L.-F. Mejia. 2004. "On the Empirics of Sudden Stops: The Relevance of Balance-sheet Effects." *NBER Working Paper Series*, No. 10520.

Calvo, G. A., A. Izquierdo, and R. Loo-Kung. 2006. "Relative Price Volatility under Sudden Stops: The Relevance of Balance Sheet Effects." *Journal of International Economics* 69: 231–254.

Cavallo, E., A. Powell, M. Pedemonte, and P. Tavella. 2015. "A New Taxonomy of Sudden Stops: Which Sudden Stops Should Countries Be Most Concerned About?" *Journal of International Money and Finance* 51: 47–70.

Cavallo, E. A., and J. A. Frankel. 2008. "Does Openness to Trade Make Countries More Vulnerable to Sudden Stops, or Less? Using Gravity to Establish Causality." *Journal of International Money and Finance* 27: 1430–1452.

Cerutti, E., S. Claessens, and D. Puy. 2019. "Push Factors and Capital Flows to Emerging Markets: Why Knowing Your Lender Matters More than Fundamentals." *Journal of International Economics* 119: 133–149.

Chari, A., K. D. Stedman, and C. Lundblad. 2017. "Taper Tantrums: Qe, Its Aftermath and Emerging Market Capital Flows".*NBER Working Paper Series,* No. 23474

Chen, N.-F., R. Roll, and S. A. Ross. 1986. "Economic Forces and the Stock Market." *Journal of Business* 59: 383–403.

Converse, N. 2018. "Uncertainty, Capital Flows, and Maturity Mismatch." *Journal of International Money and Finance* 88: 260–275.

Dahlhaus, T., and G. Vasishtha. 2020. "Monetary Policy News in the US: Effects on Emerging Market Capital Flows." *Journal of International Money and Finance* 109: 102251.

Devarajan, S., and D. S. Go. 1998. "The Simplest Dynamic General-Equilibrium Model of an Open Economy." *Journal of Policy Modeling* 20: 677–714.

Devereux, M. B., and A. Sutherland. 2009. "A Portfolio Model of Capital Flows to Emerging Markets." *Journal of Development Economics* 89: 181–193.

Duca, J., and J. Muellbauer. 2014. "Tobin LIVES: Integrating Evolving Credit Market Architecture into Flow-of-Funds Based Macro-Models." In *A Flow-of-funds Perspective on the Financial Crisis*, edited by Winkler B., A. van Riet and P. Bull. Palgrave Macmillan, London: Palgrave Studies in Economics and Banking.

Eggertsson, G. B., and P. Krugman. 2012. "Debt, Deleveraging, and the Liquidity Trap: A Fisher-Minsky-Koo Approach." *Quarterly Journal of Economics* 127: 1469–1513.

Eichengreen, B., and P. Gupta. 2016. "Managing Sudden Stops." *World Bank Policy Research Working Paper Series*, No. 7639.

Eller, M., F. Huber, and H. Schuberth. 2020. "How Important are Global Factors for Understanding the Dynamics of International Capital Flows?" *Journal of International Money and Finance* 109: 102221.

Faulkner, D., C. Loewald, and K. Makrelov. 2013. "Achieving Higher Growth and Employment: Policy Options for South Africa." *South Africa Reserve Bank Working Paper Series*, No. 13/03.

Fedderke, J. W., and W. Liu. 2002. "Modelling the Determinants of Capital Flows and Capital Flight: With an Application to South African Data from 1960 to 1995." *Economic Modelling* 19: 419–444.

Forbes, K. J., and F. E. Warnock. 2012. "Capital Flow Waves: Surges, Stops, Flight, and Retrenchment." *Journal of International Economics* 88: 235–251.

Fornaro, L. 2015. "Financial Crises and Exchange Rate Policy." *Journal of International Economics* 95: 202–215.

Frankel, J., B. Smit, and F. Sturzenegger. 2008. "South Africa: Macroeconomic Challenges after a Decade of Success." *Economics of Transition* 16: 639–677.

Ghosh, A. R., J. D. Ostry, and M. S. Qureshi. 2016. "When Do Capital Inflow Surges End in Tears?" *The American Economic Review* 106: 581–585.

Godley, W. 1996. "Money, Finance and National Income Determination: An Integrated Approach." *Levy Economics Institute, The, Economics Working Paper Archive*, No. 167.

Godley, W., and M. Lavoie. 2012. *Monetary Economics: An Integrated Approach to Credit, Money, Income, Production and Wealth*. Palgrave Macmillan, London.

Goldstein, M., and P. Turner. 1996. "Banking Crises in Emerging Economies: Origins and Policy Options." *BIS Economic Paper*, No. 46.

Grandmont, J.-M. 1998. "Expectations Formation and Stability of Large Socioeconomic Systems." *Econometrica* 66: 741–781.

Harris, L. 1979. "Catastrophy Theory, Utility Theory and Animal Spirit Expectations." *Australian Economic Papers* 18: 268–282.

Hassan, S. 2015. "Speculative Flows, Exchange Rate Volatility and Monetary Policy: The South African Experience." *South Africa Reserve Bank Working Paper Series*, No. WP/15/02.

Hommes, C. 2011. "The Heterogeneous Expectations Hypothesis: Some Evidence from the Lab." *Journal of Economic Dynamics & Control* 35: 1–24.

IMF. 2010. "Global Liquidity Expansion: Effects on Receiving Economies and Policy Response Options." *Global Financial Stability Report*. Washington, DC: International Monetary Fund: pp. 119–151.

IMF. 2013a. "Global Impact and Challenges of Unconventional Monetary Policies." *IMF Policy Paper*.

IMF. 2013b. "South Africa 2013 Article 4 Concultations." *International Monetary Fund Country Reports*, No. 13/303.

Joyce, J. P., and M. Nabar. 2009. "Sudden Stops, Banking Crises and Investment Collapses in Emerging Markets." *Journal of Development Economics* 90: 314–322.

Kreuser, F., R. Burger, and N. Rankin. 2015. "The Elasticity of Substitution and Labour Displacing Technical Change in Post-Apartheid South Africa." *Wider Working Paper Series*, No. 2015/101.

Magud, N. E., and E. R. Vesperoni. 2015. "Exchange Rate Flexibility and Credit during Capital Inflow Reversals: Purgatory ... Not Paradise." *Journal of International Money and Finance* 55: 88–110.

Makrelov, K., C. Arndt, R. Davies, and L. Harris. 2020. "Balance Sheet Changes and the Impact of Financial Sector Risk-taking on Fiscal Multipliers." *Economic Modelling* 87: 322–343.

McQuade, P., and M. Schmitz. 2017. "The Great Moderation in International Capital Flows: A Global Phenomenon?" *Journal of International Money and Finance* 73: 188–212.

Mendoza, E., V. Quadrini, and J. Ríos-Rull. 2009. "Financial Integration, Financial Development, and Global Imbalances." *Journal of Political Economy* 117: 371–416.

Mendoza, E. G. 2006. "Lessons from the Debt-Deflation Theory of Sudden Stops." *American Economic Review* 96: 411–416.

Mishkin, F. S. 1996. "Understanding Financial Crises: A Developing Country Perspective." *NBER Working Paper Series,* No. 5600.

Mishkin, F. S. 1999. "International Capital Movements, Financial Volatility and Financial Instability." *NBER Working Paper Series*, No. 6390.

NPC. 2013. *Natonal Development Plan 2030*. Pretoria: National Planning Commission. In, edited by.

Ottonello, P. 2013. "Optimal Exchange Rate Policy under Collateral Constraints and Wage Rigidity." *Mimeo, Columbia University.*

Rajan, R. G. 2005. "Has Financial Development Made the World Riskier?" *NBER Working Paper Series*, No. 11728.

Reinhart, C. M., and V. R. Reinhart. 2008. "Capital Flow Bonanzas: An Encompassing View of the past and Present." *NBER Working Paper Series*, No. 14321.

Rey, H. 2015. "Dilemma Not Trilemma: The Global Financial Cycle and Monetary Policy Independence." *NBER Working Paper Series*, No. 21162.

Roos, M. W. M., and W. J. Luhan. 2013. "Information, Learning and Expectations in an Experimental Model Economy." *Economica* 80: 513–531.

Sá, F., P. Towbin, and T. Wieladek. 2011. "Low Interest Rates and Housing Booms: The Role of Capital Inflows, Monetary Policy and Financial Innovation." *Bank of England Working Paper Series*, No. 411.

Saikkonen, L. 2015. "Estimation of Substitution and Transformation Elasticities for South African Trade." *Wider Working Paper Series*, No. 2015/104.

Sarno, L., I. Tsiakas, and B. Ulloa. 2016. "What Drives International Portfolio Flows?" *Journal of International Money and Finance* 60: 53–72.

Sawyer, M. 2013. "What Is Financialization?" *International Journal of Political Economy* 42: 5–18.

Sawyer, M. 2016. "Confronting Financialisation." In *Financial Liberalisation*, edited by Arestis P., and M. Sawyer. International Papers in Political Economy, Palgrave Macmillan, Cham.

Sawyer, M. and M. Passarella. 2017. "The Monetary Circuit in the Age of Financialisation: A Stock-Flow Consistent Model with a Twofold Banking Sector." *Metroeconomica* 68(2): 321–353.

Simon, H. A. 1955. "A Behavioral Model of Rational Choice." *The Quarterly Journal of Economics* 69: 99–118.

Simon, H. A. 1982. *Models of Bounded Rationality: Empirically Grounded Economic Reason*. Cambridge, Massachusetts: MIT press.

Smit, B., C. Grobler, and C. Nel. 2014. "Sudden Stops and Current Account Reversals: Potential Macroeconomic Consequences for South Africa." *South African Journal of Economics* 82: 616–627.

Tille, C., and E. Van Wincoop. 2010. "International Capital Flows." *Journal of International Economics* 80: 157–175.

Tobin, J. 1982. "Money and Finance in the Macroeconomic Process." *Journal of Money, Credit & Banking (Ohio State University Press)* 14: 171–204.

Wesso, G. E. R. A. L. D. R. 2001. "The Dynamics of Capital Flows in South Africa: An Empirical Investigation." *South African Reserve Bank Quarterly Bulletin* 2: 59–77.

Woodford, M. 2010. "Financial Intermediation and Macroeconomic Analysis." *Journal of Economic Perspectives* 24: 21–44.

Annexure A

Symbol	Description	Value
$\alpha_1^{gap,l}\ \alpha_1^{gap,k}$	Responsiveness of the demand for factors of production to real loan rates	0.45 0.40
$\alpha_2^{gap,l}\ \alpha_2^{gap,k}$	Responsiveness of the demand for factors of production to aggregate demand	0.35 0.3
$\alpha^{mgov,cd}$	Demand for cash and deposits as share of income: government sector	0.13
$\alpha^{mhhd,cd}$	Demand for cash and deposits as share of income: household	0.34
$\alpha^{nfin,cd}$	Demand for cash and deposits as share of income: non-financial sector	1.16
$\alpha^{cb,l}$	Demand for loans as share of income: Reserve Bank	12.7
$\alpha^{fin,l}$	Demand for loans as share of income: financial sector	0.31
$\alpha^{mgov,l}$	Demand for loans as share of income: government sector	0.25
$\alpha^{mhhd,l}$	Demand for loans as share of income: household sector	0.61
$\alpha^{mrow,l}$	Demand for loans as share of GDP: foreign sector	0.12
$\alpha^{nfin,l}$	Demand for loans as share of income: non-financial sector	1.26
α^p	Production function shift parameter (base year)	0.84
α^m	Import function shift parameter	1.32
α^t	Export function shift parameter	5.55
β^{repo}	Responsiveness of the financial sector reserve ratio to changes in the repo rate	15.3
β_1^{repo}	Responsiveness of the supply of loans to changes in the repo rate	−0.7
β_2^{repo}	Taylor Rule coefficient on inflation	2.0
β_3^{repo}	Taylor Rule coefficient on the output gap	0.3
β^{sa}	Responsiveness of the financial sector reserve ratio to changes in the growth rate of financial assets	−0.1

Description	Symbol	Value
Responsiveness of the demand for loans to changes in the real borrowing costs	μ^l	−5
Responsiveness of equity prices to inflation expectations	μ_1^{peq}	0.55
Responsiveness of equity prices to money supply	μ_2^{peq}	0.2
Responsiveness of equity prices to economic activity	μ_3^{peq}	0.33
Coefficient in the cash and deposit rate equation	μ_2^{rcd}	0.6
Tobin demand coefficient showing the steady-state share of bonds in total wealth (fin: financial sector, mrow: foreign sector)	$\lambda_{b,0}^{fin}\ \lambda_{b,0}^{mrow}$	0.22 0.18
Tobin demand coefficient showing the steady-state share of equities in total wealth	$\lambda_{eq,0}^{fin}\ \lambda_{eq,0}^{mrow}$	0.52 0.73
Tobin demand coefficient showing the steady-state share of cash and deposits in total wealth	$\lambda_{cd,0}^{fin}\ \lambda_{cd,0}^{mrow}$	0.26 0.1
Tobin demand coefficient: responsiveness of bond to changes in the bond return	$\lambda_{b,1}^{fin}\ \lambda_{b,1}^{mrow}$	1.99 1.99
Tobin demand coefficient: responsiveness of bond to changes in the equity return	$\lambda_{b,2}^{fin}\ \lambda_{b,2}^{mrow}$	−1.6 −1.6
Tobin demand coefficient: responsiveness of bond to changes in the cash and deposit return	$\lambda_{b,3}^{fin}\ \lambda_{b,3}^{mrow}$	−0.2 −0.2
Tobin demand coefficient: responsiveness of bond to changes in transactional demand for money	$\lambda_{b,4}^{fin}\ \lambda_{b,4}^{mrow}$	0 0
Tobin demand coefficient: responsiveness of equity to changes in the bond return	$\lambda_{eq,1}^{fin}\ \lambda_{eq,1}^{mrow}$	−1.3 −1.3
Tobin demand coefficient: responsiveness of equity to changes in the equity return	$\lambda_{eq,2}^{fin}\ \lambda_{eq,2}^{mrow}$	1.6 1.6
Tobin demand coefficient: responsiveness of equity to changes in the cash and deposit return	$\lambda_{eq,3}^{fin}\ \lambda_{eq,3}^{mrow}$	−0.6 −0.6
Tobin demand coefficient: responsiveness of equity to changes in transactional demand for money	$\lambda_{eq,4}^{fin}\ \lambda_{eq,4}^{mrow}$	−0.001 −0.001
Tobin demand coefficient: responsiveness of cash and deposit to changes in the bond return	$\lambda_{cd,1}^{fin}\ \lambda_{cd,1}^{mrow}$	−0.66 −0.66
Tobin demand coefficient: responsiveness of cash and deposit to changes in the equity return	$\lambda_{cd,2}^{fin}\ \lambda_{cd,2}^{mrow}$	−0.04 −0.04
Tobin demand coefficient: responsiveness of cash and deposit to changes in the cash and deposit return	$\lambda_{cd,3}^{fin}\ \lambda_{cd,3}^{mrow}$	0.7 0.7

(Continued)

(Continued).

Symbol	Description	Value	Symbol	Description	Value
γ_1^l	Steady-state growth rate of investment	1.04	$\lambda_{cd,4}^{fin}\,\lambda_{cd,4}^{mrow}$	Tobin demand coefficient: responsiveness of cash and deposit to changes in transactional demand for money	0.001 0.001
γ_2^l	Responsiveness of investment by the non-financial sector to the Tobins Q term	0.14	σ^{cd}	Growth of cash and deposit liabilities of the Reserve Bank coefficient	0.027
γ_3^l	Responsiveness of investment to the real loan rate	−0.9	grg	Government consumption growth rate	0.026
δ	Capital depreciation rate	0.5	inf	Inflation target (steady-state inflation)	0.06
δ^p	Production function share parameter	0.0001	inta	Quantity of aggregate intermediate input per output	0.06
δ^q	Import function share parameter	0.03	ivat	Quantity of aggregate output per value added	0.44
δ^t	Export function share parameter	0.99	$ld^{mgov}\,ld^{mrow}\,ld^{nfin}$	Share of wealth provided as loans for the non-financial, government, and foreign sectors	0.421 0.065 0.385
θ^p	Responsiveness of price expectation to deviations of expected prices from actual prices in the previous period	0.2	lmrow	Share of foreign loan and equity liability as percentage of GDP	0.18
θ_1^{pq}	Responsiveness of prices to changes in the output gap	1.5	$mpsbar^{fin}\,mpsbar^{nfin}$	Steady-state savings rate	0.07 0.44
θ_2^{pq}	Responsiveness of prices to changes in import prices	0.5	$oipprm^{fin}\,oipprm^{mhhd}\,oipprm^{mgov}\,oipprm^{mrow}\,oipprm^{nfin}$	Other income paid as share of GDP	0.15 0.04 0.01 0.001 0.01
θ^{prce}	Responsiveness of price expectations to the output gap	0.3	$shifoint^{fin}\,shifoint^{mhhd}\,shifoint^{mgov}\,shifoint^{mrow}\,shifoint^{nfin}$	Share of other income received	0.23 0.68 0.01 0.03 0.06
ρ^{lab}	Labour force growth	0.01	$shifsoc^{fin}\,shifsoc^{mhhd}\,shifsoc^{mgov}$	Share of social contributions received	0.51 0.46 0.03
ρ^p	Production function substitution elasticity	−1	$soc\_par^{fin}\,soc\_par^{mhhd}\,soc\_par^{mgov}\,soc\_par^{nfin}$	Social contributions paid as a share of GDP	0.05 0.09 0.03 0.004
ρ^q	Import function substitution elasticity	1	ss^{peq}	Steady-state growth in equity prices	0.03
ρ^{repo}	Interest rate smoothing coefficient	0.7	ta	Activity tax rate	0.009
ρ^t	Export function substitution elasticity	11	td	Personal direct tax rate	0.009
μ^{cd}	Responsiveness of the demand for cash and deposits to changes in the real return	0.2	tm	Import tariff rate	0.027
μ^{fw}	Target real growth rate for household wealth	0.15	ts	Sales tax rate	0.038

Do public banks reduce monetary policy power? Evidence from Brazil based on state dependent local projections (2000–2018)

Nikolas Passos ⓘ and André de Melo Modenesi ⓘ

ABSTRACT

We test the hypothesis that public banks reduce monetary policy power. Previous studies have shown that companies with access to government-driven credit present smaller fall in investment and production after a contractionary monetary policy shock. Nevertheless, these studies are based on microeconomic data and ignore macroeconomic effects, especially the cost-push effects, of monetary policy. We employ state-dependent local projections to compare monetary policy power – the sensibility of inflation to changes in policy interest rate – between periods of high credit of public banks and periods of high credit of private banks. We do not find evidence that monetary policy is less powerful in periods of high credit of public banks. Even though periods of high credit of public banks present a lower effect over output, those periods present less persistent price puzzles than periods of high private credit. Robustness of results is enhanced by performing several tests. We attribute our results to lower flexibility in interest rates of credit from public banks, what leads to lower transmission in financial costs, lower reduction in capital stock and lower variation in the exchange rate.

1. Introduction

The surveillance of contemporary macroeconomic regimes by Central Banks has relied on strong and highly flexible financial markets. Some have argued that more flexible financial markets are usually associated with stronger impacts of monetary policy on economic activity (Borio 1996; Garriga, Kydland, and Šustek 2017). In countries where companies and households are indebted in flexible rate loans, increases in interest rates are expected to further reduce the aggregate demand. In this case, the credit-channel of monetary policy (Bernanke and Gertler 1995) would be stronger in economies with more flexible financial markets.

Nevertheless, the flexibility of credit rates may have undesirable outcomes for investment and financial stability, given the uncertain character of investment decisions, which requires long term, committed, and patient finance, especially in innovation-intensive sectors (Mazzucato and Penna 2016). A financial system subject to highly volatile interest rates may harm investment planning and increase financial instability (Minsky 2008,

239). When balance sheets are filled with short-term and flexible rate debts, the effects of monetary policy over activity may increase instability (Fazzari and Minsky 1984, 112).

It could be argued that the flexibility of credit rates divergently affects two goals of Central Banks, namely, the control of inflation, and the promotion of financial stability. On the one side, Central Banks are expected to affect credit markets and economic activity in order to meet inflation targets. On the other, Central Banks are required to assure financial stability, which can be achieved through stable credit sources. Those goals should be conciliated to promote a functional macroeconomic regime (Baker 2018).

In Brazil, a broad network of public banks has sought to maintain low credit rates and lengthen credit terms, especially for agriculture, housing, and capital goods (Torres Filho 2009; Pazarbasioglu-Dutz et al. 2017). The intervention also protects real economy from exogenous shocks, such as international financial crisis. The response of public banks to the 2008 crisis fostered the recovery of Brazilian economy, although it was criticized for reducing monetary policy power (Bonomo, Brito, and Martins 2015; Segura-Ubiergo 2012). Critics argued that, once the credit granted by public banks present a less pronounced response to monetary policy, a higher change in the policy rate is required for a desired change in output (Arida 2005). This would imply a higher equilibrium basic interest rate (Barboza 2015). This argument was central in recent financial reforms that reduced government intervention in credit markets. Among the reforms, the main referential rate for the Brazilian National Development Bank (BNDES) was pegged to inflation plus the spread of a government five-year bond (Palludeto, Antonio, and Zanchetta Borghi 2020, 15).[1] The reforms aimed to bring BNDES benchmark rate closer to market levels and more responsive to Central Bank basic rate. In 2020, during the coronavirus lockdown, Public Banks were reactivated to avoid a liquidity crisis. The recurrent reliance on public banks to provide liquidity when private banks retract brings to forefront the necessity of coordinating financial stability policies with monetary policy.

This paper tests the hypothesis that the increased role of public banks reduces the power of monetary policy. We apply state-dependent local projections (Jordà 2005; Jordà, Schularick, and Taylor 2019) to compare estimates between states of high-credit of public banks with states of high-credit of private banks. Our estimations contribute to the growing literature on state-dependent effects of monetary policy (Tenreyro and Thwaites 2016; Alpanda and Zubairy 2018; Jordà, Schularick, and Taylor 2019).

We find that in periods of high indebtedness (of both private and public banks) the price puzzles are more pronounced, *i.e.*, price levels may initially rise more after an increase in interest rates. In credit cycles led by private banks, such price puzzles are more persistent. Once the credit of private banks is more flexible, the stronger credit channel is curtailed by financial costs increases after the monetary contraction. Therefore, we conclude that policies promoted by public banks, that increase the maturity of loans and the stability of credit rates, may enhance overall monetary policy goals in times of financial fragility.

In addition to the introduction, this work is divided into three sections. The second section discusses the coordination issue between public banks and monetary policy, analyzing the Brazilian debate. The third section presents the methodology of state-dependent local projections, explaining its usage for regimes of high credit of public and private banks. In the fourth section, we present our results, showing that the estimates do not support the hypothesis that regimes of high credit of public banks compromise the power of monetary policy. Even though public banks make production more resilient to monetary policy shocks, the power of monetary policy – measured in terms of inflation – is no less in periods of high public credit when compared with periods of high private credit.

2. Public banks and monetary policy

Interest rates in Brazil are historically high when compared to countries with similar economic structure. Several authors suggested that such high level could be explained by government intervention in credit markets (Arida 2005; Barboza 2015; Bonomo and Martins 2016). Persio Arida (2005), former president of both Brazilian Central Bank and Brazilian National Development Bank, has argued that public banks were responsible for reducing the power of monetary policy. Once credit rates charged by public banks were less responsive to the policy rate, borrowers from public banks would be protected from monetary tightening. Therefore, after a hike in interest rate, companies' capital expenditure and households borrowing from public banks would be reduced less than if they had borrowed only at private banks. For those authors, the Central Bank would have to further increase the policy rate to reduce excess demand and stabilize rising inflation (Segura-Ubiergo 2012; Barboza 2015).

Credit rates earmarked by government are in fact significantly less responsive to changes in the policy rate (SELIC) (Bonomo, Brito, and Lazzarini 2018). As the correlations in the Table 1 show, the average interest rates on free fund and earmarked loans are highly correlated with the policy rate (68% and 65%, respectively). In contrast, the correlation with SELIC is lower with BNDES rates and with earmarked spread (28% and 0.03%, respectively). The lower responsiveness of the credit spread of earmarked credit would curtail the bank-lending channel of monetary policy (Bernanke and Gertler 1995), requiring the Central Bank to further increase the policy rate (Bonomo and Martins 2016).

The lower responsiveness of earmarked spread and BNDES rates reflects governmental protection of priority sectors through the stabilization of credit rates. In the aftermath of the 2008 subprime crisis, Brazilian Government intensified the action of public banks to prevent international financial instability from affecting domestic production (Barros et al. 2018; Griffith-Jones et al. 2018). Although the intervention sustained the stability of financial markets, its continuity was criticized as soon as the initial effects of the crisis ceased for reducing monetary policy power (Segura-Ubiergo 2012; Bonomo, Brito, and Martins 2015).

Table 1. Correlations between basic interest rate (SELIC) and average rates and spreads for free and earmarked credits in Brazil (2007–2018).

	S5ELIC	Free Rates	Earmarked Rates	Free Spread	Earmarked Spread	BNDES
SELIC	1					
Free Rates	0,68	1				
Earmarked Rates	0,65	0,91	1			
Free Spread	0,47	0,96	0,88	1		
Earmarked Spread	0,03	0,6	0,75	0,72	1	
Average rate of BNDES	0,28	0,78	0,9	0,86	0,92	1

Data Source: Brazilian Central Bank. Authors' own elaboration.

Current empirical testing of the effects of public banks on monetary policy is limited to microeconomic studies at firm and industry level. Bonomo and Martins (2016) show that after a contractionary shock, companies with access to credit from public banks present a smaller reduction in the volume of new loans and employment. In addition, these companies are less sensitive to external shocks, such as rising international risk. Corroborating such evidence, Perdigão (2018) shows that there is a lower response to monetary policy in industrial sectors with a larger share of earmarked credit to total credit.

Monetary policy power, by definition, is a macroeconomic phenomenon, which firm-level databases empirical studies are unable to measure (Kashyap and Stein 2000). Castro (2019) questions the validity of the arguments of Bonomo and Martins (2016) due to the presentation of only microeconomic evidence. In fact, aggregate effects of monetary policy are the sum of each firm's microeconomic effects plus the external effects of all firms acting simultaneously, what alters input and final goods prices. Cross-section estimates omit external effects and are not necessarily informative about inflation outcomes.

Analyzing inflation responses disregarding external effects can hide the intensification of price puzzles. Positive variations in prices following contractionary shocks – known as price puzzles – are well documented in the economic literature (Ravenna and Walsh 2006; Ramey 2016). Price puzzles appear in most VAR estimations in the form of an initial increase in prices after a monetary policy shock (Ramey 2016, 99). The effects of financial costs were pointed as an explanation for this hump-shaped estimations in VAR models: the hump in prices would stem from increased financial costs passed on to consumers by firms setting prices as a markup on costs (Lima and Setterfield 2010; Castro 2019). If reductions on prices arising from demand contraction do not offset the rise in the financial cost, inflation is increased. The existence of price puzzles alters the optimal monetary policy problem because any shock implies a trade-off between stabilizing inflation and stabilizing the output gap (Cardim de Carvalho 2005; Ravenna and Walsh 2006; Martins et al. 2017).

Interestingly, the lower flexibility of credit rates, promoted by public banks, has divergent effects for the traditional credit channel of monetary policy and for the increase in financial costs.[2] Since interest charged on earmarked credit is less responsive to the policy rate, the cost channel is less relevant for firms with greater access to public banks' credit. Thus, earmarked credit would have a dubious effect on monetary policy. The access to public banks' credit results in lower output reduction following contractionary shocks. On the other hand, access to public bank credit reduces the intensity of cost shocks that could lead to price puzzles. Consequently, in the absence of public banks, output may respond more and inflation respond less to monetary policy changes (Castro 2018; Silva, Paes, and Bezerra 2018). In this case, public banks would be beneficial both for the financial stability and for the inflation control.

3. Methodology and empirical strategies

The extent that aforementioned mechanisms hold are a matter of empirical testing. As highlighted before, former studies are incapable of studying inflation, since they deal only with microeconomic data. Time-series approaches have been the main methodological tool for understanding the effects of monetary policy over activity and inflation at once.

Estimates of local projections have been increasingly used in the monetary policy literature. A recurring result of this literature is the fact that monetary shocks during periods of growth are more potent than those in periods of recession (Tenreyro and Thwaites 2016; Santoro et al. 2014; Jordà, Schularick, and Taylor 2019). Tenreyro and Thwaites (2016, 59) conclude that in the American economy contractionary shocks have greater effects on output than monetary easing. In stark contrast, Chen, Li, and Tillmann (2019) conclude that in China output and inflation responses are higher after a monetary easing. This result is attributed to the presence of public companies, which would suffer less from monetary contractions and benefit more in times of loosening. Chen, Li, and Tillmann (2019) results highlight the fact that the response of diverse government agencies is highly relevant for the conduct of monetary policy.

Beyond studies of cycles of monetary policy in depressions and recoveries, a recent literature has applied local projections to different monetary responses in a credit cycle. Jordà, Schularick, and Taylor (2019) show that the product responds more to monetary policy in periods of high credit to households, while inflation responds more in periods of low credit. Alpanda and Zubairy (2018, p.12) conclude that a monetary easing is more effective for economic activity in periods of low mortgage debt, but in these periods there is a worsening of the price puzzle.

As we can see, the literature of state-dependent local projections questions the linear relation between excess demand and inflation predicted by new-Keynesian models (Santoro et al. 2014). In fact, this relation would be subject to distinct phases in the credit cycle and also to the nature of the credit cycle, for instance, if it is held by households or by companies (Jordà, Schularick, and Taylor 2019).

3.1. Local projections

Local Projections were initially proposed by Jordà (2005) as a way to avoid bias due to poor specifications in Vector Auto-Regression (VAR) models. In VAR models, if the number of lags is not equal to that of the data generating process, estimated errors of impulse response functions are accumulated at each time horizon (Ramey 2016, 83). As an alternative, local projections make direct predictions from a specific regression for each time horizon. Instead of iterating one step further from a previously estimated model as in VAR models, local projections predict values of a variable using a direct forecast for a specific horizon (Ramey 2016, 84). For instance, a linear local projection can be estimated by Ordinary Least Squares (OLS), according to the Equation (1):

$$y_{t+h} = \beta_{t+h} shock_t + \theta_{t+h} x_t + u_t \qquad (1)$$

The coefficient β estimates the average effect of an exogenous *shock* on the variable y over the time horizon t + h. The matrix θ contains the coefficients of the control variables x_t. Therefore, an OLS estimate is generated for each time horizon h after the shock. Control variables ensure that the effects of the shock on the variable of interest are only caused by the shock itself. Controls may include trends, shock-correlated variables, y-lags, and seasonal dummies, for example. The method is robust to non-stationary or cointegration in the data (Ahmed and Cassou, 2016). Nevertheless, the u_t error terms are serially correlated, requiring the usage of Newey-West corrections in standard errors (Jordà 2005).

In order to identify contemporary causal relationships between the variable of interest y_{t+h} and shocks, one has to adopt one of the identification options shared with the VAR literature. The most common identification methods include Cholesky decompositions, structural identifications, narrative methods, high-frequency identification, instrumental variables, signal restrictions. In our estimates, we employ Cholesky decompositions for the estimation of shocks, that are subsequently adopted in the local projections. Such a specification is commonly used in monetary policy studies (Christiano, Eichenbaum, and Evans 1999; Jordà 2005; Ramey 2016) and has presented similar results to narrative methods (Alpanda and Zubairy 2018).

In small sample estimations or where the number of lags is poorly specified, local projections are a viable alternative to VAR impulse-response function estimates (Brugnolini 2018; Barnichon and Brownlees 2019). Even though estimations from local projections may in some cases be less accurate than VAR estimates (Kilian and Kim 2011; Ramey 2016), it presents several advantages for the question at hand. Firstly, local projections can be estimated by OLS, not requiring more complex methods used in high order VARs (Jordà 2005). Secondly, local projections are robust to the poor specification of the actual data generating process. Since it is not necessary to define a priori the functional form of the data generating process, the estimates are reliable even if the equation system that best describes the interactions of variables is not known (Jordà 2005). Thirdly, estimates with local projections may allow to save parameters and increase degrees of freedom (Tenreyro and Thwaites 2016, 47).

3.2. Local projections for credit states

Monetary policy studies with linear local projections analyze the average response of output and inflation to a monetary shock, regardless of the credit conditions of the economy (see for instance Carcel et al. 2018). As a hypothesis of this work, we test if different credit state influences the response of inflation and output to monetary policy. Moreover, we test whether the states have different outcomes for monetary policy in the cases where the credit cycle is led by public or private banks.[3]

Local projections are adopted because they allow the simple incorporation of different states. We can investigate the effects of exogenous shocks under different credit states without a priori defining how the economy moves from one state to another (Ramey and Zubairy 2018). Estimations are performed following the specification in Equation (2)[4]:

$$ y_{t+h} = F(z_{t-1})\left(\beta_{t+h}^{H}shock_t + \theta_h^H x_t\right) + (1 - F(z_{t-1}))\left(\beta_{t+h}^{L}shock_t + \theta_h^L x_t\right) + u_t \quad (2) $$

As in the linear model, the coefficients β_h estimate the average effect of a shock on the variable y for the time $t + h$. In a state-dependent local projection, the coefficient β^i depends on the credit state, defined by the function $F(z_{t-1})$ and indicated by the upper index for high (H) or low (L). Following Auerbach and Gorodnichenko (2012) the state variable is lagged, so there is no contemporary feedback between the state and the shocks. The $F(z_{t-1})$ function returns a regime definition representing the probability of the economy to be in a high or low credit state. To obtain non-discrete probabilities as values of the function $F(z_{t-1})$, the difference between the state variable and its trend is transformed according to the following logistic function.[5]:

$$F(z_t) = \frac{e^{-\gamma z_t}}{(1 + e^{-\gamma z_t})} \tag{3}$$

Function (3) approaches 1 when the economy is most likely to be in a high credit regime. Alternatively, it approaches 0 when the regime is more likely to be low-credit. The logistic function smoothes state changes to prevent that slight variations in z_t result in a sudden (and discrete) regime change. The parameter γ defines the intensity of the smoothing: higher values of γ mean that $F(z_t)$ stays longer near the limits [0,1], bringing the model closer to a discrete setting.

Credit states are identified following the financial literature on the 'credit gap' (Drehmann and Tsatsaronis 2014). Accordingly, high credit states are defined as those in which the credit balance to GDP is larger than its long-term trend. By using the credit series as a ratio of GDP, we intend to control for changes in price levels and changes in economic conditions (Alpanda and Zubairy 2018). Thus, the difference between Credit/GDP and its trend defines regimes in local projections represented by the z_t in Equation (2).

The credit gap has been recommended by the Bank for international settlements as an indicator of macroeconomic financial instability. Periods where the credit gap is positive are characterized by banking stress and are good predictors of financial crises (Jordà, Schularick, and Taylor 2016). Drehmann and Tsatsaronis (2014, 58) argue that, conceptually, the credit gap synthesizes financial instability in line with the work of Minsky (2008). High credit periods are thus important episodes to investigate the effects of the monetary policy credit channel. In these periods, interest rate shocks are more likely to affect investing and spending decisions, given the greater commitment of income to loan repayments. As we argue, credit cycles of public banks are assumed to soften this effect as there is lower response of borrowing costs to movements in basic rate.

We test whether in periods of high credit of public banks the monetary policy has lower power in comparison with periods of high credit of private banks. The two upper graphs in Figure 1 show the credit state variable of public and private banks (% GDP), as well as their dotted trend. The periods in which the credit stock is superior to the trend is considered a high credit state. The lower graphs show the state indicator, where value 1 is assigned to periods when the credit balance is greater than trend.

In the estimations, we consider separately the state variables for public and private banks, and then compare the results by examining the monetary policy multipliers. Accordingly, first we analyze how monetary policy behaves during high and low credit states led by public banks. Then, we move on to the credit cycles led by private banks. Finally, monetary policy multipliers inspired by Jordà, Schularick, and Taylor (2019) are calculated in order to compare monetary policy power in the two different cycles.

3.3. Data

The basic estimates of this work include output, inflation, and interest rate variables, as well as the credit state variable of public and private banks. For inflation data, we use the series of the National Consumer Price Index (IPCA), the index used for Central Bank's inflation target. For Gross Domestic Product data, we use Getúlio Vargas Foundation (FGV) monthly series, which is compatible with Brazilian Institute of

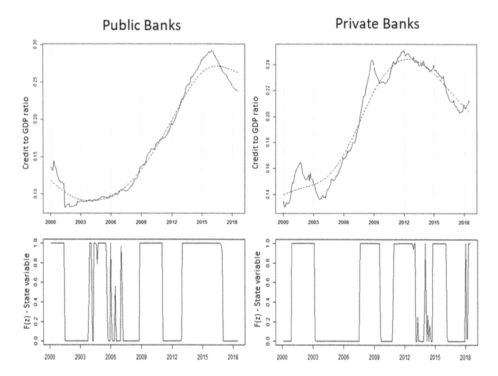

Figure 1. State variable: credit stock of public and private banks as a proportion of GDP.

Geography and Statistics (IBGE) quarterly GDP estimates. For monetary policy shocks, we employ the policy rate series (Selic), published by the Central Bank. For state variables, we use the Balances of Credit Operations of Financial Institutions under Public Control and Private Control, both disclosed by the Central Bank. The main estimates are made for the period between January 2000 and July 2018, resulting in 223 monthly observations. The lower limit of January 2000 is given by the availability of data from the FGV GDP Monitor. As the inflation targeting regime was adopted in Brazil in June 1999, the sample covers the entire period, except for the initial 6 months, due to the unavailability of monthly GDP data for the period. Further data details can be found in the Annex.

4. Results

In the next two sections, we present the estimations of the local projections first for public banks and then to private banks credit cycles. After, we compare the responses (of inflation, GDP and the policy rate) calculating the monetary policy multipliers. We adopted 12 lags, as recommended by the AIC criterion. Additionally, we added a monthly dummy to prevent the effects captured from reflecting seasonality.

The display of separate IRFs for public and private banks is useful because it highlights the different effects of monetary policy across credit states within each of the cases. From this separate exposition we can conclude that the credit cycle is relevant for monetary policy either when it is led by public or private banks.

4.1. Credit regimes of public banks

Figure 2 presents the impulse response functions for a one standard deviation policy rate shock, depending on the credit state of public banks. The first column presents the response of inflation, output and of policy rates itself. We can compare the responses during high-credit state (solid black line) with the low-credit state (dotted red line). The confidence intervals (95%) for high-credit state is represented by the solid thin lines and for the low-credit state by red-shaded area.

Inflation's response presents a remarkable price puzzle by the fifth month. This price puzzle is more relevant in the period of high credit than low credit. From the sixth month onwards, monetary policy has a significant negative effect on inflation in the high-credit regime, which is maintained until the 15th month. In the low-credit regime, however, monetary policy has no significant negative effect on prices. In this regime, after the sixth month the shock becomes not significant in almost all periods.

The effects on GDP are in line with the hypothesis that in the period of higher indebtedness the effects of monetary policy are stronger. The negative effect on GDP occurs from the first month and intensifies in the following months. Until the eighth month, the trajectories of high and low credit are similar. After the eighth month, there is a greater distance between the two estimates, with the product having larger losses in the state of high credit. Note that the effects on the product do not dissipate until the 15th month. Effects of monetary policy over periods of more than 2 years are also noted by Jordà, Schularick, and Taylor (2019) and Tenreyro and Thwaites (2016).

In both regimes, positive monetary policy shocks are followed by successive increases in policy rate. In the high-credit regime, successive increases are greater than those of the low-credit regime. During the high-credit regime, although the shocks decay, they do not converge to zero within the presented 15-month horizon. Under the low-credit regime, basic interest rate stops increasing after 1 year.

Interestingly, for the three variables, the high-credit regime has greater confidence intervals than the low credit. This reflects the fact that in addition to the more intense effects on the high-credit regime the effects are less predictable.

The second column of Figure 2 presents the T statistic for the null hypothesis that $(\beta_h^a - \beta_h^b) = 0$, with the shaded confidence interval of ± 1.96. Thus, if the line is below the shaded area during a period of time, we can reject the null hypothesis that the response (of the variable of interest) is the same during both credit states, in favor of the alternative that the same response is more negative during high-credit states (at a 5% significance level). Therefore, regarding the inflation results, the difference between the two credit regimes is significant for the initial price puzzle and for the larger drop in inflation from the seventh month on. For the GDP, the high-credit period presents more significant effects in the first months after the shock and after the 11th month. Regarding the policy rate, after the fourth period, successive rate increases are generally significantly higher in the high-credit period.

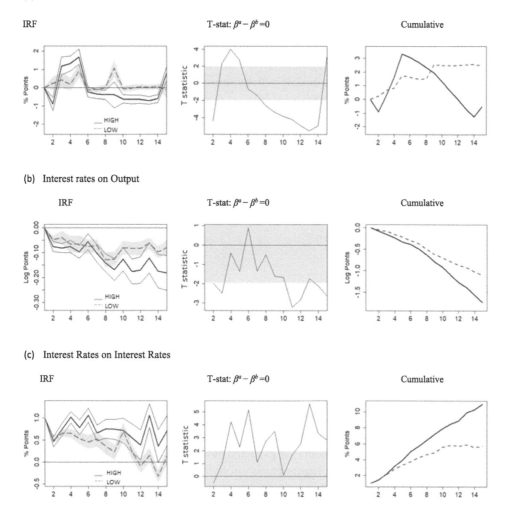

Figure 2. Impulse response functions with public banks credit as state variable. (a) Interest rate on inflation. (b) Interest rate on output. (c) Interest rate on interest rate. The first column shows the impulse response functions in the high credit (solid black) and low credit (dashed red) states, the respective confidence intervals (95%) are represented by the solid line and shaded area. The second column presents T statistics for the hypothesis that the difference between the coefficients in the two states is equal to 0, the shaded area marks the range of ± 1.96. The third column presents the cumulative impulse-response function for the state of high credit (solid) and low credit (dotted).

The third column of Figure 2 presents the accumulated result of the impulse response functions, calculated by the sum of all shocks up to a certain horizon h. For inflation, it is noted that the negative net effect only becomes relevant after the 12th month of the high-credit regime. For the GDP, there is a similar response in both regimes until the eighth month, when the monetary policy begins to have increasing effects on the high-credit regime. The high-credit regime does not indicate a stabilization of increases in basic interest rates, while in the low-credit regime such stabilization is already noted after the 11th month.

Considering only the public banks, we can conclude that in a high-credit state, the initial price puzzle is higher, the effects of monetary policy over economic activity is stronger and the cycle of increases in policy rate last longer. Without comparing with cycles led by private banks, one could conclude that higher public bank activity reduces monetary policy power. When public banks are more active, an increase of interest rate would have higher effects on economic activity, but an undesirable effect on inflation, what would lead to additional increases in interest rates. In the next section, we compare these results with the credit cycles of private banks. We note that those results are even more intense.

We should note that some of the impulse response functions show a significant effect until the 15th month. This does not mean that the effect is permanent, but it is persistent. Medium-term effects of monetary policy are not surprising and are reported in other studies based on local projections and VAR. For instance, Jordà, Schularick, and Taylor (2019) and Tenreyro and Thwaites (2016) find significant effects after more than 2 years of a monetary policy intervention.

4.2. Credit regimes of private banks

We repeat the above estimates using the private banks' credit gap as a state variable. Figure 3 presents the basic results for a standard deviation shock on policy rate over the three variables of interest, conditioned by the private bank credit regime. Although a similar price puzzle in both states is noted until the 6th month, inflation presents a stronger and longer-lasting price puzzle during periods of high private credit. From the 6th month on, there is an opposite effect to that perceived in the credit cycle of public banks: periods of low private credit have a negative inflation response and periods of high credit have a positive response. GDP is significantly more responsive in the high-credit period. Interest rate, in turn, presents the expected downward trajectory in both states. In the high-credit period, there is a higher persistence of interest rate hikes, which does not stabilize until the 15th month.

In the second column of Figure 3 we note that the differences pointed out between the two credit states are significant. In particular, the larger price puzzle, the larger fall in GDP and the persistence of interest rate increases during periods of high credit are significant. The effects are also shown in the cumulative charts in the third column. Inflation falls only after the 13th month. The large accumulated difference in GDP response, and interest rate's indistinct behavior, with the difference that in the low-credit period stabilization of policy rate increases is already noted in the 12th month.

4.3. Monetary policy multipliers

To compare estimates between public bank and private bank estimates, we calculated cumulative multipliers of the impulse-response functions, based on Jordà, Schularick, and Taylor (2019). Multipliers are calculated as the ratio of the average of GDP and of inflation response to the average of policy rate response. Interest rate normalization removes differences between states that may stem from differences in the interest rate path in each state.

(a) Interest rate on inflation

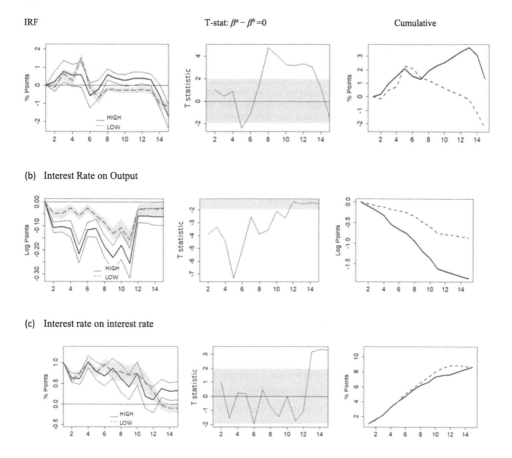

Figure 3. Impulse response functions with private banks credit as state variable. (a) Interest rate on inflation. (b) Interest rate on output. (c) Interest rate on interest rate. The first column shows the impulse response functions in the high credit (solid black) and low credit (dashed red) states, the respective confidence intervals (95%) are represented by the solid line and shaded area. The second column presents T statistics for the hypothesis that the difference between the coefficients in the two states is equal to 0, the shaded area marks the range of ± 1.96. The third column presents the cumulative impulse-response function for the state of high credit (solid) and low credit (dotted).

We can interpret the multipliers as the average change in GDP (or the average change in inflation) required to obtain the impulse response functions of Figures 2 and 3. The multiplier thus gives us the percentage of cumulative change in GDP (or percentage points in the case of inflation), given a 1 percentage point shock on policy rate in monthly values. Table 2 presents the results for the estimates based on the credit state variables of the public and private banks.

Regarding the GDP, the credit states presents negative answers. During the period of high public credit, an increase of one percentage point in the (monthly) policy rate reduces the product growth rate by 16% over the 15-month period.[6] The strongest response of the GDP occurs during the period of high private credit, and it is approximately twice as big as during the period of low private credit; and it is 30% higher than

Table 2. Monetary policy multipliers.

Public Banks	High Credit	Low Credit
Inflation	− 0,05	0,43
Output	− 0,16	− 0,19
Private Banks	High Credit	Low Credit
Inflation	0,16	− 0,27
Output	− 0,21	− 0,10

Reported values can be interpreted as how much output (or inflation) changes in percentage (percentage points) on average over the 15 months considered to construct the impulse response functions reported in Figures 2 and 3.

during the period of high public credit. This confirms the thesis that in periods of high public banks' credit, output is shielded from monetary policy shocks, partially confirming the findings of Bonomo and Martins (2016).

Negative inflation multipliers only occur during periods of high public credit and of periods of low private credit; and the effect during the low-credit period is more significant. During the period of high public banks credit, an increase of one percentage point in the (monthly) interest rate reduces the inflation rate by 0.05 percentage points over the 15-month period. In turn, in the high private bank credit regimes, the price puzzle persists throughout the period. This evidence is against the hypothesis that the monetary policy is more effective to control inflation in case the private banks led the credit cycle. It confirms the hypothesis that price puzzles are more related to periods in which there is higher credit rates flexibility (Castro 2019).

To corroborate our results, we performed robustness tests that consider: (i) alternative lags selection criteria, (ii) greater smoothing in regimes, (iii) exclusion of recessive periods, (iv) inclusion of open economy variables and (v) alternative state variables, considering free and earmarked credit stocks, instead of public and private banks. The detailed results of each test are presented in the Annex.

Inflation results are not as robust as GDP results to different specifications, but we conclude that in high credit periods price puzzles are more likely and durable. As we noted in the impulse response functions, price puzzles are considerably persistent during periods of high private credit. The periods of high public credit presented higher initial price puzzle, which reduced earlier than those in periods of high private credit. The periods of high private credit also showed the largest reductions in GDP, after rising interest rates. The periods of low private credit behave more closely than expected by new-Keynesian models, with smaller negative GDP responses and higher inflation responses.

4.4. Discussion of results

The monetary policy multipliers reveal diverging dynamics between inflation and GDP in periods of high and low credit (considering both private and public banks). The steeper fall in production does not necessarily reflect a steeper fall in inflation, as would be expected from a traditional Phillips curve. The stronger negative effect of interest rates on inflation occurred in periods of low private credit, the same period with the lowest drop in GDP. On the other hand, in the period of high private credit, the stronger response of the product was accompanied by persistent price puzzle.

GDP response confirms the hypothesis that periods of high credit from public banks show greater resilience to monetary policy than periods of high credit from private banks. We attribute this result to the higher interest rate response of private banks to basic interest rates. These results corroborate, in a macroeconomic viewpoint, the micro hypotheses tested in Bonomo and Martins (2016) and Perdigão (2018).

The vast literature on asymmetries in the transmission of monetary policy in times of recession and expansion is unable to deal simultaneously with diverging GDP and inflation outcomes (Santoro et al. 2014, 20). In general, models that apply Bernanke and Gertler's (1995) theory of financial accelerator predict analogous product and inflation asymmetries in times of recession and expansion of economic activity. Santoro et al. (2014) suggest that not only the IS curve, but also the Phillips curve would have its parameters changed over the economic cycle. In a monetary contraction, during a recessive period, there would be a simultaneous movement of supply and aggregate demand that could even lead to a price puzzle, depending on the magnitude of the changes.

In the context of Brazilian monetary policy, Castro (2019) provides an explanation for such divergence based on the rise in financial costs, given the rise in interest rates. The greater flexibility of loan terms can lead to both a contraction of the product and a sharpening of the price puzzle. Another hypothesis for the diverging results may derive from the fact that in periods of higher credit restriction, interest rate increases may result in an exchange rate devaluation, what leads to price increases (Kohlscheen 2014). The fact that in periods of higher financial instability, interest rate increases may lead to increased market fears and devaluation may ultimately cause an increase in prices. This possibility is corroborated by the robustness test that include international variables. In that test, the periods of high indebtedness are marked by devaluations following a contractionary shock, confirming the hypothesis suggested by Aghion, Bacchetta, and Banerjee (2001) and Kohlscheen (2014).

At least three factors may contribute to the divergent dynamics of price puzzle dynamics between credit cycles led by public and private banks. As we argued, following a contractionary monetary shock, lending by public banks imply lower financial cost, lower capital stock reduction and lower variation in the exchange rate. Accordingly, public banks could in fact contribute to monetary policy power.

In line with the work of Castro (2019), additional studies should formalize such hypotheses, detailing the interaction between the various mechanisms listed. Further empirical tests could provide fine-grained investigation of effects over various components of demand, for understanding whether public banks are able to protect investment.

From the point of view of the effects of interest rate shocks on demand, previous literature is practically consensual in arguing that there is greater resilience of investment to interest rate shocks, due to the existence of earmarked credit and public banks. Less responsive credit rates allow companies and households to keep expenditure levels after monetary shocks. Confirming this fact, we find that during periods of high credit of private banks the output was more responsive than in periods of high credit of public banks.

This result corroborates those presented by Bonomo and Martins (2016) and Perdigão (2018). However, it is important to emphasize that the power of monetary policy is

Table 3. Synthesis of results.

Bank Type	High Credit	
	Public	Private
Inflation	Initial/not persistent Price Puzzle	Persistent Price Puzzle
GDP	Medium	High

defined in terms of inflation, the determination of which does not depend solely on demand variations. In addition, the effects on output are also likely to apply to international shocks, i.e., government participation in the credit market also mitigates the effects of external shocks (Bonomo and Martins 2016). Considering that the economic costs of financial crises rise when the public sector does not pursue macroeconomic stabilization (Jordà, Schularick, and Taylor 2016), public banks offer an important countercyclical alternative.

5. Final remarks

The effects of monetary policy on real variables and on inflation are generally described on a linear fashion: the effect over aggregate demand are transmitted to prices regardless of economic states. This analysis is complexified when one considers the effects deriving from financial costs. If costs increase after a monetary policy shock, a sharp fall in GDP do not necessarily results in a sharp fall in prices. Estimates with state-dependent local projections empirically illustrate this paradox of monetary policy (Santoro et al. 2014; Alpanda and Zubairy 2018; Jordà, Schularick, and Taylor 2019). Our estimates are in line with this literature, which points to the need to incorporate nonlinearities in monetary policy transmission mechanisms.

We provide evidence that the power of monetary policy depends on the credit state of the economy. High credit periods showed higher price puzzle and stronger GDP responses. In turn, periods of low credit presented lower product response, but larger declines in inflation. Centrally, our work provides the first macroeconomic empirical study of the effects of public banks on monetary policy power in Brazil. Table 3 summarizes our most relevant findings:

(1) Periods of high credit of public banks presents lower – and no persistent – price puzzles. It means that monetary policy's power is stronger during this period in comparison with periods of high credit of private banks; and

(2) During periods of high credit of public banks, inflation control is associated with a less pronounced fall in GDP. It means that the sacrifice rate (the fall in GDP required to control inflation) is smaller during this period in comparison with periods of high credit of private banks.

In a few words, we do not find evidence in favor of the hypothesis that the existence of public banks reduces monetary policy's power. On the contrary: based on our results one should say that the power of monetary policy is enhanced by the existence of public banks in Brazil.

Notes

1. Formerly, the main benchmark for BNDES loans was the Long Term Interest Rate, a rate defined by the National Monetary Council and usually fixed in lower levels than the policy rate.
2. In addition to lower cost shocks, the existence of public banks could lead to less pronounced reductions in capital stock following increases in interest rates (Feijó and Sousa 2012; Modenesi and Modenesi 2012; Castro 2018). Through this channel, the short-term impacts of targeted credit on the power of monetary policy could be offset by allowing potential output to expand, reducing the impact of demand pressures on inflation.
3. Analogously, Jordà, Schularick, and Taylor (2019) compares the credit cycles defined by different debtors, whether households or companies. Authors conclude that cycles led by household debt are more influent for monetary policy interventions.
4. Estimations are performed with the *lpirfs* package, implemented in R by Adämmer (2018).
5. Function implemented by Adämmer (2018), and originally suggested by Granger and Terasvirta (1993).
6. These effects are lower than during the period of low credit of public banks, what seems to be at odds with the impulse response functions. Nevertheless, this is the case because the increases in interest rates during the period of high credit of public banks is higher. Thus, considering the higher interest rates increase in the high public credit state, GDP is more resilient to the monetary policy in the high public banks credit regime than in the low public credit regime.

Disclosure statement

No potential conflict of interest was reported by the author(s).

Funding

This work was supported by the Conselho Nacional de Desenvolvimento Científico e Tecnológico (CNPq); Fundação Carlos Chagas Filho de Amparo à Pesquisa do Estado do Rio de Janeiro (FAPERJ).

ORCID

Nikolas Passos (iD) http://orcid.org/0000-0002-4380-6392
André de Melo Modenesi (iD) http://orcid.org/0000-0001-5392-2920

References

Adämmer, P. 2018. Lpirfs: An R Package to Estimate Impulse Response Functions by Local Projections. *The R Journal* 11 (2): 421–438.
Aghion, P., P. Bacchetta, and A. Banerjee. 2001. "Currency Crises and Monetary Policy in an Economy with Credit Constraints." *European Economic Review* 45 (7): 1121–1150. doi:10.1016/S0014-2921(00)00100-8.
Ahmed, M. I., and S. P. Cassou. 2016. "Does Consumer Confidence Affect Durable Goods Spending during Bad and Good Economic Times Equally?" *Journal of Macroeconomics* 50: 86–97. doi:10.1016/j.jmacro.2016.08.008.
Alpanda, S., and S. Zubairy. 2018. "Household Debt Overhang and Transmission of Monetary Policy." *Journal of Money, Credit, and Banking* 135 (1–2): 499–526.
Arida, P. 2005. "Mecanismos compulsórios e mercado de capitais: propostas de política econômica." In *Mercado de capitais e crescimento econômico: lições internacionais, desafios brasileiros*, edited by E. L. Bacha and L. C. de Oliveira Filho, 22–45. Rio de Janeiro: Contracapa.

Auerbach, A. J., and Y. Gorodnichenko. 2012. "Measuring the Output Responses to Fiscal Policy." *American Economic Journal: Economic Policy* 4 (2): 1–27.

Baker, A. 2018. "Macroprudential Regimes and the Politics of Social Purpose." *Review of International Political Economy* 25 (3): 293–316. doi:10.1080/09692290.2018.1459780

Barboza, R. 2015. "Taxa de juros e mecanismos de transmissão da política monetária no Brasil." *Revista de Economia Política* 35 (1): 133–155. doi:10.1590/0101-31572015v35n01a08.

Barnichon, R., and C. Brownlees. 2019. "Impulse Response Estimation by Smooth Local Projections." *Review of Economics and Statistics* 101 (3): 522–530. doi:10.1162/rest_a_00778.

Barros, L., C. K. Dos Santos Silva, and R. de Freitas Oliveira. 2018. "Presença Estatal no Mercado de Crédito: o papel dos bancos públicos e do crédito direcionado na crise de 2008." *Working Paper Brazilian Central Bank No. 488.*

Bernanke, B. S., and M. Gertler. 1995. "Inside the Black Box: The Credit Channel of Monetary Policy Transmission." *Journal of Economic Perspectives* 9 (4): 27–48. doi:10.1257/jep.9.4.27.

Bonomo, M., and B. Martins. 2016. "The Impact of Government-driven Loans in the Monetary Transmission Mechanism: What Can We Learn from Firm-level Data?" *Banco Central do Brasil, Working Paper Brazilian Central Bank No 419.*

Bonomo, M., R. Brito, and S. Lazzarini. 2018. "Crédito Direcionado e Financiamento do Desenvolvimento." Chap. 14 In *Desafios da Nação,* edited by João A. N., B. Araújo, and R. Bacelette. 1 vols., 629–661. Rio de Janeiro: IPEA.

Bonomo, M., R. D. Brito, and B. Martins. 2015. "The after Crisis Government-driven Credit Expansion in Brazil: A Firm Level Analysis." *Journal of International Money and Finance* 55: 111–134. doi:10.1016/j.jimonfin.2015.02.017.

Borio, C. E. 1996. "Credit Characteristics and the Monetary Policy Transmission Mechanism in Fourteen Industrial Countries: Facts, Conjectures and Some Econometric Evidence." In *Monetary Policy in a Converging Europe,* edited by Koos A., K. Koedijk, C. Kool, and C. Winder, 77–115. Boston, MA: Springer.

Brugnolini, L. 2018. "About Local Projection Impulse Response Function Reliability." *Technical Report 6-440.* Tor Vergata University, CEIS.

Carcel, H., Luis A. Gil-Alana, and P. Wanke. 2018. "Application of Local Projections in the Monetary Policy in Brazil." *Applied Economics Letters* 25 (13): 941–944.

Cardim de Carvalho, F. J. 2005. "Uma contribuição ao debate em torno da eficácia da política monetária e algumas implicações para o caso do brasil." *Revista de Economia Política* 25 (4): 323–336.

Castro, P. 2018. "Earmarked Credit, Investment and Monetary Policy Power." Chapter 2 of PhD thesis, PUCRio, Rio de Janeiro.

Castro, P. 2019. "Earmarked Credit and Monetary Policy Power: Micro and Macro Considerations." *Working Paper Brazilian Central Bank No. 505.*

Chen, H., R. Li, and P. Tillmann. 2019. "Pushing on a String: State-owned Enterprises and Monetary Policy Transmission in China." *China Economic Review* 54: 26–40. doi:10.1016/j.chieco.2018.10.005.

Christiano, L. J., M. Eichenbaum, and C. L. Evans. 1999. "Chapter 2 Monetary Policy Shocks: What Have We Learned and to What End?" In *Handbook of Macroeconomics,* edited by John T. and M. Woodford. 1 vols., 65–148. doi:10.1016/S1574-0048(99)01005-8

Drehmann, M., and K. Tsatsaronis. 2014. "The Credit-to-GDP Gap and Countercyclical Capital Buffers: Questions and Answers." *BIS Quarterly Review March* 55–73, 2014.

Fazzari, S., and H. Minsky. 1984. "Domestic Monetary Policy: If Not Monetarism, What?" *Journal of Economic Issues* 18 (1): 101–116. doi:10.1080/00213624.1984.11504220.

Feijó, C., and A. Sousa. 2012. "A política monetária brasileira e suas recentes re-especificações: uma análise pela ótica da coordenação." In *Sistema Financeiro e Política econômica em uma era de Instabilidade,* edited by José L. O. and L. F. de Paula. 1 vols., 132–143. Rio de Janeiro: Editora Elsevier.

Garriga, C., F. E. Kydland, and R. Šustek. 2017. "Mortgages and Monetary Policy." *The Review of Financial Studies* 30 (10): 3337–3375. doi:10.1093/rfs/hhx043.

Granger, C. W., T. Terasvirta. 1993. "Modelling Non-linear Economic Relationships." *OUP Catalogue*. Oxford University Press. number 9780198773207.

Griffith-Jones, S., and J. A. Ocampo. 2018. *The Future of National Development Banks*. Oxford: Oxford University Press.

Jordà, Ò., M. Schularick, and A. M. Taylor. 2016. "Sovereigns Versus Banks: Credit, Crises, and Consequences." *Journal of the European Economic Association* 14 (1): 45–79. doi:10.1111/jeea.12144.

Jordà, Ò., M. Schularick, and A. M. Taylor. 2019. "The Effects of Quasi-random Monetary Experiments." *Journal of Monetary Economics. In Press*. doi:10.1016/j.jmoneco.2019.01.021.

Jordà, O. 2005. "Estimation and Inference of Impulse Responses by Local Projections." *American Economic Review* 95 (1): 161–182. doi:10.1257/0002828053828518.

Kashyap, A. K., and J. C. Stein. 2000. "What Do a Million Observations on Banks Say about the Transmission of Monetary Policy?" *American Economic Review* 90 (3): 407–428. doi:10.1257/aer.90.3.407.

Kilian, L., and Y. J. Kim. 2011. "How Reliable are Local Projection Estimators of Impulse Responses?" *Review of Economics and Statistics* 93 (4): 1460–1466.

Kohlscheen, E. 2014. "The Impact of Monetary Policy on the Exchange Rate: A High Frequency Exchange Rate Puzzle in Emerging Economies." *Journal of International Money and Finance* 44: 69–96. doi:10.1016/j.jimonfin.2014.01.005.

Lima, G., and M. Setterfield. 2010. "Pricing Behaviour and the Cost-Push Channel of Monetary Policy." *Review of Political Economy* 22 (1): 19–40. doi:10.1080/09538250903391863.

Martins, N. M., C. C. Pires-Alves, A. D. M. Modenesi, and K. V. B. D. S. Leite. 2017. "The Transmission Mechanism of Monetary Policy: Microeconomic Aspects of Macroeconomic Issues." *Journal of Post Keynesian Economics* 40 (3): 300–326. doi:10.1080/01603477.2017.1319249.

Mazzucato, M., and C. Penna. 2016. "Beyond Market Failures: The Market Creating and Shaping Roles of State Investment Banks." *Journal of Economic Policy Reform* 19 (4): 305–326. doi:10.1080/17487870.2016.1216416.

Minsky, H. 2008. *Stabilizing an Unstable Economy*. vols. 1. New York: McGraw-Hill.

Modenesi, A. D. M., and R. L. Modenesi. 2012. "Quinze anos de rigidez monetária no brasil pós-plano real: uma agenda de pesquisa." *Brazilian Journal of Political Economy* 32 (3): 389–411. doi:10.1590/S0101-31572012000300003.

Palludeto, A., W. Antonio, and R. A. Zanchetta Borghi. 2020. "Institutions and Development from a Historical Perspective: The Case of the Brazilian Development Bank." *Review of Political Economy* 1–19. doi:10.1080/09538259.2020.1720144.

Pazarbasioglu-Dutz, C., S. Byskov, M. Bonomo, I. Carneiro, B. Martins, and A. Perez. 2017. "Brazil Financial Intermediation Costs and Credit Allocation." *World Bank Report*.

Perdigão, B. 2018. "Essays on Monetary Economics and Banking." PhD thesis, PUCRio, Rio de.

Ramey, V. A. 2016. "Macroeconomic Shocks and Their Propagation." In *Handbook of Macroeconomics*. 2 vols., edited by J. Taylor and H. Uhlig, 71–162. Oxford: Elsevier.

Ramey, V. A., and S. Zubairy. 2018. "Government Spending Multipliers in Good Times and in Bad: Evidence from Us Historical Data." *Journal of Political Economy* 126 (2): 850–901.

Ravenna, F., and C. E. Walsh. 2006. "Optimal Monetary Policy with the Cost Channel." *Journal of Monetary Economics* 53 (2): 199–216. doi:10.1016/j.jmoneco.2005.01.004.

Santoro, E., I. Petrella, D. Pfajfar, and E. E Gaffeo. 2014. "Loss Aversion and the Asymmetric Transmission of Monetary Policy." *Journal of Monetary Economics* 68: 19–36.

Segura-Ubiergo, M. A. 2012. "The Puzzle of Brazil's High Interest Rates." *Technical Report 12-62*. International Monetary Fund.

Silva, I. É. M., N. L. Paes, and J. F. Bezerra. 2018. "Evidências de pass-through incompleto da taxa de juros, crédito direcionado e canal de custo da política monetária no brasil." *Estudos Econômicos, São Paulo* 48 (4): 559–595. doi:10.1590/0101-41614841inj.

Tenreyro, S., and G. Thwaites. 2016. "Pushing on a String: US Monetary Policy Is Less Powerful in Recessions." *American Economic Journal: Macroeconomics* 8 (4): 43–74.

Torres Filho, E. T. 2009. "Mecanismos de direcionamento do crédito, bancos de desenvolvimento e a experiência recente do BNDES." In *Ensaios Sobre Economia Financeira*, edited by F. M. Rocha Ferreira and B. B. Meirelles, 11–56. Rio de Janeiro: BNDES.

Some new insights on financialization and income inequality: evidence for the US economy, 1947–2013

Marwil J. Dávila Fernández (iD) and Lionello F. Punzo (iD)

ABSTRACT

In this article, we study the relationship between income distribution and financialization in the United States between 1947 and 2013. Financialization is introduced as a two-fold process. On the one hand, it implies an increase in the contribution of the financial sector in the composition of production. On the other hand, it is related to an increase in the importance of financial assets in terms of the composition of wealth. We take the share of financial employment as a *proxy* of the first dimension while, as to wealth composition, we make use of the share of financial assets on corporations' total assets. Applying cointegration techniques, we identify a positive long-run relationship between financialization and income inequality. Causality goes from employment to income inequality and from the latter to wealth. Nonlinear estimators suggest the existence of certain asymmetric effects such that changes in income distribution cannot be reverted by simply reverting financialization.

1. Introduction

Over the past decades, there has been increased income inequality in the United States (US) and many European countries. The causes and consequences of these trends have been the subject of a large body of scholarship, both inside and outside the social sciences. This article relates to two of the most prominent interpretations, highlighting the role of the financial sector in the broader process of institutional and structural change.

On the one hand, several scholars have pointed out that the rise and fall of unionization and a declining minimum wage account for much of the recent increase in inequality (Card and DiNardo 2002; Western and Rosenfeld 2011; Volscho and Kelly 2012). The same follows the increased ability of 'superstars' and top executives to set their pay (Dew-Becker and Gordon 2005; Gordon and Dew-Becker 2007). Financial market activity has been fueled by the deregulation of financial institutions, creating a niche of very high wages. By controlling key assets such as knowledge, teams, and clients, financiers exercise a 'hold up' power over the firm. This has resulted in an increase in the finance wage-premium gap (Godechot 2020).

In a similar vein, it has been also argued that the deregulated financial environment altogether with the rise of shareholder-value orientation have implied in higher inequality at both personal and functional income distribution (Epstein 2005; Lin and Tomaskovic-Devey 2013; Dunhaupt 2017). For instance, Stiglitz (2013) has sustained that growing inequality is mostly the result of how the market economy has been structured in the last thirty years. Not by chance, the explosion of top managerial compensation and its link with substantial cuts in top tax rates have been investigated by constructing time-series of top income shares over the long-run (Atkinson, Piketty, and Saez 2011; Piketty and Saez 2013).

On the other hand, starting from Kuznets' principle that the structure composition of the economy primarily guides changes in inequality, a second group of contributions has decomposed variations on income distribution following wage variations across US economic sectors (Conceição and Galbraith 2001; Galbraith 2012; Galbraith and Hale 2014). It is hard to parse out in a clean and precise way a separation between institutional and structural change. Still, this last strand of the literature has put more emphasis in changes in the sectoral composition of the economy. Such disaggregation has enlightened the roles played by the financial sector, the technology boom and wartime public spending, in driving the evolution of income distribution in that country.

Our contribution focuses on the US economy and combines elements of this two groups as we are interested in the implications for income inequality of the rise in relative importance of the financial sector, i.e. *financialization*. The literature on financialization has extensively documented an 'increasing role of financial motives, financial markets, financial actors, and financial institutions in the operations of the domestic and international economies' (Epstein 2005, 3). Even though a precise definition varies considerably across analyses, the shared premise is that the relative size of the financial sector has grown significantly over the past three decades.[1]

While there have been a significant number of publications in the area (see Mader, Mertens, and van der Zwan 2020), our understanding is that at least two main issues remain to be solved. First, the very term *financialization* as well as a proper way to measure it, are still unclear in economics. A second problem concerns the causal direction in the relationship between financialization and income inequality. Most of the literature has focused on causality going from the former to the latter. However, authors such as Kumhof, Rancière, and Winant (2015) and Fligstein, Hastings, and Goldstein (2017) have also discussed whether high levels of inequality may push financialization in terms of indebtedness. We argue that the correspondence might depend on the dimension of the phenomenon under analysis. In this paper, we treat financialization as a two-fold process. On the one hand, it can be understood as an increase in the contribution of the financial sector in the composition of e.g. total production. On the other hand, it can be seen as an increase in the importance of financial assets in terms of the composition of a country's wealth.

In an attempt to cope with some of the limitations of current metrics, we propose two distinct measures of financialization, one for each dimension. We use the share of financial employment on total employment as a *proxy* for the structure of production. By not relying on value-added or on the share of financial income on GDP, this metric has the advantage of avoiding several accounting problems (e.g. Nakamura 2010; Mazzucato and Shipman 2014; Pagano 2014). There is evidence indicating that the value-added of the financial sector has been overstated in National Accounts, where pure rent-seeking activities have been counted as productive (Basu and Foley 2013; Assa 2016). Moreover, we look at the structure of the economy also from the point of view of wealth by taking the share of financial assets on corporations' total assets. Such a simplification does not come without an important cost since we are not explicitly considering households and we do not differentiate financial from non-financial firms. While still a reasonable *proxy* for the relative size of finance in wealth, this choice allows us to focus the analysis on corporations, which after all are the economic units responsible for production activities.

This article looks at the relationship between income distribution and financialization in the United States between 1947 and 2013. Applying cointegration techniques, we identify a positive long-run relationship between the share of financial employment and income inequality as well as between the share of financial assets and income distribution. An increase of one percentage point in the share of employment related to financial activities increases the Gini index between 0.25 and 0.5%. On the other hand, a one percentage point increase in inequality is associated with a 1 to 1.75% greater wealth-share in the form of financial assets. At least in what concerns the traditional Gini index, causality goes from the employment dimension to inequality and from inequality to the wealth dimension.

The observed increase in financial employment has boosted income inequality which in turn has been responsible for the prevalence of wealth accumulation in the form of financial assets. This gives an extra flavor to Godechot's (2020) story. Labor in the financial sector and related business activities controls key assets of the firm, thus receiving a higher wage premium. In a context of deindustrialization, this has deepened inequality. Income which is not consumed, has not been accumulated in the form of physical assets, but rather as financial wealth. Moreover, nonlinear cointegration models suggest the existence of asymmetric effects such that changes in income distribution cannot be reverted by simple reverting financialization.

When it comes to policy implications, three results are worth stressing. First, an increase of wealth accumulation in the form of more financial assets is at least partially the shortcoming of a (*previous*) increase in inequality. Therefore, limiting financial intermediation and securitization requires the problem of a continuous deterioration of income distribution be solved. Second, income inequality is likely to continue to rise unless the share of financial employment stabilizes. Hence, coping with financialization and the associated distributive issue might require a more direct intervention in the labor market. Finally, policymakers should act as soon as possible because just reverting financialization is not capable of decreasing inequality.

The remaining of the paper is organized as follows. Giving a preliminary look at the US experience using descriptive data analysis, in the next section we further elaborate our main argument. Section 3 presents our empirical exercise on the relation between financialization and income inequality. Section 4 concludes with a discussion of the main results.

2. A brief descriptive narrative

The suffix '-ization' has been intensively used in social sciences to designate a changing weight and importance of the thing or quality preceding it. Intuitively speaking, financialization would correspond to an increase in the relative importance of finance in the economy. Thereafter, we will introduce financialization as a two-fold process character-ized by (i) an increase in the relative contribution of the financial sector in the composi-tion of production, and (ii) an increase in the importance of finance in the wealth composition, as we are interested in capturing the increase in relative importance of the financial sector in terms of both dimensions.

Karwowski, Shabani, and Stockhammer (2020) recently presented an extensive review of the empirical literature on financialization, distinguishing between two different and frequently used types of measures. In their classification, activity or flow indicators refer to financial incomes or payments relative to total income. On the other hand, vulner-ability in terms of stock indicators is taken to be outstanding debt to income. Such meters might be misleading, and for at least two reasons. On the one hand, the modern economy produces less measurable output than the traditional manufacturing, mining, and agri-culture (see, for example, Nakamura 2010; Pagano 2014). Mazzucato and Shipman (2014), e.g., have lately shown an increase in divergence between value-creation and value-added measures. They attribute it to the rise in size and influence of financial institutions and markets. Financial sector's value-added would have been overstated while pure rent-seeking activities had been counted as productive. This jeopardizes otherwise natural candidates to capture the relative weight of Financial vis-a-vis Real activity, as represented by gross financial income or financial sector's value added as shares of GDP.

As to the stock indicator, recalling how its ratio to a flow is, in fact, a time measure, the debt to GDP ratio may be a useful measure of financial fragility but it does not capture accurately the wealth dimension of financialization. This is because no reference is made to variations in relative importance of the financial sector in the economy. We may discuss whether financialization goes along with financial fragility. The role of such indicators to assess, e.g., conditions preceding a financial crisis is unquestioned. Still, in our understanding, one should make a step forward and propose an indicator devised to captures changes in wealth composition.

Given the many distortions an inadequate treatment of financial outputs might have in an analysis of income inequality, hereafter the share of financial employment in total employment will be taken as the *proxy* for the flow financialization dimension. The use of employment shares allows us to establish a link to a distinct though often conjugated process, deindustrialisation. In a sense, for certain countries at least, deindustrialization may be an integral component of globalization, technological and institutional argu-ments for increasing inequality, whilst an increase in financial employment share gets associated with a broader upswing in the share of non-manufacturing employment. If deindustrialization is defined as a relative decrease in manufacturing employment, financialization may come to be the other side of it.

In a detailed and relatively recent assessment of financialization in the United States, Krippner (2005) contrasts a more traditional perspective on long-term economic change focused on what is being produced in the economy, *with* one engaged in understanding where profits are being generated. Indicators such as the share in financial employment or in financial GDP would be in the former group. The relative importance of financial to non-financial profits, on the other hand, would reflect the latter view. As she did at the time, we are not claiming that ours or her separation is more fundamental or 'truer' given that 'how one conceptualizes structural change in the economy depends very much on one's theoretical purpose' (Krippner 2005, 176). Our motivation is much simpler. With a long tradition in economics (and other social sciences), we try to provide new empirical insights on financialization from both an employment and a wealth composition perspective, and on their relationship to income inequality. Furthermore, since we conceptualize it as a process of structural change involving changes in the relative importance of the financial sector, we want to emphasize this perspective also regarding these two dimensions.

Figure 1, on the left, shows the evolution of the share of financial employment between 1950 and 2010 with the respective *lowess* curves. A detailed description of data used in this study appears in the next section. It is possible to identify a positive trend for the whole period; one can also divide the sample before and after the eighties. While between 1950 and 1980, financial employment share increased from 7% to 10% in total employment, in the next 30 years it got closer to 20%.

Even though we do not report the full evolution of overall non-manufacturing employment, it is to be noticed how it increased from 75% in 1947 to 91.3% in 2014, i.e. with an expansion of 16.3 points. During the same period, the share of financial employment increased 11.3 points, going from 6.7% to 18%. This means that the flow dimension of financialization may be responsible for something in the order of 70% of deindustrialization, 11.3/16.3 = 0.69.

The process seems to be correlated with the recent upswing in income inequality. It is well known that the service sector, which includes financial activities, has a more significant income gap between high paying and low paying jobs, as compared to manufacturing.[2] If workers displaced from manufacturing activities used to be in the middle-income percentiles of the population and can no longer be reallocated in non-manufacturing sectors with higher or equal earnings, we shall expect income distribution to deteriorate. Labor saving technological change has reduced the demand for many blue-collar jobs while globalization is creating a global marketplace putting workers of tradable sectors in competition with comparable workers from abroad. De-industrialization has weakened workers bargaining power, producing higher inequality.

Dew-Becker and Gordon (2005) documented that, in the United States between 1966 and 2001, no quantile below the 90th percentile has experienced growth in wages commensurate with the growth of average rate of productivity. Yellen (2006) pointed

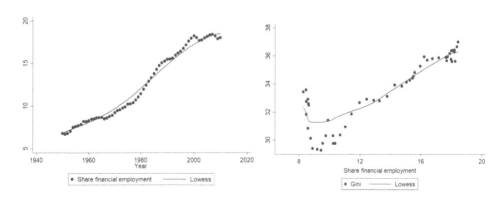

Figure 1. Share of financial employment and income inequality.

out that displaced workers have to take a pay cuts of about 17% on average, on new jobs. In the early 2000s, the size of this wage loss was the highest in at least 20 years. These considerations might also have an impact on the functional distribution of income. A reduction in the wage-share over the past decades has been reported for several countries including the United States, by e.g. Karabarbounis and Neiman (2014), Dunhaupt (2017), Hein et al. (2017), among others.

In addition, the increase in economic importance of finance is deeply related to the remunerations of corporate officers. As argued by Godechot (2020), financiers exercise a 'hold up' power over firms by controlling key assets. The threat of moving them to a competitor has increased their bargain power and wage-premium gap. Moreover, the financial sector has developed expertise in a widespread variety of rent-seeking forms, going from taking advantage of asymmetries of information to lending and abusive credit card practices (Stiglitz 2013). If workers in the financial sector are in the top-income percentiles of the population and manage to increase their wages above productivity gains, we will likely have a deterioration of income distribution.

Figure 1, on the right, depicts the positive relationship between income inequality and financial employment share. Higher levels of income inequality are associated with higher levels of employment in the sector. With a financial employment share between 8 and 10%, there is no clear correspondence between the two variables: if something, it looks negative. However, an increase in the financial employment share from 10% to almost 20% is associated with an increase of 7 points in the Gini index.

As we believe that indicators of the type debt to Income or GDP ratios are inappropriate to capture the wealth dimension of financialization, we are using the share of financial assets on corporations' total assets. Corporations' wealth is traditionally defined as net (or residual) wealth of the corporation' sector, i.e. the sum of non-financial and financial assets owned by corporations minus their debt liabilities. In practice, firms are owned either by the private sector, or the government or by the rest of the world. Hence, the corporations' value is included already in their financial assets and therefore in their net wealth. With a Tobin's Q ratio equal to one, by construction, net corporate wealth is equal to zero: the full value of corporations is already included in private and public wealth, so there is nothing to add (Alvaredo et al., 2016).

In this paper, however, we are not looking at corporations' net wealth but instead at the composition of corporations' gross wealth, i.e. the sum of non-financial and financial assets owned by firms. An increase in the share of financial assets in corporations' gross wealth does not imply that wealth has been created, but it still highlights that there are major changes underway in the economy. Reasons behind them go from an increase in financial activity and interest for financial assets per se to the reorganization of industrial activity. Such changes might point at an increase in intermediation and securitization activities, with most likely implications for income distribution.

To the extent that firms are owned by households, we only indirectly assess the correspondence of income inequality and the composition of their wealth. One should notice, nonetheless, that the latter significantly changes as we move from the middle class to the very rich. An average middle-income individual in the US has around 60% of her wealth in the form of principal residence, 16% in pension accounts, and only 12% as either equity or stocks and securities. These last two components increase to 40% in upper-income households reaching 80% for the ultra-rich (see Saez and Zucman 2016; Wolff 2017).

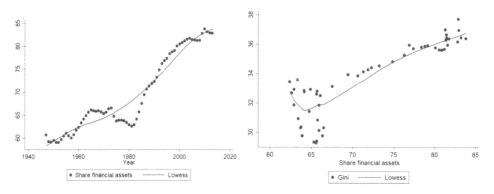

Figure 2. Share of financial assets on corporates total assets and income inequalities.

Two further reasons justify our use of the composition of corporates' gross wealth to account for the second dimension of financialization. First, we thus properly capture the composition of a wealth indicator, a feature which is vital if we are to understand financialization as an increase in relative importance of the financial sector. Next, although we do not differentiate between financial and non-financial companies, it permits us to concentrate on the basic units of production, i.e., on firms.

Figure 2, on the left, presents the evolution of the share of financial assets on corporations' total assets in the last 60 years and, on the right, the positive relationship between this *proxy* of wealth composition and inequality. The 1980s divide the sample into two periods. In this decade, there is a clear structural break that contrasts with a more continuous trajectory in financial employment. Between 1950 and 1980 the share of financial assets increased from 60 to 65%. In the next thirty years, we observe an increase to around 85% by the end of the period.

Such trajectories can be partially explained by a change in corporate governance towards shareholder-value, as observed in the past four decades. This process has fundamentally benefited asset holders by increasing the dividend payout ratio and by rising stock prices (Dunhaupt 2017; Hein et al. 2017). Once there is a sufficiently uneven distribution of financial assets favoring those in the top-income percentiles of the population, income inequality will increase as long as returns on the financial sector are higher than in the real sector.[3] The magnitude of this effect further depends on the size of finance on wealth. This hypothesis is quite plausible considering, for example, that the rate of return on fortunes increases with their size (Wade 2014). An increase in the share of financial assets from 70 to 85% is associated with an increase of 4 points in the Gini index. Still, for values below 70%, there is no clear correspondence, a pattern already identified for employment.

3. Data and empirical exercise

Our dataset is annual and comprehends the period after the Second World War. We use the Gini net index provided by the Standardized World Income Inequality Database (SWIID) available from 1960 to 2014. The SWIID is the most comprehensive dataset on income inequality at international levels, incorporating the OECD Income Distribution Database, the Socio Economic Database generated by CEDLAS, the Eurostat, the World

Bank's PovcalNet, the UN Economic Commission for Latin America and the Caribbean, and national statistic offices around the world.[4]

The share of financial assets over total corporations' assets was computed using data provided by the World Inequality Database (WID) from 1947 to 2013. Corporations' financial assets correspond to the sum between corporate equity, fund shares, offshore wealth, corporate currency, deposit banks, loans, corporate pension funds and life insurance. Corporate non-financial assets are equal to the sum of business, housing, and other non-financial assets.

Finally, the share of financial employment comes from the 10-sector database provided by the Groningen Growth and Development Centre (GGDC) from 1950 to 2010. The financial sector comprehends financial intermediation, insurance, real estate, renting, and business services as defined in the International Standard Industrial Classification (ISIC rev. 3.1). Our choice has two motivations. First, a significant amount of contributions in the financialization literature does rely on data of the so-called FIRE sector, which is close to the metric used here. Therefore, our results can be somehow compared to other existing studies in the field. Next, the GGDC 10-sector dataset is a long-run internationally comparable database on sectoral macroeconomic series for countries in all five continents. This will allow us to replicate, in our future research, the present exercise for other countries, once again with comparable results. The main disadvantage of working with such a broad concept is that we overlook intra-sectoral dynamics. In fact, while the share of financial intermediation in employment went from 3.5% in the 1970s up to 4.25% in 2010, renting and business services increased their share from 5.5% to 13.5%. Hence, it is worth saying that financial intermediation currently is no more than a quarter of financial employment.

In what follows, we converted time-series to the logarithmic form. We also make use of a set of control variables which include GDP, a human capital index, the degree of trade openness, and the size of the population. Data come from the Penn World Table (PWT 9.0).

To ascertain the existence of a long-run relationship between financialization and income inequality, we use country-specific cointegrating techniques. While still constrained by parametric assumptions, this approach overcomes several of the main shortcomings of the usual cross-country regressions. Indeed, the presence of cointegration brings a form of robustness to many of the classical empirical problems that lead to the violation of the so-called exogeneity condition for the regressors. This includes reverse causality, simultaneity, omitted variables, measurement error, among others (for a review of the super-consistency properties under cointegration, see Pedroni 2019). For example, under cointegration, an omitted variable is less likely to bias our estimates because it will either be non-stationary – in which case we would not have cointegration – or stationary – in which case the obtained coefficients do not change when we include it.[5]

Considerable attention has been paid over the past decades to testing the existence of relationships in levels between variables. Different approaches are available in the literature, for example: (i) the Engle-Granger two-step residual-based procedure; (ii) the Johansen system-based reduced rank regressions, or (iii) the Hansen instability test. Those tests, however, require series to be unequivocally non-stationary and integrated of the same order. This might be a problem when it is not known with certainty whether the underlying regressors are trend or first-difference stationary.

Ascertaining the order of integration becomes an essential precondition to establish whether the use of traditional cointegration techniques is warranted. In this respect, we performed the traditional Augmented Dickey-Fuller (ADF) and the Dickey-Fuller test with GLS detrending (DF-GLS). The share of financial employment and the Gini index were found to be integrated of order one while the share of financial assets is at least integrated of order two (see table A1 in the appendix). This posits a problem since under such conditions we cannot proceed with our exercise.

Notice, however, that from visual inspection we identified a structural break around the 1980s for the share of financial assets. Multiple breakpoint Bai-Perron test identifies several repartition breaks for Gini (1986, 1994, 2004), the share of financial employment (1959, 1970, 1981, 1993, 2002), and the share of financial assets (1961, 1988, 1998). As Perron (1989) points out, structural change and unit roots are closely related and might invalidate conventional unit root tests. An extensive literature has followed outlining unit root tests that remain valid in the presence of a break. Hence, we also performed the Dickey-Fuller test allowing for a structural break, as reported in the appendix (see table A2). Results indicate that series are at most integrated of order one, though the share of financial employment and assets might be I(0).

Hence, we make use of the Auto-Regressive Distributed Lag (ARDL) bounds testing procedure developed by Pesaran and Shin (1998) and later extended by Pesaran, Shin, and Smith (2001). This methodology has several advantages over other cointegration methods as it allows the undertaking of analysis regardless of whether the variables are a mixture of stationary, I(0), and integrated of order one, I(1), which is our case. Unless explicitly said otherwise, we included dummy variables to capture the structural break effects. We assigned a dummy variable for each indicator. They assume value 1 for years with breaks and 0 for years with no break. The underlying general ARDL (p,q) model is given by:

$$Gini_t = \alpha_0 + \sum_{i=1}^{p} \varphi_i Gini_{t-i} + \sum_{j=1}^{q} \beta_j Finan_{t-j} + \gamma X_{t-1} + \varepsilon_t \qquad (1)$$

$$Finan_t = \alpha_0 + \sum_{i=1}^{p} \varphi_i Finan_{t-i} + \sum_{j=1}^{q} \beta_j Gini_{t-j} + \gamma X_{t-1} + \varepsilon_t \qquad (2)$$

where α_0 is a constant term, φ_i is the coefficient associated with the lags of the dependent variable, β_j stands as the lag coefficient of the independent variable of interest, X is a vector of exogenous controls, γ corresponds to the associated marginal effects, and ε is the error term.In what follows, our estimation strategy consists of investigating the relationship between financialization and inequality in separate models. The analysis also includes the investigation of Granger causality and the estimation of long-run coefficients.[6]

3.1. ARDL estimations

Different ARDL models were estimated, all of them including income distribution either as dependent or explanatory variable. In order to avoid potential serial correlation issues, the order of lags was obtained using the Akaike (AIC) informational criteria. We allow for automatic lag selection imposing a maximum of four lags for dependent and independent variables. This means that, for each set of controls, we estimated 20 times the model using different lag combinations and selected the one that performs best according to AIC. Some models were found to have serial correlation, in which case, we maintained the automatic lag selection criteria but increased the maximum number of lags until we removed it.

If two series are cointegrated, this means that they have a long-term relationship, which prevents them from wandering apart without bound. Pesaran, Shin, and Smith (2001) and Narayan (2005) provided supply bounds on the critical values for the asymptotic distribution of the F-statistic. Table 1 reports our estimates of Equation (1) as well as the ARDL/Bounds cointegration test.

Independently of the dimension of financialization, the calculated F-statistic is higher than the 5% of significance critical value in almost all models indicating the existence of a cointegrating vector. That is, we identify the existence of a long-run relationship between the variables under analysis. Financialization and income distribution are related. Predictive causality can be assessed by incorporating the lagged error-correction (EC) term (Narayan and Smyth 2006; Odhiambo 2009). The estimated EC is in all cases negative, indicating convergence to the long-run equilibrium solution. Models I–IV show that between 17 and 35% of any movements into disequilibrium are corrected for within one period when financial employment is used as explanatory variable. On the other hand, when we focus on the wealth dimension of financialization, we have that this is the case from 18 to 25%, as we can see from models V–VIII.

Despite the existence of a cointegrating relationship, we identify an important difference in terms of the long-run estimated coefficients. An increase of one percentage point in the share of financial employment is related to an increase between 0.25 and 0.5% in income inequality. This effect increases in magnitude as we introduce our set of control variables, being always significant at 1%. However, when it comes to the share of financial assets in total wealth, only in model V we found a significant effect. Still, it vanishes once we control for the size of the economy and human capital, remaining not significant as we add further controls.

We proceed by presenting our estimates of Equation (2). It consists of the evaluation of the reverse causality case, i.e. changes in income inequality fueling financialization. Results are provided in Table 2. The F-statistic of the Bounds cointegration test indicates the existence of a long-run correspondence between income inequality and financialization whose strength varies depending on the dimension under analysis. For instance, in models IX to XII, we either were not able to find a cointegrating vector or it was found barely after the 5% threshold. This implies that the respective estimated coefficients should be interpreted with caution. We do find that an increase of 1 percentage point in inequality is associated with 1.5 to 2% higher financial employment. Unfortunately, we cannot guarantee the consistency of such a result because in the absence of cointegration, our regressions are subject to endogeneity problems.

On the contrary, when we use the share of financial assets as dependent variable, the cointegrating relationship becomes stronger as we include our set of controlling variables. In this case, an increase of 1 percentage point in the Gini index is associated with an increase between 1 and 1.7% in the share of wealth in financial. The estimated EC term is negative and significant, confirming the direction of Granger causality. From 6 to 8% of deviations from equilibrium are corrected for within one period.

Table 1. ARDL estimates and bounds cointegration from financialization to income inequality.

	Gini							
	I	II	III	IV	V	VI	VII	VIII
Model	ARDL(2,0)	ARDL(2,0)	ARDL(2,0)	ARDL(2,0)	ARDL(3,3)	ARDL(4,2)	ARDL(3,2)	ARDL(3,2)
Financial employment	0.261200***	0.400810***	0.495574***	0.484603***	–	–	–	–
Financial assets	–	–	–	–	0.630890***	-0.210541	0.184444	0.079571
constant	2.832504***	3.118102***	5.196375***	4.474358	0.809943**	0.715825	3.175771*	2.049150
Y	–	Yes	Yes	Yes	–	Yes	Yes	Yes
HC	–	Yes	Yes	Yes	–	Yes	Yes	Yes
Trade	–	–	Yes	Yes	–	–	Yes	Yes
Pop.	–	–	–	Yes	–	–	–	Yes
Dummy	Yes	Yes	Yes	Yes	Yes	Yes	Yes	Yes
ECM	-0.17558***	-0.24983***	-0.29778***	-0.35546***	-0.184180**	-0.2049***	-0.23889***	-0.26941***
F-statistic	5.015386***	3.906477*	5.162867***	5.208650**	2.288055	5.116271**	6.219295***	6.664440***
Adjs. R2	0.972569	0.972127	0.973943	0.973706	0.974781	0.979155	0.978394	0.978851
LM Prob.	0.8590	0.8346	0.6271	0.7628	0.1807	0.6437	0.2718	0.2467

*, **, and *** stand by 10%, 5% and 1% of significance. Models I–IV include financial employment as explanatory variable and a set of controls. Models V–VIII use instead the share of corporate's gross wealth in the form of financial assets. Control variables include output, Y; human capital, HC; trade openness, Trade; population size, Pop; and structural break dummies. The reported F-statistic corresponds to the Bounds cointegration test. The LM probabilities indicate that the null-hypothesis of no serial correlation cannot be rejected.

Table 2. ARDL estimates and bounds cointegration from inequality to income financialization.

	Share financial employment				Share financial assets			
	IX	X	XI	XII	XIII	XIV	XV	XVI
Model	ARDL(2,2)	ARDL(2,4)	ARDL(2,4)	ARDL(2,4)	ARDL(5,2)	ARDL(5,2)	ARDL(5,2)	ARDL(5,2)
Gini	12.10882	1.967499***	1.588799***	1.727915***	1.607426***	1.191080	1.659921**	1.720854*
constant	−40.91975	−1.209734	−5.702473**	−5.048661**	−1.325467	0.864526	−7.637349**	−7.72495***
Y	–	Yes	Yes	Yes	–	Yes	Yes	Yes
HC	–	Yes	Yes	Yes	–	Yes	Yes	Yes
Trade	–	–	Yes	Yes	–	–	Yes	Yes
Pop.	–	–	–	Yes	–	–	–	Yes
Dummy	Yes	Yes	Yes	Yes	Yes	Yes	Yes	Yes
ECM	0.00677***	−0.10444***	−0.15949***	−0.22936***	−0.06280***	−0.07371***	−0.08893***	−0.08756***
F-statistic	3.146310	4.954678**	3.451975	4.999800**	4.315043**	5.235913**	12.04319***	10.78514***
Adjs. R2	0.998364	0.998547	0.998575	0.998695	0.996367	0.996805	0.998069	0.998018
LM Prob.	0.9415	0.6398	0.4057	0.1217	0.3699	0.1337	0.2509	0.2569

*, **, and *** stand by 10%, 5% and 1% of significance. Models IX–XII present financial employment as dependent variable and a set of controls among which the Gini index. Models XIII–XVI explain the financial share of corporate's gross wealth as a function of income distribution. Control variables include output, Y; human capital, HC; trade openness, Trade; population size, Pop; and structural break dummies. The reported F-statistic corresponds to the Bounds cointegration test. The LM probabilities indicate that the null-hypothesis of no serial correlation cannot be rejected.

Comparing our estimates in Tables 1 and Tables 2, there is a long-run relationship between financialization and inequality but that the causal relationship is conditional to the dimension of the problem under analysis. The observed increase in financial employment seems to boost inequality while the latter is responsible for the prevalence of wealth accumulation in the form of financial assets. This gives an extra flavor to Godechot's (2020) story. Labor in the financial sector has created a niche of high wages that control key assets and clients of the firm. In this way, financiers 'hold up' power over the firm, receiving a higher wage-premium which in a context of deindustrialization deepens inequality. Income which is not consumed, has not been accumulated in the form of physical assets, but rather as financial wealth. As firms are owned by households, this last result converges with findings that show that most of the middle-income class wealth derives from their principal residence, whereas the very rich have around 80% of their wealth as financial assets and business equity (Wolff 2017).

Our results are in line with previous studies that have also documented an association between financialization and inequality (e.g. Sjoberg 2009; Dunhaupt 2014). Still, a novelty of the present exercise lies in the sequence of events, with income distribution as an intermediator between the two dimensions of financialization. Changes in the composition of production impact income distribution that amplifies this effect on the composition of wealth. It is also worth to notice that convergence to equilibrium is much faster in models I–IV than in models IX–XII. This result is expected considering that changes in stocks are supposed to happen slower than adjustments of flows.

Residuals were checked for serial correlation using the Breusch-Godfrey LM test. The probability of rejecting the null hypothesis of no serial correlation is reported in the last line of Table 1 and 2 . If residuals are correlated, the estimated coefficients would be biased and inconsistent. Given that the null hypothesis cannot be rejected, even at 10%, we conclude that our inference is valid, and our regressions are not spurious.

3.2. Exploring nonlinearities

We let down the assumption that the long-run relationship is represented as a symmetric linear combination of nonstationary regressors, and, instead, adopt the flexible NARDL estimator proposed by Shin, Yu, and Greenwood-Nimmo (2014), an extension of the ARDL framework. It pays with the advantage of being capable of simultaneously and coherently modeling asymmetries both in the underlying long-run relationship and in the patterns of dynamic adjustment. The question we want to address is: does a negative shock have the same long-run impact as a positive shock? The underlying general NARDL (p,q) model is given by:

$$Gini_t = \alpha_0 + \sum_{i=1}^{p} \varphi_i Gini_{t-i} + \sum_{j=1}^{q} \left(\beta_j^+ Finan_{t-j}^+ + \beta_j^- Finan_{t-j}^- \right) + \varepsilon_t \qquad (3)$$

$$Finan_t = \alpha_0 + \sum_{i=1}^{p} \varphi_i Finan_{t-i} + \sum_{j=1}^{q} \left(\beta_j^+ Gini_{t-j}^+ + \beta_j^- Gini_{t-j}^- \right) + \varepsilon_t \qquad (4)$$

where $Finan = Finan_0 + Finan^+ + Finan^-$ and $Gini = Gini_0 + Gini^+ + Gini^-$. The superscripts + and – indicate the partial sum processes of positive and negative changes in the respective variables:

$$Finan_t^+ = \sum_{k=1}^{t} \max(\Delta Finan_k, 0) \qquad Finan_t^- = \sum_{k=1}^{t} \min(\Delta Finan_k, 0)$$

$$Gini_t^+ = \sum_{k=1}^{t} \max(\Delta Gini_k, 0) \qquad Gini_t^- = \sum_{k=1}^{t} \min(\Delta Gini_k, 0)$$

Following our results in the previous subsection, we estimate two NARDL models. The first takes the Gini index as dependent variable and the share of financial employment as the independent one, being equivalent to model I. The second brings in income inequality as the explanatory variable of the share of financial on total corporate assets, being equivalent to model XIII. AIC informational criteria chose a NARDL (2,2) and NARDL (3,3), respectively. Table 3 reports our estimated long-run coefficients.

The calculated F-statistic is higher than the 5% of significance critical value, indicating the existence of cointegration in both models. An increase of 1% in the share of financial employment causes an increase of 0.27% in inequality while a positive shock of 1% in the latter leads to a 0.87% increase in the share of financial assets. On the other hand, negative shocks have no significant effects, emphasizing the importance of asymmetries in the underline relationship. That is, financialization in terms of employment leads to a once and for all increase in inequality that fuels the share of financial assets and cannot be easily reversed.

Figure 3 shows how coefficients converge to the long-run solution. The green and red lines correspond to positive and negative shocks, respectively. Given that the latter is not statistically significant, we concentrate in the dynamics of the former. Three years after an increase in financial employment, we observe a momentaneous reduction in income inequality. While Gini indexes are falling, there is also a reduction in the accumulation of financial assets relatively to real assets. However, as soon as distributive conditions start to deteriorate, the final and long-lasting effect on financial assets becomes positive.

Although the idea that deindustrialization is behind the rise of inequality in the US finds an echo in the literature, our exercise also brings some insights in this respect.[7] For instance, deindustrialization is often interpreted as a reduction in the share of manufacturing employment. As previously reported in this article, around 70% of this reduction corresponds to a rise in the share of financial and business services employment. To the extent that, in this case, financialization can be seen as a mirror to deindustrialization, we

Table 3. Long-run coefficients, NARDL models.

Dependent	Explanatory	Model	+	-	F-statistic
Gini	Financial employment	ARDL(2,2)	0.272***	−0.754	4.3034**
Financial assets	Gini	ARDL(3,3)	0.869***	−0.255	5.3957***

*, **, and *** stand by 10%, 5% and 1% of significance. Superscripts + and – indicate the response of the dependent variable to a positive and negative shock, respectively, in the explanatory one. The reported F-statistic corresponds to the Bounds cointegration test.

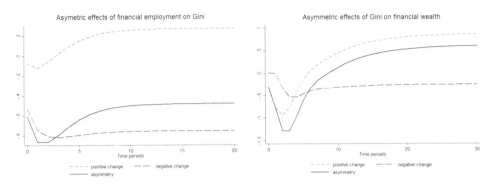

Figure 3. Asymmetric effects on the relationship between financialization and income inequality.

have here also a correspondence between the latter and the rise of income inequality in the United States.

Taking as departure point Kuznet's principle that changes in income inequality largely depend on the sectoral composition of the economy, our results indicate that the composition of wealth is to some extent guided by the composition of the flow behind it, that is, production. Wealth represents accumulation and captures the 'state of the system,' providing the basis for making choices. Its change, however, depends on the flows going into or coming out.

Even though an increase in the share of financial assets on corporates' gross wealth does not imply that wealth has been created, it does suggest that the structure of the economy is changing. In particular, it indicates an increase in financial intermediation and securitization. This process has strong ties with changes in financial employment. The observed increase in the share of financial employment has exacerbated income inequality leading to a further movement towards financial assets. The recent rise in income inequality is strongly linked with the increase in the importance of finance concerning production and wealth composition.

The reader may ask how sustainable this process is in the very long-run. It is true that the share of financial employment or financial assets cannot growth forever since, in the limit, all employment or assets would become financial. Even then, financialization, as described in this article, will come to an end because there will not be anymore an increase in relative importance of the financial sector. Still, we have provided evidence that this once and for all change in the sectoral composition of the economy has important implications for income distribution that cannot be reverted by simple reverting deindustrialization.

4. Considerations by way of conclusion

The United States and many countries in Europe have experienced increased income inequality over the past decades. There is a broad agreement among social scientists that technological change, international trade and social norms have been playing key roles in this process. Inspired by Kuznets' principle that changes in inequality are largely driven by the structure composition and the stage of development of the economy, we conceptualized financialization as a two-fold process characterized by (i) an increase in the

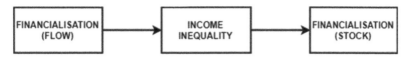

Figure 4. A summarizing diagrams.

contribution of the financial sector in terms of the composition of production, and (ii) an increase in importance of finance in terms of the composition of wealth.

Worried by the distortions that an inadequate treatment of financial outputs might have in the analysis of income inequality, we opted for the share of financial employment on total employment as a *proxy* for one of the two dimensions of financialization. Its use allows us to avoid accounting problems and to establish a link between financialization and deindustrialization as the two 'faces of the same coin,' a process of structural change. We then brought wealth composition into our discussion, through the share of financial assets on corporations' total assets. We maintain that such a wealth measure for financialization, overcomes some of the main problems with the otherwise misleading, most commonly used, Debt to GDP ratio measure.

Using cointegration techniques, we have identified a *positive* long-run relationship between the share of financial employment and income inequality, as well as *positive* relationship between the share of financial assets and income inequality. An increase of one percentage point in the flow measure of financialization increases Gini net between 0.25 and 0.5% while a one percentage point increase in inequality is associated with an increase from 1 to 1.6% of the wealth indicator. At least in what concerns the traditional Gini index, causality goes from the flow dimension to income distribution and from distribution to the stock dimension.

Our findings are summarized in Figure 4. An increase in the share of financial employment increases income inequality which in turn leads to increases in the share of financial assets on wealth, (likely) due to an expansion of financial intermediation and securitization.

In terms of policy implications, three results are worth stressing. First, as a wealth increase in the form of financial assets is at least partially the consequence of a previous increase in inequality, limiting financial intermediation and securitization requires addressing the continuous deterioration of income distribution. Second, unless the share of financial employment stabilizes, income inequality is expected to continue rising. Challenging financialization and the distributive problem might require direct interventions into labor markets. Finally, policymakers must act as soon as possible because reverting financialization may not be enough to reduce inequality.

Notes

1. The concept of financialization is particularly controversial and has been defined in many ways that often look mutually inconsistent. A more comprehensive discussion of those issues can be found in Vercelli (2013) and Karwowski, Shabani, and Stockhammer (2020). For a multisectoral approach using input-out tables, see Dávila-Fernández and Punzo

    (2020). A discussion on the determination of the rate of profit also from a multisectoral perspective can be found in Di Bucchianico (2020).

2. This does not mean we do not acknowledge that manufacture activities are also heterogeneous. The literature on complexity, for example, has stressed the relation between technology and manufacture as well as possible bridges with income inequality (e.g. Hartmann et al. 2017). Moreover, services also exhibit different patterns of growth that vary according to each country development level (Eichengreen and Gupta 2013).

3. For a preliminary discussion of these issues from a functional income distribution perspective see Dávila-Fernández et al. (2017). Recently, Piketty (2014) has suggested that the size of the gap between the rate of return on capital, r, and the economy's growth rate, g, is one of the important forces that can account for the historical magnitude and variations of wealth inequality. It is interesting to notice that if we divide Piketty's broad definition of capital between financial and non-financial assets, $r > g$ becomes $\theta i + (1 - \theta)r > g$ where i is the rate of return of financial assets, θ is the share of financial assets over total assets, and r is strictly the rate of return on capital.

4. When it comes to the finance-inequality nexus, Godechot (2016) argued that using top earning shares might be more accurate. While we agree that top incomes provide important insights, our understanding is that the most widely accepted indicator of inequality continues to be the Gini index. Using data for the top 1% income share from WID, it would be possible to perform a similar exercise to the one done in this paper. However, preliminary regressions, available under request, indicated serious serial correlation and structural break problems, invalidating the use of the time-series techniques applied here.

5. Only if an omitted variable is strongly correlated with one of the variables in the cointegration analysis we might end up with spurious estimates. For a further discussion and references on the econometric properties of the time-series approach and its advantages or disadvantages in comparison to cross-country analysis, see Gobbin and Rayp (2008).

6. As requested by the reviewer, in the appendix, we provide a detailed explanation of how the long-run parameters as well as the error correction term were calculated. Of course, a specialized reader can directly refer to Pesaran and Shin (1998) and Pesaran, Shin, and Smith (2001).

7. For example, Moller, Alderson, and Nielsen (2009) include a 'deindustrialization' variable when addressing income inequality in the United States. However, they use a within-county panel approach. On the other hand, studies like Alderson and Nielsen (2002) or Jaumotte, Lall, and Papageorgiou (2013) rely on traditional cross-country regressions.

Acknowledgements

An earlier version of this paper was presented at the Universities of Tuscany Annual Meeting, Certosa di Pontignano, Italy. We are grateful to the participants for helpful comments and suggestions. We also thank Fabio Petri, Ugo Pagano, and Fethiye Ceylan for insightful conversations and correspondence on earlier drafts of the article. Marwil Dávila gratefully acknowledges the University of Brasília, in particular, Adriana Amado and Ricardo Araújo, whose hospitality assisted his work on the completion of this paper. In addition, we thank the editor and the anonymous referee for their careful reading and indications for improvement. The usual caveats apply.

Disclosure statement

No potential conflict of interest was reported by the author(s).

References

Alderson, A., and F. Nielsen. 2002. "Globalization and the Great U-Turn: Income Inequality Trends in 16 OECD Countries." *American Journal of Sociology* 107 (5): 1244–1299. doi:10.1086/341329.

Alvaredo, F., A. Atkinson, L. Chancel, T. Piketty, E. Saez, and G. Zucman. 2016.. "Distributional National Accounts (DINA) Guidelines: Concepts and Methods Used in WID.world." WID. world Working Paper.

Assa, J. 2016. *The Financialization of GDP*. Routledge: New York.

Atkinson, A., T. Piketty, and E. Saez. 2011. "Top Incomes in the Long Run of History." *Journal of Economic Literature* 49 (1): 3–71. doi:10.1257/jel.49.1.3.

Basu, D., and D. Foley. 2013. "Dynamics of Output and Employment in the US Economy." *Cambridge Journal of Economics* 37 (5): 1077–1106. doi:10.1093/cje/bes088.

Card, D., and J. DiNardo. 2002. "Skill-Biased Technological Change and Rising Wage Inequality: Some Problems and Puzzles." *Journal of Labor Economics* 20 (4): 733–783. doi:10.1086/342055.

Conceição, P., and J. Galbraith. 2001. "Toward a New Kuznets Hypothesis: Theory and Evidence on Growth and Inequality." In *Inequality and Industrial Change: A Global View*, edited by J. Galbraith and M. Berner, 139–160. Cambridge: Cambridge University Press.

Dávila-Fernández, M., J. Oreiro, L. Punzo, and S. Bimonte. 2017. "Capital in the Twenty First Century: Re-interpreting the Fundamental Contradiction of Capitalism." *Journal of Post Keynesian Economics* 40 (2): 168–182. doi:10.1080/01603477.2016.1248978.

Dávila-Fernández, M., and L. Punzo. 2020. "Financialisation as Structural Change: Measuring the Financial Content of 'Things'." *Economic Systems Research* 32 (1): 98–120. doi:10.1080/09535314.2019.1643294.

Dew-Becker, I., and R. Gordon. 2005. "Where Did the Productivity Growth Go? Inflation Dynamics and the Distribution of Income." *Brookings Papers on Economic Activity* 2005 (2): 67–127. doi:10.1353/eca.2006.0004.

Di Bucchianico, S. 2020. "The Impact of Financialization on the Rate of Profit." *Review of Political Economy*. Forthcomingdoi:10.1080/09538259.2020.1835109.

Dunhaupt, P. 2014. "An Empirical Assessment of the Contribution of Financialization and Corporate Governance to the Rise of Income Inequality." Working Paper, Institute for International Political Economy (IPE), 41.

Dunhaupt, P. 2017. "Determinants of Labour's Income Share in the Era of Financialization." *Cambridge Journal of Economics* 41 (1): 283–306. doi:10.1093/cje/bew023.

Eichengreen, B., and P. Gupta. 2013. "The Two Waves of Service-Sector Growth." *Oxford Economic Papers* 65 (1): 96–123. doi:10.1093/oep/gpr059.

Epstein, G. 2005. "Introduction: Financialization and the World Economy." In *Financialization of the World Economy*, edited by G. Epstein, 3–16. Cheltenham: Edward Elgar.

Fligstein, N., O. Hastings, and A. Goldstein. 2017. "Keeping up with the Joneses: How Households Fared in the Era of High Income Inequality and the Housing Price Bubble, 1999–2007." *Socius: Sociological Research for a Dynamic World* 3: 2378023117722330. doi:10.1177/2378023117722330.

Galbraith, J. 2012. *Inequality and Instability: A Study of the World Economy Just before the Crisis*. New York: Oxford University press.

Galbraith, J., and T. Hale. 2014. "The Evolution of Economic Inequality in the United States, 1969–2012: Evidence from Data on Inter-industrial Earnings and Inter-regional Incomes." *World Economic Review* 3: 1–19.

Gobbin, N., and G. Rayp. 2008. "Different Ways of Looking at Old Issues: A Time-Series Approach to Inequality and Growth." *Applied Economics* 40 (7): 885–895. doi:10.1080/00036840600771106.

Godechot, O. 2016. "Financialization Is Marketization! A Study of the Respective Impacts of Various Dimensions of Financialization on the Increase in Global Inequality." *Sociological Science* 3: 495–519. doi:10.15195/v3.a22.

Godechot, O. 2020. "Financialization and the Increase in Inequality." In *The Routledge International Handbook of Financialization*, edited by P. Mader, D. Mertens, and N. van der Zwan, 413–424. London: Routledge.

Gordon, R., and I. Dew-Becker. 2007. "Selected Issues in the Rise of Income Inequality." *Brooking Papers on Economic Activity* 2: 169–190.

Granger, C. 1969. "Investigating Causal Relations by Econometric Models and Cross- Spectral Methods." *Econometrica* 37 (3): 424–438. doi:10.2307/1912791.

Hartmann, D., M. Guevara, C. Jara-Figueroa, M. Aristarán, and C. Hidalgo. 2017. "Linking Economic Complexity, Institutions, and Income Inequality." *World Development* 93: 75–93. doi:10.1016/j.worlddev.2016.12.020.

Hein, E., P. Dunhaupt, A. Alfageme, and M. Kulesza. 2017. "Financialization and Distribution from a Kaleckian Perspective: The United States, the United Kingdom, and Sweden Compared before and after the Crisis." *International Journal of Political Economy* 46 (4): 233–266. doi:10.1080/08911916.2017.1407735.

Jaumotte, F., S. Lall, and C. Papageorgiou. 2013. "Rising Income Inequality: Technology, or Trade and Financial Globalization?" *IMF Economic Review* 61 (2): 271–309. doi:10.1057/imfer.2013.7.

Karabarbounis, L., and B. Neiman. 2014. "The Global Decline of the Labor Share." *Quarterly Journal of Economics* 129 (1): 61–103. doi:10.1093/qje/qjt032.

Karwowski, E., M. Shabani, and E. Stockhammer. 2020. "Dimensions and Determinants of Financialisation: Comparing OECD Countries since 1997. A Cross-country Study." *New Political Economy* 25 (6): 957–977. doi:10.1080/13563467.2019.1664446.

Krippner, G. 2005. "The Financialization of the American Economy." *Socio-Economic Review* 3 (2): 173–208. doi:10.1093/SER/mwi008.

Kumhof, M., R. Rancière, and P. Winant. 2015. "Inequality, Leverage, and Crises." *American Economic Review* 105 (3): 1217–1245. doi:10.1257/aer.20110683.

Lin, K., and D. Tomaskovic-Devey. 2013. "Financialization and U.S. Income Inequality, 1970–2008." *American Journal of Sociology* 118 (5): 1284–1329. doi:10.1086/669499.

Mader, P., D. Mertens, and N. van der Zwan. 2020. *The Routledge International Handbook of Financialization.* London: Routledge.

Mazzucato, M., and A. Shipman. 2014. "Accounting for Productive Investment and Value Creation." *Industrial and Corporate Change* 23 (4): 1059–1085. doi:10.1093/icc/dtt037.

Moller, S., A. Alderson, and F. Nielsen. 2009. "Changing Patterns of Income Inequality in U.S. Counties, 1970–2000." *American Journal of Sociology* 114 (4): 1037–1101. doi:10.1086/595943.

Nakamura, L. 2010. "Intangible Assets and National Income Accounting." *Review of Income and Wealth* 56 (1): 135–155. doi:10.1111/j.1475-4991.2010.00390.x.

Narayan, P. 2005. "The Saving and Investment Nexus for China: Evidence from Cointegration Tests." *Applied Economics* 37 (17): 1979–1990. doi:10.1080/00036840500278103.

Narayan, P., and R. Smyth. 2006. "Higher Education, Real Income and Real Investment in China: Evidence from Granger Causality Tests." *Education Economics* 14 (1): 107–125. doi:10.1080/09645290500481931.

Odhiambo, N. 2009. "Energy Consumption and Economic Growth Nexus in Tanzania: An ARDL Bounds Testing Approach." *Energy Policy* 37 (2): 617–622. doi:10.1016/j.enpol.2008.09.077.

Pagano, U. 2014. "The Crisis of Intellectual Monopoly Capitalism." *Cambridge Journal of Economics* 38 (6): 1409–1429. doi:10.1093/cje/beu025.

Pedroni, P. 2019. "Panel Cointegration Techniques and Open Challenges." In *Panel Data Econometrics: Theory,* edited by M. Tsionas, 251–288. Amsterdam: Elsevier.

Perron, P. 1989. "The Great Crash, the Oil Price Shock, and the Unit Root Hypothesis." *Econometrica* 57 (6): 1361–1401. doi:10.2307/1913712.

Pesaran, M., and Y. Shin. 1998. "An Autoregressive Distributed-lag Modelling Approach to Cointegration Analysis." *Econometric Society Monographs* 31: 371–413.

Pesaran, M., Y. Shin, and R. Smith. 2001. "Bounds Testing Approaches to the Analysis of Level Relationships." *Journal of Applied Econometrics* 16 (3): 289–326. doi:10.1002/jae.616.

Piketty, T. 2014. *Capital in the Twenty-First Century.* Cambridge Massachusetts: Harvard University Press.

Piketty, T., and E. Saez. 2013. "Top Incomes and the Great Recession: Recent Evolutions and Policy Implications." *IMF Economic Review* 61 (3): 456–478. doi:10.1057/imfer.2013.14.

Saez, E., and G. Zucman. 2016. "Wealth Inequality in the United States since 1913: Evidence from Capitalized Income Tax Data." *Quarterly Journal of Economics* 131 (2): 519–579. doi:10.1093/qje/qjw004.

Shin, Y., B. Yu, and M. Greenwood-Nimmo. 2014. "Modelling Asymmetric Cointegration and Dynamic Multipliers in a Nonlinear ARDL Framework." In *Festschrift in Honor of Peter Schmidt,* edited by R. Sickles and W. Horrace, 281–314. New York: Springer.

Sjoberg, O. 2009. "Corporate Governance and Earnings Inequality in the OECD Countries 1979–2000." *European Sociological Review* 25 (5): 519–533.

Stiglitz, J. 2013. *The Price of Inequality: How Today's Divided Society Endangers Our Future.* New York: W.W. Norton & Company.

van der Zwan, N. 2014. "Making Sense of Financialization." *Socio-Economic Review* 12 (1): 99–129. doi:10.1093/ser/mwt020.

Vercelli, A. 2013. "Financialization in a Long-Run Perspective." *International Journal of Political Economy* 42 (4): 19–46. doi:10.2753/IJP0891-1916420402.

Volscho, T., and N. Kelly. 2012. "The Rise of the Super-Rich: Power Resources, Taxes, Financial Markets, and the Dynamics of the Top 1 Percent, 1949 to 2008." *American Sociological Review* 77 (5): 679–699. doi:10.1177/0003122412458508.

Wade, R. 2014. "The Piketty Phenomenon and the Future of Inequality." *Real World Economics Review* 69: 2–17.

Western, B., and J. Rosenfeld. 2011. "Unions, Norms, and the Rise in U.S. Wage Inequality." *American Sociological Review* 76 (4): 513–537. doi:10.1177/0003122411414817.

Wolff, E. 2017. "Household Wealth Trends in the United States, 1962–2013: What Happened over the Great Recession?" *RSF Journal of the Social Sciences* 2 (6): 24–43.

Yellen, J. 2006. "Economic Inequality in the United States." Speech to the Center for the Study of Democracy, Economics of Governance Lecture. Irvine: University of California.

Determinants of social outreach of microfinance institutions

Shakil Quayes and George Joseph

ABSTRACT

The paper analyzes the determinants of social outreach of microfinance institutions (MFIs), using three measures of outreach – depth of outreach, breadth of outreach, and outreach to women, and its possible complementarity with financial performance. We use an unbalanced panel of 1,219 MFIs over a period of 20 years to investigate the effect of firm-specific characteristics and the impact of prevailing legal system on social outreach of MFIs. Rejecting the notion of tradeoff between financial performance and social outreach, our empirical results show that better financial performance has a positive association with social outreach. Furthermore, we observe that a common law legal system is more conducive in facilitating social outreach, and MFIs operating under common law legal system achieve better depth of outreach, breadth of outreach, and outreach to women, than MFIs under code law legal system and mixed law legal system. While we had expected nonprofit MFIs to exhibit better social outreach than for profit MFIs, we found empirical evidence of such only in the case of outreach to women. Finally, we find empirical evidence that unregulated MFIs achieve better social outreach than regulated MFIs.

1. Introduction

One of the many problems faced by the large number of households trapped in the vicious cycle of poverty is their limited access to credit. Over the last few decades, microfinance has emerged as an option of accessing credit, allowing poor households to pursue entrepreneurship and consumption smoothing. The primary mission of the microfinance industry is to extend credit to the poor who have no access to credit from formal financial institutions, and an important yardstick of measuring the success of microfinance institutions (MFIs) is their level of social outreach. While MFIs have expanded their scale and scope of credit operation, they have also had to navigate the path to financial sustainability without curtailing their primary goal of social outreach. Quayes (2020) shows empirical evidence of a general increase in the depth of outreach over the last two decades, and no reduction in the outreach to women borrowers.

We can argue that an increased emphasis on financial sustainability of MFIs, that are current or previous beneficiaries of subsidies from donors, could possibly reduce the extent of their social outreach. In other words, there could be a tradeoff between outreach of MFIs and their financial performance. On the other hand, it has been claimed that the lending models utilized by the microfinance industry for extending credit to poor borrowers, has resulted in consistently high repayment rate and low administrative and monitoring costs. Hence, greater outreach may actually complement financial performance instead of adversely affecting it. While some early studies focusing on empirical analyses with small data sets showed evidence of such a tradeoff, recent studies utilizing larger samples of MFIs have failed to show evidence of such a tradeoff or have demonstrated a positive association between financial performance and social outreach.

This study also analyzes the possible impact of prevailing legal system on the social outreach of MFIs. Legal systems can facilitate the functioning of regulatory environment through a long tradition of institutions. Following widely accepted norms based on historic perspective of legal traditions, we categorize the legal systems into code law system, common law system, and mixed law system. Under the code law system, which primarily emerges from early Roman and French Napoleonic systems, governments typically exercise political influence on regulatory systems. In the case of accounting standards and governance, code law countries prescribe regulations and the government enforces the law by imposing civil and criminal penalties on violators. In contrast, common law system with its English origin evolves in the private sector over time through legal precedents, and accepted practices eventually assume the status of the law.

The rest of the paper is structured as follows. Section 2 provides a brief review of the relevant literature; Section 3 describes the data and variables; Section 4 details the model and the empirical results; and Section 5 provides summary, conclusions, and policy implications.

2. Literature review

Sharma and Zeller (1997) and Zeller (1998) find a negative correlation between repayment and poverty level because poor borrowers are more susceptible to economic shocks. This may result in a tradeoff between depth of outreach and sustainability. Morduch (1999, 2000) was skeptical that financial sustainability would ensure better outreach, since he discounted the likelihood of a positive impact of high recovery rate on financial sustainability of MFIs. Navajas et al. (2000) and Cull, Demirguc-Kunt, and Morduch (2007) claimed that MFIs maintained sustainability by allocating loans to households near the poverty line without extending credit to the poorest households. Navajas et al. (2000), Schreiner (2002), Rhyne (1998) and Von Pischke (1998) based their argument of a tradeoff between financial sustainability and depth of outreach on higher transaction costs of smaller loans, while Olivares-Polanco (2005) demonstrated empirical evidence of such a tradeoff. According to Manos and Yaron (2009), tradeoff may exist in the short-run, but both outreach and financial sustainability may improve over time due to economies of scale and innovations. Paxton (2003) and Hartarska and Nadolnyak (2007) provide arguments against the notion of a tradeoff, and Gonzalez and Rosenberg (2006) fail to find any evidence of a tradeoff. Finally, Quayes (2012, 2015) found empirical evidence of a positive association between financial performance and outreach.

Kar (2012) showed a negative effect of equity on depth of outreach, but no statistically significant effect on breadth of outreach or on outreach to women. Barry and Tacneng

(2014) found that nonprofit non-governmental organizations have better social outreach than for profit MFIs. Abdullah and Quayes (2016) report evidence of a positive association between financial performance and outreach to women. Hartarska (2005) emphasizes the importance of governance on the performance of MFIs, but Hartarska and Nadolnyak (2007) did not find that regulation of MFIs leads to better social outreach. Finally, Staschen and Nelson (2013) argued that the regulatory framework helps in removing demand and supply side barriers.

We also address the impact of legal system on the social outreach of MFIs by taking into account the legal institutional environments under which they operate (La Porta et al. 1997a). MFIs portray institutional rationality (Friedland and Alford, 1991) as a blend of development mission that seeks to uplift the poor, and a banking framework within which it maintains its social and fiduciary obligations (Battilana and Dorado 2010). They operate with divergent goals that compete for priority; financial performance is associated with validity in financial markets and social outreach is associated with its network between individuals and organizations (Almandoz 2012). As to the role of women in the legal system, Chakrabarty and Bass (2014) argue that the absence of institutional support for women entrepreneurs may perpetuate the cycle of poverty. While analyzing the factors that influence the level of financial disclosure and amount of social outreach information online, Gutiérrez-Nieto, Fuertes-Callén, and Serrano-Cinca (2008) stress the importance of regional transparency requirements, which depended on the prevailing legal system. They report that for profit MFIs disclose more financial information whereas nonprofit MFIs disclose better social information.

According to La Porta et al. (1997b) there is a linkage between institutional factors and the legal system prevailing in the country, which also affect the organizational structure and governance mechanism of institutions. Quayes and Joseph (2017) show that MFIs operating in code law systems exhibit better financial performance than MFIs operating in common law legal systems. The current study investigates the influence of legal institutions on the social outreach of MFIs. La Porta, Lopez-de-silanes, and Shleifer (2008) observe that sources of variation in the market environment may reside in the legal systems.

3. Description of data and variables

The study utilizes a panel of 1,219 MFIs over 20 years from 1999 to 2018, collected from MIX Market data, freely available from the World Bank Data Catalogue. We collected two variables, an MFI's non-profit status, and its being subject to regulation, from the MIX website platform through its subscription service. The MIX Market contains an extensive database of financial information for a large number of MFIs across the world. The database converts variables expressed in local currency into US dollars.

The focus of this study is to analyze the impact of firm-specific characteristics of MFIs, and the prevailing legal system of the country, on the social outreach of MFIs. To this end, we use three measures of social outreach – depth of outreach, breadth of outreach, and outreach to women. To our knowledge, the formal definitions of depth and breadth of outreach can be attributed to Navajas et al. (2000), reiterated by Schreiner (2002). Depth of outreach is a measure of dispersion of credit access with respect to the level of poverty; the poorer the borrower, the greater is the depth of outreach. Breadth of outreach measures the extension of credit; larger number of borrowers implies greater breadth of outreach. While poor households in developing countries have limited access to credit, the extent of access to credit for women is even worse. The majority of borrowers in the microfinance industry are women, and it helps to mitigate the dearth of access to credit for women. We define the proportion of female borrowers in an MFI's portfolio as a measure of outreach to women.

Average loan balance per borrower (ALB) represents the size of loans allocated by an MFI; it is the size of the loan when originated. A smaller ALB implies disbursement of smaller loans indicating better depth of outreach. ALB is divided by the per capita gross national income of the respective country to normalize the variation in income across countries. Following prior studies, we use average loan balance per borrower adjusted for GNI (ALBG) to measure the depth of outreach. We measure breadth of outreach by the number of active borrowers (NB) for the MFI. Finally, the proportion of borrowers that are women (WBR) represents the outreach to women by an MFI.

Following Quayes and Joseph (2017), we define three categories of legal systems: (i) legal system based on common law, (ii) legal system based on code law, and (iii) legal system based on a mixture of common law and code law. We use two dummy variables for our estimation model, (i) Code law (CDL) = 1 if legal system is based on code law and 0 otherwise, and (ii) Mixed Law (MXL) = 1 if legal system is a mix of common law and code law and 0 otherwise. Common law (CML) legal system is the baseline in our empirical model. We hypothesize that MFIs operating under a common law legal system have a better social outreach than MFIs operating under a code law legal system or those under a mixed law legal system.

Size of the firm, measured by gross loan portfolio (LP), will have a negative effect on depth of outreach if an increase in portfolio size is due to diversification of portfolio by expanding into larger loans but it will not have any effect if it is simply due to geographic expansion. The same argument holds for the effect of size on outreach to women. On the other hand, an increase in size will have a positive association with breadth of outreach.

Intuitively, a larger level of equity (EQ) should have a positive impact on social outreach, since greater equity may require MFIs to commit to specific outreach targets. Therefore, a larger level of total equity should have a positive impact on depth of outreach, breadth of outreach and outreach to women.

Smaller loans are usually associated with a larger per dollar loan expense. An increase in the total expense ratio (ER) should have a favorable effect on depth of outreach. In other words, smaller loan sizes would be associated with higher expense ratio. An increase in higher expense ratio should be associated with greater breadth of outreach. The effect of expense ratio on outreach to women is ambiguous and we do not have any *a priori* assumption about its association with outreach to women.

Loan loss reserve ratio (LR), defined as loan loss reserve as a fraction of loan portfolio, is an indicator of anticipated loss from defaulting loans, and an increase in LR may result in MFIs curtailing their overall social outreach.

There has been a general trend of for all MFIs to attain financial self-sufficiency. It has allowed them to wean themselves off donor subsidies and chart a path that can ensure a steady source of funds. Financial performance has developed into one of the essential goals even for MFIs that have operated as nonprofit organizations. We use rate of return on assets (ROA) defined as net operating income (less taxes) divided by average level of assets as our measure of financial performance. Following Quayes (2012, 2015), Quayes and Hasan (2014), we use two additional measures of financial performance to check for robustness of our results. These two measures are profit margin rate (PMR) defined as net operating income divided by total financial revenue, and operational self-sufficiency (OSS) defined by total financial revenue divided by the sum of financial expense, loan loss provision expense, and operating expense. We expect to show that financial performance has a positive association with depth of outreach, breadth of outreach, and outreach to women.

In an attempt to compare the level of disclosure transparency between nonprofit and for profit entities who provide public goods, Goodell, Goyal, and Hasan (2020) observed that MFIs are an excellent example, since there are a large number of both nonprofit and for profit MFIs. Their analysis revealed that for profit MFIs exhibit better financial transparency than nonprofit MFIs. More than half of all the MFIs in our sample are nonprofit organizations; we use the dummy variable nonprofit (NP) which is equal to 1 if the MFI is a nonprofit organization and zero otherwise. The variable will allow us to identify whether nonprofit MFIs achieve better social outreach than for profit MFIs. We expect nonprofit MFIs to achieve better social outreach in terms of depth of outreach, breadth of outreach, and outreach to women, than for profit MFIs.

Fifty eight percent of all the MFIs in our sample fall under the purview of some regulatory authority in their respective countries. Adherence to regulation is costly and it may constrain the flexibility of MFIs' outreach efforts. We define a dummy variable regulation (RG) as equal to 1 if MFI is regulated and zero otherwise. We expect unregulated MFIs to achieve better depth of outreach, better breadth of outreach, and better outreach to women.

A careful reading of the descriptive statistics reported in Table 1 indicates that MFIs operating in common law legal system achieve better social outreach than MFIs operating in either code law legal system or mixed law legal system, in terms of depth of outreach, breadth of outreach, and outreach to women. The mean of average loan balance per borrower divided by per capita GNI (ALB) of MFIs operating in common law legal system is 0.271 in comparison to 0.696 for MFIs in code law system and 0.784 for MFIs in mixed law system. Mean ALB for MFIs in common law system is lower than the mean ALB for MFIs in both code law system and mixed law system, and the differences are statistically significant at one percent level. The mean of ALB for MFIs from mixed law system is greater than the mean of ALB for MFIs from code law system, and this difference is statistically significant at one percent level.

The average of the number of borrowers (NAB) for MFIs operating in common law system is 245,572 while the average NAB is 52,320 for MFIs operating in code system and 47,281 for MFIs in mixed law system. The mean NAB for MFIs operating in common law system is greater than the mean of NAB of MFIs operating in both code law system and mixed law system, and the differences are statistically significant at the one percent level. While the mean of NAB for MFIs under code law system is greater than the mean of NAB under mixed law system but this difference is not statistically significant.

The proportion of female borrowers (WBR) is an important measure of outreach. The average WBR for MFIs operating in common law system is 87.58% while the average WBR for MFIs operating under code law system is 59.21% and the average WBR for MFIs in mixed law countries is 61.79%. Clearly, the average WBR for MFIs operating under common law system is greater than the mean WBR for MFIs operating in code law system and mixed law system, and the mean of WBR for MFIs operating in mixed law

Table 1. Descriptive statistics.

Variable	All MFIs	Common[a]	Code[b]	Mixed[c]
Average Loan Balance (ALB)	0.6264	0.2713	0.6960	0.7842
		(0.0000)	(0.0000)	(0.0000)
Number of Borrowers (NAB)	99.9168	345.5718	52.3197	47.2812
		(0.0000)	(0.4127)	(0.0000)
Women Borrowers (WBR)	0.6521	0.8758	0.5921	0.6179
		(0.0000)	(0.0000)	(0.0000)
Gross Loan Portfolio (LP)	53.9936	60.7106	52.1953	52.9635
		(0.3104)	(0.8902)	(0.1319)
Total Equity (EQ)	13.6487	17.4035	13.0587	12.7310
		(0.0465)	(0.7821)	(0.0016)
Total Expense Ratio (ER)	0.2540	0.1988	0.2960	0.2403
		(0.0000)	(0.0000)	(0.0000)
Loan Loss Reserves (LR)	0.0259	0.0086	0.0254	0.0328
		(0.0004)	(0.0130)	(0.1969)
Nonprofit Status (NP)	0.5307	0.5560	0.4399	0.5960
		(0.0000)	(0.0000)	(0.0074)
Regulated (RG)	0.5818	0.7720	0.5183	0.5631
		(0.0000)	(0.0001)	(0.0000)
Return on Assets (ROA)	0.0191	0.0127	0.0237	0.0178
		(0.0028)	(0.0195)	(0.0445)
Profit Margin Rate (PMR)	0.0601	0.0307	0.0806	0.0542
		(0.0221)	(0.1504)	(0.3562)
Operational Self Sufficiency (OSS)	1.1876	1.1677	1.1997	1.1850
		(0.0124)	(0.0140	(0.3615)
Sample Size	8,817	1,500	3,305	4,012

NAB reported in thousands.
LP and EQ reported in millions of USD.
[a]p-value in parentheses for testing$H_0 : \mu_{Common} = \mu_{Code}$; $H_a : \mu_{Common} \neq \mu_{Code}$
[b]p-value in parentheses for testing$H_0 : \mu_{Code} = \mu_{Mixed}$;$H_a : \mu_{Code} \neq \mu_{Mixed}$
[c]p-value in parentheses for testing$H_0 : \mu_{Mixed} = \mu_{Common}$;$H_a : \mu_{Mixed} \neq \mu_{Common}$

system is greater than the average WBR of MFIs operating in code law system. All three differences are statistically significant at the 1% level.

A majority of MFIs operating in code law countries are for profit whereas the majority of MFIs operating in common law or mixed law countries are nonprofit organizations. More than 75% of MFIs operating in the common law system are regulated, whereas about half of the MFIs in code law system, and 56% of the MFIs in mixed law system are regulated. Finally, MFIs operating in code law countries exhibit a better return on assets than MFIs operating in common law countries and mixed law countries, and MFIs operating in mixed law countries show a better ROA than MFIs in operating in common law countries. All these differences are statistically significant at the five percent level or less. Finally, the MFIs operating in code law systems exhibit better financial performance (ROA) than MFIs operating in common law system or mixed law systems, and MFIs operating in mixed law system show better financial performance than MFIs operating in common law system.

4. Model and empirical results

The focus of this study is to ascertain the factors that determine the social outreach of MFIs. Equation (1) models the social outreach of MFIs as a function of the firm-specific characteristics of an MFI and the country's legal system:

$$SO_{it} = \alpha_i + \beta'M_{it} + \lambda'LS_i + \varepsilon_{it} \qquad (1)$$

where, SO measures social outreach, M is a vector of MFI-specific variables and LS represents the prevailing legal system in the country. To delineate the impact of firm characteristics and the legal system on the social outreach of MFIs based on Equation (1), we estimate the following empirical panel model described by Equation (2).

$$SO_{it} = \alpha_i + \beta_1 LP_{it} + \beta_2 EQ_{it} + \beta_3 ER_{it} + \beta_4 LR_{it} + \beta_5 NP_{it}$$
$$+ \beta_6 RG_{it} + \beta_7 FP_{it} + \lambda_1 CDL_i + \lambda_2 MXL_i + \varepsilon_{it} \qquad (2)$$

where SO = Social Performance measured by (i) average loan balance per borrower divided by per capital GNI; (ii) number of borrowers; and (iii) proportion of female borrowers.

LP = Log of gross loan portfolio
EQ = Total equity
ER = Total expense ratio
LR = Loan loss reserve ratio
NP = 1 if MFI is not for profit and 0 otherwise
RG = 1 if MFI is regulated and 0 otherwise
FP = Financial performance measured by rate of return on assets (ROA)
CDL = 1 if legal system is based on code law and 0 otherwise
MXL = 1 if legal system is a mix of common law and code law and 0 otherwise

We have an unbalanced panel of 1,219 MFIs over a period of 20 years. Nonprofit status, regulation, and legal system are all time-invariant dummy variables. As such, a fixed effect panel model would not be able to capture the effect of these variables. We estimated a random effects panel model, and we estimated a Hausman–Taylor panel model in order to account for possible endogeneity between outreach and financial performance. Table 2 reports the estimates of the random effects panel regression model described in Equation (2), and we present the corresponding estimates from the Hausman–Taylor panel regression model in Table 3. In each of the two tables, the three columns represent the three dependent variables – depth of outreach, breadth of outreach, and outreach to women, respectively.

The results in Table 2 show that higher rate of return on assets is positively associated with greater breadth of outreach and an increase in the fraction of women borrowers. We can also see that an increase in ROA results in a decrease in average loan balance per borrower, which implies better depth of outreach. The estimated coefficients are all statistically significant at the 1% level. We see similar results (not reported) when we use profit margin rate or operational self-sufficiency as a measure of financial performance of MFIs. This shows clear evidence of a positive association between financial performance and all the three measures of social outreach.

To address possible endogeneity between social outreach and financial sustainability, we estimate a Hausman–Taylor model, and the estimated coefficients from this model reported in Table 3 show a positive association between each of the three measures of social outreach of an MFI and its financial performance. The results are significant at the 1% level.

Table 2. Social outreach: random effects panel regression.

Variable	Depth	Breadth	Women
Gross Loan Portfolio (LP)	0.1088*	0.6586*	−0.0127*
	13.62	88.73	−6.27
Total Equity (EQ)	−0.0422*	0.0094	0.0044**
	−5.04	1.20	2.06
Total Expense Ratio (ER)	−1.1112*	0.9525*	0.1524*
	−17.16	15.80	9.36
Loan Loss Reserve Ratio (LR)	0.0327*	−0.0236*	0.0017*
	3.60	−2.79	0.75
Nonprofit Status (NP)	−0.0581	−0.0448	0.0366*
	−0.93	−0.73	2.71
Regulated (RG)	0.7788*	−0.2002*	−0.1003*
	12.15	−3.20	−7.25
Code Law (CDL)	0.7213*	−2.0008*	−0.3426*
	8.39	−23.84	−18.48
Mixed Law (MXL)	0.8700*	−1.4117*	−0.3049*
	10.71	−17.79	−17.41
Return on Assets (ROA)	−0.4876*	0.7369*	0.1736*
	−6.90	11.22	9.70
Constant	−3.1355*	0.0713	1.0417*
	−25.03	0.59	36.19
Number of MFIs	1,219	1,219	1,219
Number of Observations	8,817	8,817	8,817
Overall R^2	0.2680	0.7298	0.2415
χ^2	1,082.91	28,192.07	639.89

t-statistics are reported within parentheses.
*,**: Indicates statistical significance at the 0.01 level, and 0.05 level.

Table 3. Social outreach: Hausman–Taylor panel regression.

Variable	Depth	Breadth	Women
Gross Loan Portfolio (LP)	0.1084*	0.6538*	−0.0130*
	13.85	90.23	−6.54
Total Equity (EQ)	−0.0401*	0.0051	0.0051**
	−4.88	0.67	2.43
Total Expense Ratio (ER)	−1.0173*	0.8600*	0.1340*
	−15.86	14.48	8.23
Loan Loss Reserve Ratio (LR)	0.0324*	−0.0237*	0.0018
	3.68	−2.91	0.82
Nonprofit Status (NP)	−0.0371	−0.0906	0.0358***
	−0.43	−1.05	1.89
Regulated (RG)	0.7736*	−0.2150*	−0.0944*
	8.67	−2.43	−4.86
Code Law (CDL)	0.7548*	−2.0805*	−0.3280*
	6.32	−17.60	−12.64
Mixed Law (MXL)	0.9435*	−1.5898*	−0.2946*
	8.38	−14.25	−12.03
Return on Assets (ROA)	−0.4233*	0.7031*	0.1598*
	(−6.05)	(10.88)	(8.95)
Constant	−3.2386*	0.3710**	1.0281*
	−20.10	2.36	28.05
Number of MFIs	1,219	1,219	1,219
Number of Observations	8,817	8,817	8,817
χ^2	906.70	28,082.04	382.78

t-statistics are reported within parentheses.
*,**,***: Indicates statistical significance at the 0.01 level, 0.05 level, and 0.10 level.

In both panel models, the estimated coefficients for the dummy variables CDL and MXL (reported in column one) are positive and statistically significant at the 1% level, indicating that MFIs operating in common law system have better depth of outreach than

MFIs operating in both code law and mixed law systems. The estimated coefficients for MXL are larger than that of CDL in both panel models, and the difference between the two coefficients are statistically significant at five percent level. Our empirical results show that unregulated MFIs achieve better depth of outreach than regulated MFIs, but the estimated coefficient for nonprofit is not statistically significant.

The second columns of Tables 2 and 3 report the estimated panel regressions for breadth of outreach. The estimated coefficients for dummy variables CDL and MXL and the difference between the estimated coefficients of CDL and MXL are significant at the 1% level for both the panel models. This indicates that MFIs operating in common law system have better breadth of outreach than MFIs operating in either code law system or mixed law system, and MFIs operating in mixed law system have better breadth of outreach than MFIs in code law system.

We present the estimated coefficients of outreach to women in the third columns of Tables 2 and 3. In the case of outreach to women, the estimated coefficients for CDL and MXL are statistically significant for both panel models, indicating that MFIs operating in common law system exhibit better outreach to women than the MFIs operating in code law system and mixed law system. MFIs operating in mixed law system have better outreach to women than MFIs operating in code law system; the difference between the estimated coefficients of CDL and MXL are statistically significant, at the 1% level for the random effects panel model and at the 10% level for the Hausman-Taylor panel model.

Nonprofit MFIs have better outreach to women than for profit MFIs; the estimated coefficients are statistically significant at the 1% level for the random effects model and at the 10% level for the Hausman–Taylor model. Finally, unregulated MFIs provide better outreach for all three measures of outreach than regulated MFIs; the estimated coefficients are statistically significant at 1% for both panel models.

The empirical results consistently demonstrate that MFIs operating under common law system exhibit better social outreach than MFIs operating under code law system or MFIs operating under mixed law system. The source of the better social outreach outcome in common law system possibly stems from better protection accorded to shareholders. This protection allows donors to exert better control over the governing board in setting out their mission goals and more effectively dictate MFI management in achieving better social outreach. On the other hand, the institutional environment under code law legal system achieves better financial performance (Quayes and Joseph 2017). Note that MFIs in common law legal system achieve better social outreach despite the fact that more than 75% of the MFIs are under the purview of regulation in this system.

5. Conclusion

This study attempted to identify the factors that have an impact on the social outreach of MFIs. In doing so, we looked at three different measures of outreach – depth of outreach, breadth of outreach, and outreach to women. We accounted for factors such as financial performance of an MFI, its nonprofit status, regulatory environment and the prevailing legal system in the country. We found strong empirical evidence of a positive association between all three measures of outreach and financial performance. Next, we demonstrate that MFIs in the common law legal system have better depth of outreach, greater breadth of outreach, and better outreach to women, than MFIs operating in code law legal system and mixed law system. We observe this phenomenon consistently from our results in the random effects panel models and the Hausman–Taylor panel models. We believe that our results are robust to the different estimation models and choice of measurement for outreach. While we had expected nonprofit MFIs to achieve better social outreach than for profit MFIs, we found empirical evidence of such only in the case of outreach to women. The empirical evidence that unregulated MFIs achieve better social outreach than regulated MFIs has a somewhat discouraging policy implication and should be a topic of more extensive future research.

Disclosure statement

No potential conflict of interest was reported by the author(s).

References

Abdullah, S., and S. Quayes. 2016. "Do Women Borrowers Augment Financial Performance of MFIs?" *Applied Economics* 48 (57): 5593–5604. doi:10.1080/00036846.2016.1181831.

Almandoz, J. 2012. "Arriving at the Starting Line: The Impact of Community and Financial Logics on New Banking Ventures." *Academy of Management Journal* 55 (6): 1381–1406. doi:10.5465/amj.2011.0361.

Barry, T. A., and R. Tacneng. 2014. "The Impact of Governance and Institutional Quality on MFI Outreach and Financial Performance in Sub-Saharan Africa." *World Development* 58: 1–20. doi:10.1016/j.worlddev.2013.12.006.

Battilana, J., and S. Dorado. 2010. "Building Sustainable Hybrid Organizations: The Case of Commercial Microfinance Organizations." *Academy of Management Journal* 53 (6): 1419–1440. doi:10.5465/amj.2010.57318391.

Chakrabarty, S., and A. Erin Bass. 2014. "Institutionalizing Ethics in Institutional Voids: Building Positive Ethical Strength to Serve Women Microfinance Borrowers in Negative Contexts." *Journal of Business Ethics* 119 (4): 529–542. doi:10.1007/s10551-013-1833-9.

Cull, R., A. Demirguc-Kunt,and J. Morduch. 2007. "Financial Performance and Outreach: A Global Analysis of Leading Microbanks." *Economic Journal* 117: F107–F133.

Friedland, R., and R. Robert Alford. 1991. "Bringing Society Back In: Symbols, Practices, and Institutional Contradictions." In *The New Institutionalism in Organizational Analysis*, edited by W. Powell and P. DiMaggio, 232–266. Chicago: University of Chicago Press.

Gonzalez, A., and R. Rosenberg. 2006. "The State of Microfinance: Outreach, Profitability and Poverty." Paper presented at the World Bank Conference on Access to Finance, Washington, DC.

Goodell, J. W., A. Goyal, and I. Hasan. 2020. "Comparing Financial Transparency between For-profit and Nonprofit Suppliers of Public Goods: Evidence from Microfinance." *Journal of International Financial Markets, Institutions and Money* 64: 101146. doi:10.1016/j.intfin.2019.101146.

Gutiérrez-Nieto, B., Y. Fuertes-Callén, and C. Serrano-Cinca. 2008. "Internet Reporting in Microfinance Institutions." *Online Information Review* 32 (3): 415–436. doi:10.1108/14684520810889709.

Hartarska, V. 2005. "Governance and Performance of Microfinance Institutions in Central and Eastern Europe and the Newly Independent States." *World Development* 33 (10): 1627–1643. doi:10.1016/j.worlddev.2005.06.001.

Hartarska, V., and D. Nadolnyak. 2007. "Do Regulated Microfinance Institutions Achieve Better Sustainability and Outreach? Cross-country Evidence." *Applied Economics* 39 (10): 1207–1222. doi:10.1080/00036840500461840.

Kar, A. K. 2012. "Does Capital and Financing Structure Have Any Relevance to the Performance of Microfinance Institutions?" *International Review of Applied Economics* 26 (3): 329–348. doi:10.1080/02692171.2011.580267.

La Porta, R., F. Lopez-de-silanes, A. Shleifer, and R. W. Vishny. 1997a. "Legal Determinants of External Finance." *The Journal of Finance* 52 (3): 1131–1150. doi:10.1111/j.1540-6261.1997. tb02727.x.

La Porta, R., F. Lopez-de-silanes, A. Shleifer, and R. W. Vishny. 1997b. "Trust in Large Organizations." *American Economic Review* 87 (2): 333–338.

La Porta, R. L., F. Lopez-de-silanes, and A. Shleifer. 2008. "The Economic Consequences of Legal Origins." *Journal of Economic Literature* 46 (2): 285–332. doi:10.1257/jel.46.2.285.

Manos, R., and J. Yaron. 2009. "Key Issues in Assessing the Performance of Microfinance Institutions." *Canadian Journal of Development Studies* 29: 101–122.

Morduch, J. 1999. "The Microfinance Promise." *Journal of Economic Literature* 37: 1569–1614.

Morduch, J. 2000. "The Microfinance Schism." *World Development* 28: 617–629.

Navajas, S., M. Schreiner, R. L. Meyer, C. Gonzalez-Vega, and J. Rodriguez-Meza. 2000. "Microcredit and the Poorest of the Poor: Theory and Evidence from Bolivia." *World Development* 28 (2): 333–346. doi:10.1016/S0305-750X(99)00121-7.

Olivares-Polanco, F. 2005. "Commercializing Microfinance and Deepening Outreach: Empirical Evidence from Latin America." *Journal of Microfinance* 7: 47–69.

Paxton, J. 2003. "A Poverty Outreach Index and Its Application to Microfinance." *Economics Bulletin* 9: 1–10.

Quayes, S. 2012. "Depth of Outreach and Financial Sustainability of Microfinance Institutions." *Applied Economics* 44 (26): 3421–3433. doi:10.1080/00036846.2011.577016.

Quayes, S. 2015. "Outreach and Performance of Microfinance Institutions: A Panel Analysis." *Applied Economics* 47 (18): 1909–1925. doi:10.1080/00036846.2014.1002891.

Quayes, S. 2020. "An Analysis of the Mission Drift in Microfinance." *Applied Economics Letters* 2020. doi:10.1080/13504851.2020.1813240.

Quayes, S., and G. Joseph. 2017. "Legal Systems and Performance of Microfinance Institutions." *International Review of Applied Economics* 31 (3): 304–317. doi:10.1080/02692171.2016.1249832.

Quayes, S., and T. Hasan. 2014. "Financial Disclosure and Performance of Microfinance Institutions." *Journal of Accounting and Organizational Change* 10 (3): 314–337. doi:10.1108/ JAOC-12-2011-0067.

Rhyne, F. 1998. "The Yin and Yang of Microfinance: Reaching the Poor and Sustainability." *Microbank Bulletin* 2: 6–8.

Schreiner, M. 2002. "Aspects of Outreach: A Framework for Discussion of the Social Benefits of Microfinance." *Journal of International Development* 14 (5): 591–603. doi:10.1002/jid.908.

Sharma, M., and M. Zeller. 1997. "Repayment Performance in Group Based Credit Programs in Bangladesh: An Empirical Analysis." *World Development* 25: 1731–1742.

Staschen, S., and C. Nelson. 2013. "The Role of Government and Industry in Financial Inclusion." In *The New Microfinance Handbook a Financial Market System Perspective*, edited by J. Ledgerwood, J. Earne, and C. Nelson, 71–96. Washington, DC: World Bank.

Von Pischke, J. D. 1998. "Measuring the Tradeoff between Outreach and Sustainability of Micro-Enterprise Lenders." *Journal of International Development* 8: 225–239.

Zeller, M. 1998. "Determinants of Repayment Performance in Credit Groups: The Role of Program Design, Intragroup Risk Pooling, and Social Cohesion." *Economic Development and Cultural Change* 46: 599–620.

Rethinking growth and inequality in the US: what is the role of measurement of GDP?

Remzi Baris Tercioglu

ABSTRACT

Five sectors have increased their contribution to US GDP growth since 1973: professional-business services (PBS), finance, information, healthcare, and arts-entertainment. Among these, however, finance, healthcare, and PBS have questionable foundations for being regarded as final consumption of households. Contra published National Income and Product Accounts, treating expenditures on finance, healthcare, and PBS as intermediate consumption reveals a significantly different picture of US economic growth, including i) a deeper slowdown of real output growth since 1973; ii) a more moderate rise in consumption share since 1980; and iii) a sharper decline in labor share, defined as the compensation of employees over GDP since 1985. Through a measurement approach, this paper thus contributes to the literature on secular stagnation and rising inequality in the United States.

1. Introduction

The US economy grew rapidly until 1980, lifting all lifeboats including those of working-class Americans. Since then, however, two fundamental changes have occurred. First, growth has slowed down, thus creating concerns about a secular decline in the economy's production capacity. Second, median income and wages have stagnated even though output and productivity growth have only slowed down. Rising distributional inequalities have been used to explain these changes, but in connection with the secular decline in growth, a question has arisen as to whether we have been measuring real output growth correctly.

In response to this question, Stiglitz (2012) points to rising rent-seeking in growing sectors of the economy as a possible distortion of growth statistics. Foley (2013) discusses the questionable foundations of post-1980 economic growth given the shift in the American economy from manufacturing to finance and information-based services, which give individuals an opportunity to generate income without necessarily contributing to production. Mazzucato (2018) similarly mentions the widening gap between income generation and value creation, and Stiglitz (2019b) argues that rising wealth extraction is a result of predatory lending by financiers and of excess prices within the health and other sectors.

Basu and Foley (2013) investigate the weak correlation between GDP and employment growth in recent business cycles. Their analysis decomposes the US value added into measurable industries, which have independent output measures, and imputed components (finance, real estate, professional services, management, healthcare, education, government, and other services) for which national accounts impute a value added based on their revenues less expenses. The authors observe that the measurable part, unlike the recent GDP, correlates strongly with current employment trends. This finding draws attention to the treatment of rapidly growing services in national accounts as a possible explanation for the decoupling of GDP from median income and wages seen in recent decades. Assa (2015) treats the output of financial and real estate services as intermediate consumption of the economy and arrives at a measure that correlates better with employment and median income than does GDP since 1980.

Following this line of thinking, I explore the impact that treating financial, healthcare, and professional-business service (PBS) expenditures as intermediate consumption of the economy has on US growth rate and income distribution from 1947 to 2017. These sectors in particular are significant because they have grown more rapidly than has the economy in recent decades, and they have questionable foundations for being regarded as final consumption of households. To distinguish between the standard published figures of the National Income and Product Accounts (NIPAs) and my own estimates, throughout this paper I refer to the former as conventional measures of income and output and to the latter as alternate measures of income and output.

In contrast to conventional measures of income and output, my approach reveals a deeper slowdown in growth since 1973, a more moderate rise in consumption share after 1980, and a sharper decline in labor share, defined as the compensation of employees (CE) over GDP, from 1985 to 2015. In terms of real output, the alternate measure's annual growth rate is 40 basis points lower than the conventional measure over 1973–2017, whereas the decline over 1947–1973 is 20 basis points. On the expenditures side, the alternate consumption share rises by 6 percentage points from 1980 to 2016, while the conventional share increases by 9 percentage points. On the income side, the alternate labor share declines by 5 percentage points from 1985 to 2015, while the conventional share falls by 2 percentage points. Given such differences, this paper shows the limitations of employing the conventional measurement approach to capture the depth of the growth and distribution problems of the post-1980 US economy.

To determine which sectors to analyze, I used a double criterion. First, I decomposed the 1947–1973 and the 1974–2018 average annual real GDP growth rate into industries and determined which sectors have increased their contribution to growth after 1973. I chose 1973 as the breaking point because the conventional output growth has decoupled from real wages and median income after 1973. If the post-1973 economy had grown more slowly than what the conventional measures of output indicate, stagnant median income and wages in recent decades would be less puzzling. Still, that alone provides insufficient grounds for excluding the output of the rapidly growing sectors from GDP. So, second, among the sectors previously identified, I focused on the industries whose output has questionable foundations for being treated as the final product.

Figure 1 shows the distribution of the US value added in 1947, 1973, and 2017. Over 70 years, real estate, PBS, finance, healthcare, information, arts-entertainment, and education increased their share of GDP, but PBS, finance, and healthcare grew mainly

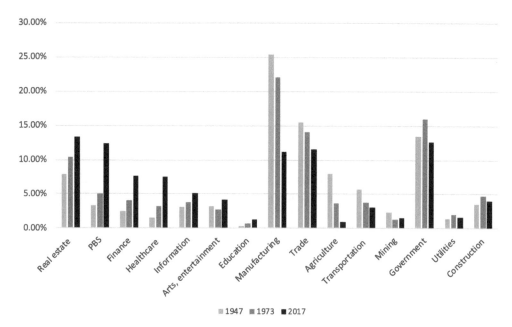

Figure 1. Value added shares of industries in 1947, 1973, and 2017. Abbreviation: PBS, professional-business services (Source: Use Tables of the Bureau of Economic Analysis before redefinitions at producers' prices).

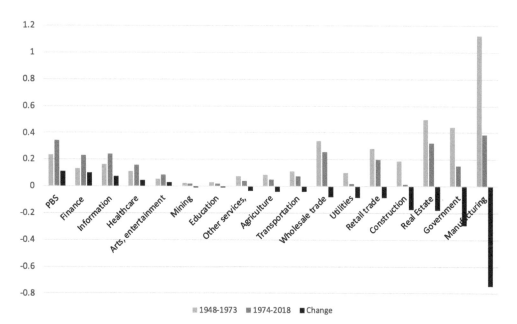

Figure 2. Contributions to percent change in real GDP by industry (Source: GDP-by-industry data of the Bureau of Economic Analysis).

after 1973. Figure 2 shows each industry's average annual contribution to real GDP growth for 1948–1973 and 1974–2018 as well as the change between these two periods. Before 1973, manufacturing, real estate, government, wholesale, and retail trade were the most significant contributors to growth; after which, their share declined significantly. By contrast, PBS, finance, information, healthcare, and arts-entertainment have increased their contribution since 1973.

Pointing to deficiencies in the conventional measures, Assa (2016) argues that financial services are better treated as intermediate consumption because households pay financial fees to consume other goods and services. With credit and debit card fees, for example, the point of departure is the good or service that is bought, and the credit and debit card fee becomes the cost of the final use. By comparison, national accounts treat mortgage points as intermediate consumption of the housing sector. Also controversial among national income accountants are traditional banking services, which constitute financial intermediation between lenders and borrowers. This issue is known as the banking problem, and Mazzucato (2018, 105) sees it as the tension between banks' decreasing role in stimulating growth and their increasing ability to generate income. Assa (2016) offers a solution to the controversy by treating all financial services as intermediate consumption of the economy.

One can make a similar argument for healthcare and PBS. Healthcare spending comprises 'necessary overhead costs of a complex industrial nation-state' (Nordhaus and Tobin 1972, 7), and its burden on the US economy has been increasing in recent decades. Nordhaus and Tobin (1972) shift public health expenditures to intermediate consumption and private healthcare to investment. However, the invariance principle (Vanoli 2005, 155) dictates that its treatment should not change under different institutional frameworks. Even though they discuss healthcare under defensive expenditures, which are generally defined as intermediate consumption, Stiglitz, Sen, and Fitoussi (2010) recommend treating healthcare spending as fixed asset formation. They argue that health expenditures are maintenance costs of human capital. However, maintenance cost formulation does not directly imply investment because the System of National Accounts (SNA) 93 defines only major improvements that increase capacity as investment expenditures, whereas ordinary maintenance is treated as an input cost of production (on National Accounts, I.-S. W. G. and of the European Communities, C 1993, 13). Foley (2013), by contrast, argues that healthcare spending is the cost of capitalist production, which creates environmental problems, stress, and poor working conditions. The case for accounting fees or attorney costs is quite similar. Expenditures on these services reflect the institutional costs of the US legal system and the complex tax code, and they are not direct sources of satisfaction for households. As such, I treat the consumer-related component of PBS in addition to expenditures on finance and healthcare as intermediate consumption in this paper to provide an account of secular stagnation and the recent decline in labor share.

Addressing secular stagnation, Summers (2015) points to demand-side problems by arguing that excess savings over investment and zero lower bound on nominal interest rates continuously push output below full employment. By contrast, Gordon (2015) explains secular stagnation using productivity slowdown, which depresses potential output. From a distributional perspective, Kiefer et al. (2020) point to a decline in potential output growth after 2000. Challenging the technological slowdown argument, Eichengreen (2014) notes the importance of investing in infrastructure, education, and training in response to secular stagnation. Using a measurement approach, I treat finance, healthcare, and PBS as intermediate consumption and find a deeper growth slowdown after 1973 that precedes the Great Recession. Therefore, if finance, healthcare, and PBS grow faster than the economy in subsequent decades, then the slowdown might turn into stagnation. Because the conventional measures of income and output fail to capture this result, they do not allow for an adequate response to secular stagnation. My findings based on the alternate measures of income and output, by contrast, suggest pro-growth policies at the industry level. Understanding these findings in light of the recent decline in labor share of GDP requires a brief examination of rising inequality as one of the causes of secular stagnation.

By redistributing income from lower- to upper-class individuals, who save more of their income, rising inequality depresses aggregate demand. Among the main drivers of inequality in the United States is the declining labor share of income. Karabarbounis and Neiman (2014) and Elsby, Hobijn, and Şahin (2013) document a decline of approximately 5–6 percentage points in US labor share since 1980. However, Elsby, Hobijn, and Şahin (2013) mention that approximately one-third of the decline is artificial due to the incorrect treatment of self-employment income. To them, the main driver of this decline is the share of GDP attributed to the CE, which has declined by 4 percentage points since 2000. Bridgman (2018) investigates possible problems with the denominator in measuring labor share. After taking out depreciation and taxes on production less subsidies from GDP, the author reports the same decline (4 points) since 2000 but argues that it is within the historical range of the labor share. The alternate measure of income points to a sharper decline in the CE share of GDP (6 points) post-2000 and suggests lowering the cost of finance, healthcare, and PBS on wage income in addition to labor-supporting policies. If we are to bring into view an adequate account of the decline in labor share and address not only secular stagnation, but also growing inequality in the United States, then an intervention is needed. This paper contributes to the literature on precisely these issues by introducing alternate measures of income and output to correct for the distortion in conventional measures. Without such adjustment to measures of GDP, the US economy will appear to be growing in ways that both belie real wage stagnation and shroud the increasing inequality among households.

Analyzing the national income accounting conventions and the rationale for treating finance, healthcare, and PBS output as intermediate consumption, Section 2 explains the methodology for the alternate measures of income and output. In Section 3, I present the results of shifting finance, healthcare, and PBS to intermediate consumption via discussion of the alternate growth rates, distribution of expenditures, and income before and after the mid-1970s. Section 4 details the importance of the results and their macro-economic policy implications, and Section 5 concludes this analysis with the significance of the alternate measures of income and output.

2. Methodology

2.1. National income accounting conventions

The NIPAs track the market value of newly produced final goods and services within the borders of the United States over a year based on their GDP measurement. GDP is calculated through three theoretically equivalent methods: the expenditures approach, the income approach, and the value added approach. For the expenditures approach, final expenditures are measured as the sum of personal consumption (C), private investment (I), net exports (NX), and government expenditures (G).

$$GDP = C + I + NX + G$$

GDP measured by income is the sum of CE (paid), gross operating surplus (GOS) (profits, rents, interest, proprietors' income, and depreciation), net taxes (NT) (taxes on production and imports less subsidies), and statistical discrepancy (SD).

$$GDP = CE + GOS + NT + SD$$

The industry value added is calculated by subtracting intermediate consumption (IC) of firms from the market value of their output (O). Gross value added (GVA) is defined as the sum total of value added of each industry (i).

$$GVA = \sum_i (O_i - IC_i)$$

Adding taxes on products less subsidies (T) to GVA yields GDP based on the value added.

$$GDP = GVA + T = \sum_i (O_i - IC_i) + T$$

Because GDP is measured at current prices, its growth rate includes a percent change in prices and a percent change in the quantity of production. For growth purposes, the critical variable is the real GDP, which tracks the volume of goods and services produced in a year. Real GDP is the result of deflating nominal GDP by the GDP deflator.

Suppose industry i is using all output of industry j. In that case, the market value of the product will be reported under IC_i and subtracted from O_i, thus leaving the value added of industry i as its net product. Expenditures on commodity j will not be included in GDP because it is an intermediate product consumed by industry i. Counting both expenditures on commodity i and j in the final product would be double-counting. If commodity j were classified as a final product instead, the market value of the expenditures would be added to GDP.

Final products are distinguished from intermediate inputs according to their direct contribution to households' current and future consumption. Kuznets (1951), the founder of US national accounts, emphasizes the link between the final product and social welfare as follows:

> The term 'net' implies that products are distinguished with reference to some set of goals, whose satisfaction is treated as a positive contribution. If by social welfare we mean a positive contribution to some socially determined set of goals, it is clear that 'net product' is an approximation to net additions to social welfare. I don't mean to imply that national income can be an accurate measure of social welfare, but it must be viewed as an approximation to it, since any measure of net product is an approximation to it. (Kuznets 1951, 179)

Eisner (1988) supports Kuznets: 'Our accounts may better seek to measure not welfare itself but the nation's output of final goods and services, which are presumed to contribute to welfare' (p. 1617). However, in conventional national accounts, the boundary separating final from intermediate expenditures depends on 'convention rather than on any well-reasoned philosophy of life. But the convention is a fickle thing, and what is regarded as a final product at one time may be regarded as an intermediate one at another' (Studenski 1961, 188).

In national accounts, government services were initially treated as intermediate consumption, then as partially final–partially intermediate, and eventually as final expenditures. The treatment of finance, which I discuss in detail below (see Section 2.2.2), is quite similar to that of government services. Finance was initially placed under intermediate consumption of firms; however, since the changes in SNA 93, the household-related part has been treated as final expenditures. The most recent example of this shift occurred in 2013 when, following the advice of SNA 2008, expenditures for intellectual property were reclassified. Before 2013, research and development as well as entertainment originals were treated as intermediate consumption of firms and subtracted from their value added. However, since 2013, the Bureau of Economic Analysis (BEA) has measured these expenditures under investment in intellectual property products and added them to GDP.[1]

Measuring only final products in GDP is crucial because, otherwise, our national accounts become prone to the double-counting problem. However, defining the final product is a practical question for the NIPAs, one that concerns the input cost of firms without a welfare dimension for households. If households are willing to pay for a good or service, our national accounts tend to interpret money paid as the basis of satisfaction of a particular need or want. This is a problem particularly for the treatment of finance, healthcare, and PBS under final consumption in the NIPAs, as Foley (2013) argues:

> Capitalist production creates health problems through environmental degradation and stresses and dangers of production environments. It is not clear that there is any inherent human need for legal or financial services, which seem rather to be needs produced by the social relations of capitalism itself. (Foley 2013, 3)

2.2. Healthcare, finance, and PBS output as intermediate consumption

Foley (2013) argues that finance and legal services are not genuine consumption and that health deterioration is a negative externality. Following this line of thinking, I argue that expenditures on finance, PBS, and healthcare are better treated as intermediate consumption because they are a means to the end of economic welfare. This treatment does not imply that these services are unnecessary. A properly functioning financial system is vital for funding productive investment and growth. Similarly, PBS are essential for the organization of economic production. As the ongoing global pandemic has shown, a good state of health is also a socially necessary condition for production. However, treating these services as final products is double-counting because they already contribute to economic growth indirectly, as I argue in Sections 2.2.1–2.2.3 (see below).

2.2.1. Healthcare

In the NIPAs, the health sector consists of five parts: healthcare (hospitals, doctors' and physicians' offices), health insurance, health-related expenditures on goods (drugs and medical equipment), health-related private and public fixed investment (equipment, structures, intellectual property products), and government expenditures on health.

Here, I discuss expenditures on durable and nondurable health-related goods and healthcare services, including government expenditures on health. I do not adjust healthcare fixed investment because it is not an intermediate consumption. I treat health insurance as intermediate consumption under finance and insurance in Section 2.2.2.[2]

Figure 3 shows the evolution of expenditures on different components of the health sector since 1947. Health expenditures composed approximately 5% of GDP until 1970, and this percentage doubled until the mid-1990s. After a short stabilization in the second half of the 1990s, it started rising again and reached 18% of GDP in 2017. The most significant expenditures are health-related services, including hospitals, doctors' offices, laboratories, nursing, and residential care services. Medical equipment and pharmaceutical expenditures are the second-most prominent component of the health sector, and their cost has been rising since the mid-1990s.

Currently, most healthcare spending is reported under personal consumption expenditures and added to GDP in the NIPAs. A small part of health services, such as work-related medical tests, is under the intermediate consumption of firms.[3] Employer contributions for health and life insurance are reported under the CE (supplements to wages and salaries) and imputed to personal consumption expenditures.[4] Government expenditures on healthcare are also imputed to households' final consumption expenditures. SNA 93 suggests treating health spending as final consumption because of the short- and long-term benefits to consumers; however, it generates regrettable expenditures from the household perspective.

In their search for a measure of economic welfare, Nordhaus and Tobin (1972) treat public health expenses as instrumental expenditures due to their collective nature and then shift private healthcare to investment.[5] However, this treatment violates the invariance principle, which requires equal treatment under different institutions. For example, under this treatment, a shift from private to public provision of healthcare would depress GDP. Thus, to satisfy the invariance principle, I extend the overhead cost argument of

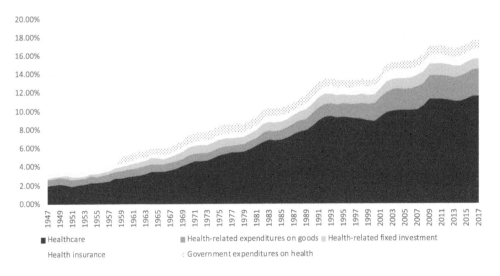

Figure 3. Health expenditures of the US as a percent of GDP, 1947–2017 (Source: National Income and Product Accounts).

Nordhaus and Tobin (1972) to private healthcare because healthcare is the cost of the net product regardless of who pays for it.

Even though they discuss healthcare under defensive expenditures, which are generally defined as intermediate consumption, Stiglitz, Sen, and Fitoussi (2010) recommend treating expenditures on healthcare as fixed asset formation. They argue that health expenditures are maintenance costs of human capital. Although keeping track of society's overall health is understandable because a good state of health is an indispensable part of well-being, treating healthcare expenditures as an investment is misleading: Health spending does not increase an economy's productive capacity, but simply prevents it from declining. The maintenance cost formulation by Stiglitz, Sen, and Fitoussi (2010) does not directly imply investment because SNA 93 defines only major improvements that increase capacity as investment expenditures. In contrast, ordinary maintenance is treated as an input cost of production (on National Accounts, I.-S. W. G. and of the European Communities, C 1993, 13). Major improvements, such as longevity, are results of research and development expenditures on health. These are already treated under investment, and my treatment here does not change this. Rather, I argue that healthcare services fit better under the ordinary maintenance category, thus they should be treated as intermediate consumption.

My discussion concerns the treatment of healthcare as intermediate consumption in the ideal case when prices are competitively determined. Currently, however, the US health system charges excessive prices for care, insurance, and drugs. Stiglitz, Sen, and Fitoussi (2010) indicate a lack of transparency in the pricing of health services and identify rising rent-seeking by insurance and drug companies. Mazzucato (2018, 208) also mentions the rising number of 'drugs with little or no therapeutic value.' Such market imperfections are increasing the distortion of the current treatment of healthcare costs as final consumption. Shifting health spending to investment will only change where in GDP these costs, including the excess cost coming from market power, are recorded. By contrast, treating health spending as intermediate consumption will report these expenditures, whether normal or excess, as costs of the net product. This shift allows the alternate measures to address the rising burden of healthcare better than does the conventional approach.

2.2.2. Finance and insurance

There has been a long controversy among national income accountants about how to treat financial services in the NIPAs. Assa (2015) describes three types of financial income and how each is treated differently in national accounts. The first income type is capital gains, which are left out of GDP because they are not productive. For example, profits from the valuation of a bank's assets are not recorded as profits. The second type of income is interest income from financial intermediation services, now known as financial intermediation services indirectly measured. Intermediation services are treated as intermediate consumption of firms and subtracted from their value added. The third type is income generated from fee-based financial services that is treated as a final product (Assa 2015, 3). Although there is no controversy about how to treat capital gains, the treatment of intermediation and fee-based income has been controversial.

According to Christophers (2011), intermediation services had not been counted in banks' output in many advanced economies until the SNA 68 because they are not explicitly priced.[6] When banks are treated like other businesses, their value added becomes negligible or even negative (Christophers 2011, 124) because the fee-based income of banks is generally lower than their input costs. This discrepancy is known as the banking problem, and national income accountants have tried to overcome it by imputing the value of financial intermediation services to banks' output. Without that, the financial sector would appear as a value extractor instead of a value adder in the NIPAs. As a result, some banking services receive unique treatment:

> BEA refers to these services as "financial services furnished without payment" or as "implicitly priced services." For example, banks may provide some services—such as processing of checks, disbursing or transferring funds when needed, protecting deposit funds, and investment services—without charging explicit fees. To account for such services, the NIPAs include an imputation of the value of these services because they are omitted by the standard measure of output based on revenue from fees or prices, which is used for most industries. (Hood 2013, 8)

The difference between interest rates on loans and deposits times volume of non-bank capital is defined as the intermediation income, and this amount is imputed to banks' output. The imputed amount was 'assumed to be fully consumed by financial and non-financial companies, so none made it through into final output' (Mazzucato 2018, 107). However, with the SNA 93, intermediation services began to be distributed to households and firms. When firms use intermediation services, the market value of the service is still treated as intermediate consumption and subtracted from their value added. However, when households use these services, the service's market value is imputed to personal consumption expenditures. Mazzucato (2018) argues that, due to this treatment, GDP appeared to increase even further with the household debt boom of the 2000s.

By contrast, Assa (2015) argues that intermediation services do not produce final use-value for households; therefore, imputed intermediation services should be subtracted from personal consumption expenditures just as they are deducted from the value added of firms. People use bank loans, not as a source of satisfaction, but because they cannot finance out of savings their expenditures. In addition to intermediation services, Assa (2015) discusses the inconsistent treatment of fee-based financial services in national income accounts. According to Assa (2015), people do not consume fee-based services per se; rather, they pay for these services as a means for consuming other goods and services.

> From a theoretical point of view, however, the non-FISIM [financial intermediation services indirectly measured] part of financial services, that is, the fee-based income in the GDP-by-output approach, is as problematic as interest-based income. Finance, in its various manifestations, ultimately involves the transfer of money. Unlike other commodities, money has no use value, only an exchange value. In fact it is exchange value par excellence. (Assa 2015, 6)

Following Assa (2015), I treat financial services as intermediate consumption of the economy; however, I do not adjust housing services because, as sheltering, they produce final use-value for households. Under finance, I subtract from GDP expenditures on Federal Reserve banks, credit intermediation, and related activities; securities, commodity contracts, and investments; as well as insurance carriers and related activities and funds, trusts, and other financial vehicles. I shift under real estate some services such as brokerage and property management to finance because they are intermediate in reaching housing services.

Figure 4 shows the evolution of financial and professional services expenditures over 1947–2017. Expenditures on financial services were approximately 2–3% of GDP until 1980. Roughly half of such expenditures were on insurance, whose share of GDP has moved within the 1–2% band over the 70 years tracked. However, imputed intermediation services and fee-based expenses multiplied over the 1980s and the 1990s. The GDP share of financial expenditures reached 7% in 2000 and stabilized around this level over the 2000s.

2.2.3. Professional-business services

PBS comprise legal services, design, research and technical services, administration, and waste management services in the NIPAs. PBS are a mixed bag, and firms use most of their output as intermediate consumption (73% in 2016), adding the remaining part to GDP (27% in 2016). Among final expenditures on PBS, investment composes the most significant part (57.4% in 2016) as a result of expenditures on research and development and fixed investment on computer systems. I do not adjust private and public investments on PBS because they generate future income streams; thus, they are not intermediate consumption. Some PBS services (computer systems design, management of companies and enterprises) are used only by firms, and they do not appear under final expenditures. For households, legal services constitute the most significant part of personal consumption expenditures on PBS (40% in 2016). Here, I focus mainly on this part of PBS, which was approximately 1.7% of GDP in 2017.

PBS are not direct sources of satisfaction for households. Kuznets (1953) defines these services as 'occupational expenses,' which are the expenditures made to earn a living.

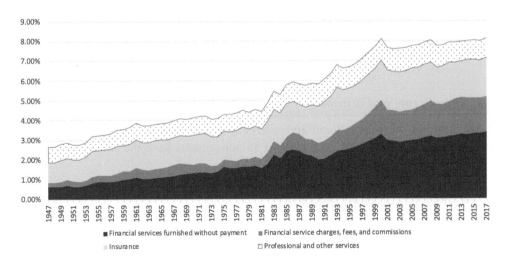

Figure 4. Financial and professional services as a percent of GDP, 1947–2017 (Source: National Income and Product Accounts).

Kuznets (1953) uses trade union dues and work-related expenses as examples, and one can extend this definition to include legal and accounting fees. Consider fees paid to accountants for filing tax returns. This process is connected to earning income, and people spend time and money on it because of the complicated US tax code. This expenditure is hardly adding to the satisfaction of consumers. The same argument could be made for fees paid to attorneys and other litigation costs.[7] People do not consume these services per se, but people pay for them to obtain a particular outcome. If fees are paid in return for a service, which is related to production or consumption, then they become a cost of the economic outcome of the legal process. For example, if an immigrant is seeking a US work permit, then she will need to hire a lawyer to defend her legal status. The money she pays to lawyers will, thus, function as an intermediate consumption. Even when lawsuits are not directly economic (for example, child custody cases resulting from the increase in divorces within modern societies), legal fees can still be considered socially regrettable expenditures for producing the value added of modern economies.[8]

3. Results of shifting sectors

While shifting finance, healthcare, and PBS expenditures to intermediate consumption, I apply a consistent adjustment to the three approaches for measuring GDP. I deduct final expenditures on finance and healthcare at current prices from consumption, investment, net exports, and government spending. For PBS, I subtract only consumption expenditures and net exports. Then, I deflate the alternate measure of final expenditures by the GDP deflator to get the alternate real GDP. In Section 3.1, I report the alternate real output growth and distribution of expenditures using the Input-Output Use Tables of the BEA. For GDP by income, I use data from the Consumer Expenditure Survey of the US Bureau of Labor Statistics (BLS) to distribute the same deduction to the CE and GOS. In Section 3.2, I provide the alternate income distribution over quintiles together with the functional distribution of income. In the Appendix, I reestimate GDP by the value added by distributing the same deduction to the value added of industries based on their shares in gross value added so that value added shares remain constant. In this way, the alternate national accounts preserve the equivalence of the three methods for measuring GDP.

3.1. Growth rate and distribution of expenditures

Table 1 shows the conventional and alternate annual real output growth rates over 1947–1973, 1973–2000, and 2000–2017. PBS adjustments have the smallest impact on

Table 1. Conventional and alternate measures of annual real output growth rates for 1947–1973, 1973–2000, and 2000–2017.

	1947–1973	1973–2000	2000–2017
Real GDP	3.96	3.10	1.85
Real GDP-PBS	3.94	3.09	1.85
Real GDP-Finance	3.90	2.99	1.83
Real GDP-Health	3.84	2.86	1.56
Real GDP-(PBS+Finance+Health)	3.76	2.69	1.50

output growth over each period. The effect of adjusting finance increases from 1947–1973 to 1973–2000, when financial services grew rapidly. Healthcare adjustments have the highest impact on the decrease over each period; recently, the magnitude of this decline has been rising. Total adjustments (shown in the last row of Table 1) reduce real output growth by 20 basis points over 1947–1973, by 41 points over 1973–2000, and by 35 points after 2000. Given that real GDP growth declined to 1.85% after 2000, 35 basis points correspond to almost one-fifth of the conventional growth rate.[9]

Table 2 shows the conventional and alternate shares of final expenditures in 1947, 1980, and 2016. In conventional accounts, consumption is the largest part of GDP in 1947, 1980, and 2016. Its share of total expenditures declines from 65% to 60% over 1947–1980 and rises to 69% in 2016. Private investment and government expenditures make 15% and 16%, respectively, of GDP in 1947. Their shares of GDP increase significantly until 1980, after which they decline to almost 1947 levels. The most significant difference is in net exports. The US economy had a trade surplus of approximately 4% of GDP in 1947, a trade balance in 1980, and a 3% deficit in 2016.

In the alternate accounts, consumption share declines from 63% in 1947 to 55% in 1980; then, owing to increasing consumerism after 1980, it rises back to 61% in 2016. The rise after 1980 is more moderate than in the conventional accounts. The 2016 level of consumption share in the alternate accounts is still below the 1947 level, showing that consumerism has brought lower consumption in recent years than it did 70 years ago when rising financial, healthcare, and PBS expenditures were treated as intermediate consumption. The alternate investment share rises from 16% in 1947 to 21% in 1980 and stays at that level in 2016. The share of government spending also does not change significantly in the alternate accounts over 1980–2016. Unlike government spending, which remains an essential part of final expenditures despite attempts to downside the government, net exports share declines from 5% in 1947 to −5% in 2016. In sum, by taking out financial products and professional services from exports, the alternate measures reveal a sharp decrease in the international competitiveness of the US economy.

3.2. Distribution of income

Here, I estimate how much of financial, healthcare, and PBS expenditures are financed out of the CE and GOS. I then recalculate GDP by income. To estimate sources of expenditures, I turn to the Consumer Expenditure Survey of the BLS as a valuable data source. Annual survey data was first collected in 1984, and current surveys are conducted

Table 2. Conventional and alternate shares of expenditures in 1947, 1980, and 2016.

	Share of 1947 GDP	Share of 1980 GDP	Share of 2016 GDP
Consumption	65	60	69
Alternate consumption	63	55	61
Investment	15	19	16
Alternate investment	16	21	21
Net exports	4	0	−3
Alternate net exports	5	0	−5
Government	16	21	18
Alternate government	17	24	23

with more than 130,000 consumer units. Using survey data from 1985–2015, I adjust the income distribution over quintiles and functional distribution of income.

3.2.1. Distribution of income over quintiles

BLS Consumer Expenditure Survey data provide three types of information. The first is the means of aggregate expenditures of quintiles together with before-tax incomes. Healthcare has a separate entry in the survey, but financial services and PBS do not. I approximate those costs using the miscellaneous expenditures category.[10] Figure 5 shows the distribution of aggregate healthcare (H) and finance-PBS (FPBS) spending over quintiles for 1985, 2000, and 2015.

Figure 5 shows that the share of the bottom 60% in aggregate healthcare expenditures declines, whereas the share of the top 20% rises continuously over 1985–2015, particularly after 2000. In 2015, the percentage of the top 20% in aggregate healthcare expenditures was 33%, and the bottom 60% is 43%. For FPBS, the distribution of spending is even more unequal. In 2015, the top 20% made 39% of aggregate expenditures, which is more than the share for the bottom 60% (38%).

The second type of information available in the survey data is the distribution of before-tax income over quintiles. Using the expenditure shares shown in Figure 5, I deduct aggregate healthcare and FPBS expenditures from each quintile's income.

$$alternate(y_i) = \frac{Y_i - (h_i * H) - (fpbs_i * FPBS)}{GDP - H - FPBS} \; \forall i = 1, ..., 5,$$

where Y_i is the before-tax income of quintile i, y_i is the income share of quintile i, h_i is the share of quintile i in aggregate health spending, and $fpbs_i$ is the share of quintile i in total FPBS spending. Providing a comparison between the conventional and the

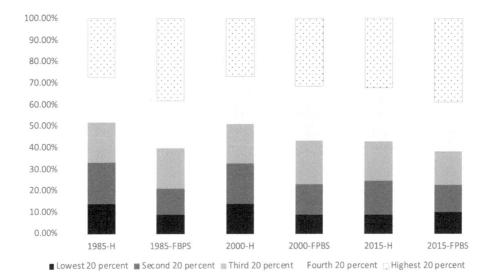

Figure 5. Distribution of healthcare (H) and finance-PBS (FPBS) expenditures over quintiles in 1985, 2000, and 2015 (Source: Consumer Expenditure Survey data).

alternate distributions of income, Figure 6 shows Lorenz curves with Gini coefficients in 1985 and 2015.

As Figure 6 indicates, the conventional income inequality over quintiles increased from 1985 to 2015 owing to a rising share of the top 20%. The conventional Gini coefficient rises from 0.39 to 0.44. In the alternate measures, the Lorenz curves shift outward, and distributions become even more unequal. The alternate Gini coefficient increases from 0.43 to 0.51. Although high-income households account for more of the aggregate healthcare, finance, and PBS expenditures, low-income households are more impacted by adjustments because the deducted amount is high relative to their income. Results, thus, indicate that rising healthcare, finance, and PBS costs had a more severe impact on low-income American households than on those with incomes in the top 20%.

3.2.2. Functional distribution of income
The third type of information available in the survey data is the decomposition of each quintile's income to its sources. One problem with analyzing both the Consumer Expenditure Surveys and the NIPAs is that the income categories used by each do not match. I take the NIPA decomposition of income (CE, GOS, and NT) as the benchmark. To this end, I reorganize the survey data income sources to make them compatible with the NIPA definitions of CE and GOS.

I start with money income before taxes. I subtract other income (scholarships, fellowships not based on working), social security (private and government), and other transfers (food stamps, public assistance) from before-tax income to arrive at income before taxes and transfers. I add unemployment and workers' compensation to wages and salaries to arrive at the CE. Lastly, I add self-employment income to interest, dividends, and rental income to arrive at the GOS.[11]

Via this methodology, I decompose before-tax and -transfer income of each quintile to the CE and GOS. One should be careful in interpreting this decomposition as

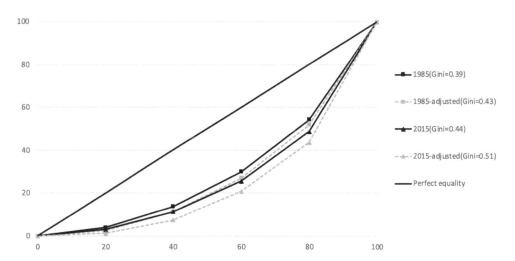

Figure 6. Lorenz curves for the conventional and the alternate income distributions in 1985 and 2015 (Source: Consumer Expenditure Survey tables of the US Bureau of Labor Statistics and Use Tables of the Bureau of Economic Analysis).

a capital-labor split because self-employment income is shifted totally to GOS instead of being distributed to labor and capital according to their shares.[12] The CE share here is less than the labor share of income. Moreover, GOS includes depreciation in addition to profits, interest, rent, and self-employment income; thus, its share is much more than the capital share.

Based on the decomposition of income, I calculate the CE and GOS shares of healthcare expenditures as follows:

$$CE^{health} = \frac{\sum_{i=1}^{5} h_i w_i y_i}{\sum_{i=1}^{5} h_i w_i y_i + \sum_{i=1}^{5} h_i (1 - w_i) y_i}$$

$$GOS^{health} = \frac{\sum_{i=1}^{5} h_i (1 - w_i) y_i}{\sum_{i=1}^{5} h_i w_i y_i + \sum_{i=1}^{5} h_i (1 - w_i) y_i}$$

where w_i represents the share of the CE in the income of quintile i. I similarly calculate the CE and GOS share of FPBS expenditures. Table 3 shows these shares for 1985, 2000, and 2015.

As shown in Table 3, almost all expenditures on healthcare services and FPBS are financed out of the CE. However, one can also observe a trend of rising GOS share in funding these services over 1985–2015. This is not surprising because the profit-type income of those in the top income percentiles has risen since the 1980s, and this group spends more on healthcare, finance, and PBS (see Figure 5). Despite this trend, the share of the CE does not fall below 90% over the sample.

The alternate CE and GOS shares of GDP can be calculated as follows:

$$alternate(CE) = \frac{CE - (CE^{health} * H) - (CE^{fpbs} * FPBS)}{GDP - H - FPBS}$$

$$alternate(GOS) = \frac{GOS - (GOS^{health} * H) - (GOS^{fpbs} * FPBS)}{GDP - H - FPBS}$$

where H indicates total health expenditures. I calculate these ratios for five years, starting with 1985.[13] Figure 7 shows the conventional and alternate GDP by income. The alternate CE share is lower, and it declines more rapidly than the conventional share because rising healthcare, finance, and PBS expenditures are largely financed out of wage income. This happens even though CE shares in Table 3 decline because increasing costs of healthcare, finance, and PBS since 2000 dominate the opposite effect.[14]

Table 3. Shares of compensations of empoyess (CE) and gross operating surplus (GOS) in expenditures on healthcare, and finance and professional-business services (FPBS) in 1985, 2000, and 2015.

	1985	2000	2015
CE share of healthcare expenditures (CE^{health})	99.3	93.2	91.4
GOS share of healthcare expenditures (GOS^{health})	0.7	6.8	8.6
CE share of FPBS expenditures (CE^{fpbs})	95.6	92.9	91.1
GOS share of FPBS expenditures (GOS^{fpbs})	4.4	7.1	8.9

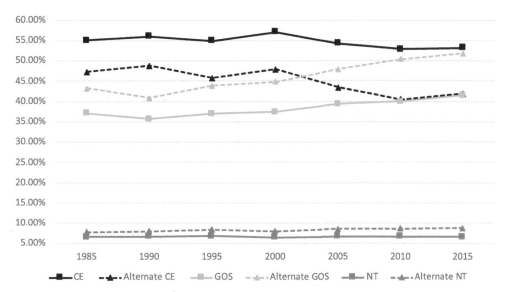

Figure 7. The conventional and alternate functional distribution of income over 1985–2015. Abbreviations: CE, compensation of employees; GOS, gross operating surplus; NT, net taxes (Source: Consumer Expenditure Survey tables of the US Bureau of Labor Statistics and Use Tables of the Bureau of Economic Analysis).

The alternate CE share declines by 6 percentage points after 2000, whereas the conventional share falls by 4 points. My finding indicates a sharper decline than do the estimates of Karabarbounis and Neiman (2014) and Elsby, Hobijn, and Şahin (2013). Bridgman (2018) argues that a decline in gross labor share does not necessarily imply capital gains, given rising depreciation (in connection with high depreciation rates of information technologies) and taxes on production. I repeat my calculations by subtracting first depreciation and then depreciation and taxes from the denominator. However, I still find a decline above 5 percentage points after 2000, which is 2 percentage points more than the conventional share. In addition to the absolute decrease, my findings point to a sharper relative decline than that of the conventional share because the initial alternate CE shares are lower than the conventional shares.

To complete the analysis, I discuss the conventional and alternate measures of productivity and average compensation over 1985–2015 shown in Figure 8.[15] The conventional measures of productivity and average compensation grow together from 1985 to 2015. In the alternate measures, average compensation stagnates after 2000 even though productivity continues to grow, thus creating a productivity gap.

The standard theory assumes a constant wage share of output; therefore, productivity gains must appear in wages. However, this is not the case, as the alternate measures in Figure 8 indicate. Stiglitz (2012) points to rising rent-seeking activities after the 1980s to explain this puzzle. Monopolies in finance, healthcare, and high-tech industries increased their prices at the expense of all others and did not share productivity gains with their workers. To explain this decline, Hein (2015) argues that rising financialization has decreased the wage share of output by changing sectoral composition, increasing overheads and profit claims, and weakening labor unions. Mazzucato (2018) points to cash

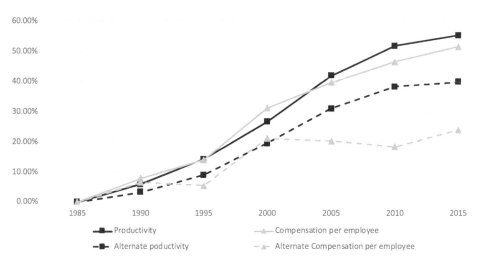

Figure 8. The conventional and alternate measures of productivity and compensation per employee with cumulative growth rates indicated on the vertical axis (Source: Consumer Expenditure Survey tables of the US Bureau of Labor Statistics and Use Tables of the Bureau of Economic Analysis).

flows arising from productivity gains used in share buybacks to increase shareholder value at the expense of stakeholder value. Taylor and Ömer (2018) point to wage repression in stagnant and leading sectors of the economy as a tool of class struggle while giving credit to market power–based explanations. These findings are compatible with wage-repression theories, as average compensation stagnates after taking out finance, healthcare, and PBS even though labor productivity continues to grow.

3.2.3. Possible sources of bias

Problematically, the survey data do not extend to the top incomes, and they depend on what people say about their income and its sources. Moreover, the surveys do not ask questions about capital gains and stock options, which constitute a critical portion of top incomes (Pressman 2015, 73). Therefore, one could argue that there is an underestimation bias in the share of capital income in finance, healthcare, and PBS expenditures reported above. It may be a concern primarily for decomposition of the top 1% of income.

To test for this bias, I compare my results with the data of Piketty (2014), who uses individual income tax returns to measure inequality. Excluding unrealized capital gains, gifts, and some government transfers to limit income to taxable income, Piketty (2014) defines total income as gross adjusted income plus adjustments to income determined by the US tax code (Pressman 2015, 57). Showing how the coverage of total income in tax returns changes over time, Pressman (2015) also discusses tax evasion issues, changing tax laws, and changes in corporate and individual tax systems that may impact how business owners fill their tax returns and report their income. Keeping these limitations in mind, I repeat my estimation from Section 3.2.2 using Piketty (2014) data to investigate possible sources of bias.

Piketty (2014) gives a decomposition of top US incomes for 2007 in the appendix of his seminal book *Capital in the Twenty-First Century*. For comparison, I form four

Table 4. Comparing shares using consumer expenditure survey and piketty data for 2007.

	Consumer Expenditure Survey data	Estimations based on Piketty data
CE^{health}	93.9	95.0
GOS^{health}	6.1	5.0
CE^{fpbs}	92.8	94.3
GOS^{fpbs}	7.2	5.7

income groups: low (bottom 60%), middle (60%–90%), high (90%–99%), and top 1%. Dividing the highest incomes into the top 1% and the top 9% is to account accurately for the rising capital share of incomes. For the low-income CE and GOS shares, I again use the survey data of the lowest three quintiles and assume that the decomposition of income of the fourth quintile represents middle-income shares.

For both the top- and high-income CE and GOS (capital income + mixed-income) shares, I use Piketty data. For the income distribution, I again use Piketty data from the World Top Incomes Database for each income group. As before, I take expenditure shares from the Consumer Expenditure Survey data. I estimate the high-income health-care and FPBS expenditures by using Engel curves derived from the survey data. Engel curves describe how spending on a good changes as income level rises. For healthcare and FPBS, curves are upward sloping, and they are almost linear. Therefore, I approximate Engel curves by linear trend lines and estimate expenditure shares of top incomes by entering mean expenditures of high incomes from the World Top Incomes Database. I summarize the results for 2007 in Table 4.

As shown in Table 4, results obtained using the Consumer Expenditure Survey and Piketty data are quite similar. The CE shares with Piketty data are even slightly higher than those obtained with the Consumer Expenditure Survey data. Thus, the under-reporting problem also affects the CE because of the high compensations of the top 1%.

4. Discussion

The findings of this paper have three macroeconomic policy implications. First, cheaper finance, healthcare, and PBS are better for economic growth. Governments can reduce the cost of these services on the net product and shift more resources to industries that directly contribute to the final product. Second, contra what the conventional measure of output indicates, the US economy has a deeper growth problem that precedes the Great Recession. As a result, industry-level pro-growth policies are suggested to generate more income while keeping an eye on resource and environmental constraints. Third, the finding of a lower and more rapidly declining labor share in the early 2000s signals growing distributional inequalities more strongly than indicated by the conventional labor share. In addition to policies that will increase the bargaining power of labor, it is more urgent to counter this trend by lowering the cost of finance, healthcare, and PBS on wage earners.

The current national accounts were born during the Second World War, just after the Great Depression (Temin 2020). Governments initially wanted to measure how much of the output had melted down during the Depression (Stiglitz 2019a, 16). The idea behind official national accounts was to provide governments a reliable estimate to measure the effects of the Depression and government policies for recovery. With the rise of Keynesian economics, governments took an active role in shaping their economies and steering growth. In the absence of economic welfare indicators in the national accounts, maximizing GDP growth became the single policy objective, and policymakers started to care less about what we put in GDP as long as GDP continued to grow.

However, how we measure GDP matters because it shapes how we perceive our economies and how we make policy decisions about them, as summarized in Figure 9. The conventional measurement of financial, healthcare, and PBS output under the final product justifies excess-prices of these services because they bring a faster GDP growth. As these services expanded after 1980, policymakers tended to believe that they were becoming more productive and adding more value to our economies. This interpretation gave rise to incorrect incentives that encouraged rent-seeking activities such as predatory lending of banks, record drug prices, and expensive lawyer fees. According to Stiglitz (2012), rent-seeking in growing industries has distorted US growth figures and income distribution and increased the living costs of ordinary Americans.

Price formation is quite opaque for finance, healthcare, and PBS, and there is much room for exercising market power. Government policies (or their lack thereof) have significant impacts on the price of finance, healthcare, and PBS, as the rising burden of these services under neoliberal policies shows. The alternate measure of the output reports their rising cost as intermediate consumption of the economy and suggest policies to lower their prices. Cheaper finance, healthcare, and PBS will not only decrease the living costs of wage earners, but also reduce the economy-wide cost of production and free more resources for other uses. In short, lowering the costs of these services will be beneficial for economic growth. Employment of this policy is particularly important for healthcare, given the public debate about how to make it more affordable and more accessible. To better guide policymakers toward the pro-growth implications of cheaper healthcare, the alternate measure of output and income can be included in the dashboard of indicators discussed by Stiglitz (2019a).

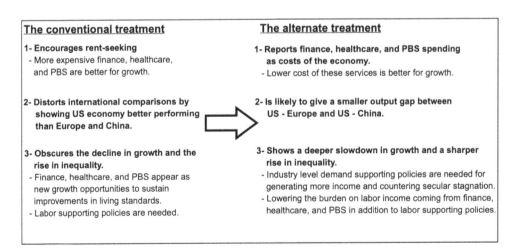

Figure 9. How the conventional treatment distorts our perception of the economy and policy implications of the alternate measurement.

Another advantage of the alternate measure of output and income pertains to international comparisons. The dominance of finance, healthcare, and PBS in GDP is more of a US phenomenon. Because European countries are less financialized and spend less of their national income on healthcare than does the United States, the conventional measure of their output is less prone to distortion. Therefore, the relatively better performance of the US economy compared with that of Europe may be a statistical mirage. The alternate measure of output is also likely to yield a smaller per capita output gap between the United States and China, thus telling a different story about the pace of global convergence. The alternative approach may explain the rising political tension between these two countries better than do conventional measures.

The conventional measure of output also distorts how we perceive the future growth possibilities and constraints of the US economy. The structural shift of the economy from manufacturing to finance-like services appears as a change in sources of growth, preserving the income-generation capacity of the US economy. From this lens, expanding finance, healthcare, and PBS among rapidly growing services becomes a natural tendency of the market moving towards new growth opportunities. However, the benefits of this expansion for ordinary Americans are quite questionable, as they become new constraints for producing the rest of the value added.

This paper puts into question the argument that the post-industrial US economy can sustain improvements in living standards by simply shifting to more sophisticated services. The finding of a deeper growth slowdown supports the emphasis that Foley (2013) placed on the limitations of the so-called new economy in overcoming resource and environmental constraints of capitalist production. With the alternate measure of output, rising financial, healthcare, and PBS costs become drains on aggregate demand. Secular stagnation points to long-term demand problems behind the decline in the US growth and provides a more plausible framework for addressing the situation. There is a need for active, sector-based demand supporting policies to prevent secular stagnation. Given low inflation rates, there is more room for policies such as greater investment in infrastructure, education, and green technologies to increase short-run growth and improve the long-run trajectory of the economy through positive feedbacks on labor productivity.

Another impact of the conventional measurement is on how we perceive the magnitude of distributional inequalities and what we can do about them. The conventional labor share has declined since 2000 by around 4 percentage points. Elsby, Hobijn, and Şahin (2013) discuss increasing import competitiveness to address the globalization behind the decline in labor share, whereas Taylor and Ömer (2018) point to wage repression as a tool for class struggle. The main driver is $r > g$ for Piketty (2014), where r is the rate of return on capital, which has been above the growth rate (g) in advanced economies since 1980. With a rising capital/income ratio and elasticity of substitution between labor and capital higher than one, Piketty (2014) predicts that r will remain above g in advanced economies. The alternate measure of income shows a deeper structural inequality between labor and capital resembling the levels attained in the 19th-century European countries discussed by Piketty (2014). The difference results from the rising finance, healthcare, and PBS costs of wage earners that work like a levy on labor income. These costs, particularly for healthcare given an aging population, are expected to rise in the coming decades. Under the current institutional framework, they will eat more of the labor income and depress living standards. Even if the decline in the conventional labor share is tamed by strengthening the bargaining power of labor, the alternate labor share will still decline as long as finance, healthcare, and PBS costs increase. The US economy urgently needs to lower this levy on labor income while introducing labor-supporting policies.

5. Conclusion

In the 1980s, Keynesian welfare state policies shifted to a neoliberal policy agenda in the United States (and other Western economies). These policies aimed to increase profit rates by attacking labor unions, deregulating labor markets, and cutting government spending on healthcare and other social benefits. Changes in the sectoral structure of the economy from manufacturing to services have accompanied this policy shift. As a result, growth rates have remained high within the ever-expanding healthcare, finance, and PBS sectors, whereas median income and real wages have stagnated. Treating these services as costs of the economy points to a significant decline in the US economy's income-generation capacity since the mid-1970s, thereby increasing the burden of neoliberal policies on society.

The alternate measure of final expenditures has a different distribution from that of the conventional GDP. Compared with the conventional share, the alternate consumption share displays a more moderate rise over 1980–2016. In 2016, it is below its 1947 level, showing that consumerism has brought lower consumption once finance, healthcare, and PBS have been taken out. The alternate investment share stays constant over 1980–2016, and the alternate government share rises to almost one-fourth, which implies more space for fiscal policy. The alternate share of net exports declines to –5% in 2016, showing the depth of the US trade deficit.

On the income side, the alternate CE share of output is lower and declines more rapidly than the conventional share because almost all expenditures on finance, healthcare, and PBS are financed out of the CE. The decline in the CE share is in line with the lower consumption share as the propensity to consume out of CE is likely to be higher than the propensity to consume out of GOS. Even though the alternate labor productivity continues to grow, the alternate CE per employee stagnates after 2000, thus creating a productivity gap. Current treatment of the NIPAs not only overestimates growth and consumption, but also obscures the sharper wage-profit inequality. As a result, American capitalism appears to be better performing and less unequal. Bringing these discrepancies into relief, this paper shows how the alternative measures of income and output provide a more accurate picture of the US economy by reclassifying finance, healthcare, and PBS expenditures to intermediate consumption.

Notes

1. The BEA-added costs for research and development as well as entertainment originals ($560 billion in 2013) increased GDP by 3.6% in July 2013 (Goodwin et al. 2015, 113). This adjustment had minimal impact on the growth rate in 2013 because past expenditures on intellectual property products were also added to previous years' GDP.
2. I include social assistance under healthcare to correspond with its classification in Input-Output Tables over 1947–1962. However, this is not a universal convention, and other countries report social services separately from healthcare.

3. Costs of work-related medical tests covered by employers are reported as business expenses and subtracted from their value added. However, if employees pay for such tests, their costs are treated as final consumption of households (Assa 2015, 14).

4. This imputation constituted approximately one-fourth of all healthcare expenditures in 2016.

5. Instrumental expenditures are 'necessary overhead costs of a complex industrial nation-state,' (Nordhaus and Tobin 1972, 7) and they are treated as intermediate consumption.

6. Christophers (2011) defines the treatment of finance by the SNA 68 as implicitly productive because intermediation services are imputed to banks' output but subtracted from the rest of their value added either by defining a nominal sector whose only input is intermediation services (as is done in the United Kingdom) or by decreasing the value added of users of intermediation services (as done in France). The United States followed the suggestion by the SNA 53 of imputing the banking sector's wages and profits to banks' output. However, the 'US case is very much an anomaly' (Christophers 2011, 130).

7. Legal services can also be described under transaction costs. Coase (1995) discusses the expenses that firms incur in using markets instead of internal resources for inputs of production. Fees paid to lawyers for preparing contracts fit well to this definition. Financial fees may also be included because they are the costs of firms' decisions to use external finance instead of retained earnings. Rising legal and financial service fees, thus transaction costs, in the United States after 1980 is in line with the rent-seeking explanation of Stiglitz (2012).

8. Interestingly, William Petty, who conducted the first national income estimations, treated 'great professions' as facilitators of the production and maintenance costs of the social order (Mazzucato 2018, 25).

9. When I deflate deductions by the personal consumption expenditures deflator and then subtract from real GDP, results do not change significantly. When I use each expenditure category's price deflator, the decline is 21 basis points over 1947–1973, 14 points over 1973–2000, and 18 points after 2000. However, using sectoral price deflators are not without problems because real expenditures are not additive.

10. Under miscellaneous expenditures, BLS collects safety deposit box rentals, funerals, cemetery lots, union dues, and occupational expenses in addition to financial and PBS expenditures.

11. In the NIPAs, proprietors' income is reported under net operating surplus, and I follow the same methodology. Instead of net operating surplus, GOS is preferred because I aim to adjust GDP by income, which includes depreciation.

12. Elsby, Hobijn, and Şahin (2013) mention several problems in imputing self-employment income to labor and capital because there is no observable measure that splits returns to labor and capital. I prefer to use the CE, which is an unambiguous return to labor, as labor income because its decline has driven the change in labor share in recent decades [(Elsby, Hobijn, and Şahin 2013, 16).

13. Expenditure survey data were published for 1960–1961 and 1972–1973 and then annually after 1984. However, the quintile distribution of expenditures is available only after 1984.

14. I do not adjust NT, and they increase slightly owing to the denominator effect.

15. I use the GDP deflator to deflate productivity and compensations. When I use the personal consumption expenditures deflator, results do not change significantly.

Disclosure statement

No potential conflict of interest was reported by the author(s).

References

Assa, J. (2015). "Financial Output as Economic Input: Resolving the Inconsistent Treatment of Financial Services in the National Accounts." *Working Paper 01/2015, Department of Economics, The New School for Social Research.*

Assa, J. 2016. *The Financialization of GDP: Implications for Economic Theory and Policy*. London and New York: Routledge.

Basu, D., and D. K. Foley. 2013. "Dynamics of Output and Employment in the Us Economy." *Cambridge Journal of Economics* 37 (5): 1077–1106.

Bridgman, B. 2018. "Is Labor's Loss Capital's Gain? Gross versus Net Labor Shares." *Macroeconomic Dynamics* 22 (8): 2070–2087.

Christophers, B. 2011. "Making Finance Productive." *Economy and Society* 40 (1): 112–140.

Coase, R. H. 1995. "The Nature of the Firm." In *Essential Readings in Economics* (pp. 37–54). London: Palgrave.

Eichengreen, B. 2014. "Secular Stagnation: A Review of the Issues." In C. Teulings & R. Baldwin (Eds.), *Secular Stagnation: Facts, Causes and Cures* (pp. 41–46). London: Centre for Economic Policy Research-CEPR.

Eisner, R. 1988. "Extended Accounts for National Income and Product." *Journal of Economic Literature* 26 (4): 1611–1684.

Elsby, M. W., B. Hobijn, and A. Şahin. 2013. "The Decline of the Us Labor Share." *Brookings Papers on Economic Activity* (2013 (2): 1–63.

Foley, D. K. 2013. "Rethinking Financial Capitalism and the "Information" Economy." *Review of Radical Political Economics* 45 (3): 257–268.

Goodwin, N., J. M. Harris, J. A. Nelson, B. Roach, and M. Torras. 2015. *Macroeconomics in Context*. London and New York: Routledge.

Gordon, R. J. 2015. "Secular Stagnation: A Supply-side View." *American Economic Review* 105 (5): 54–59.

Hein, E. 2015. "Finance-dominated Capitalism and Re-distribution of Income: A Kaleckian Perspective." *Cambridge Journal of Economics* 39 (3): 907–934.

Hood, K. K. 2013. "Measuring the Services of Commercial Banks in the National Income and Products Accounts." *Survey of Current Business* 93: 8–19.

Karabarbounis, L., and B. Neiman. 2014. "The Global Decline of the Labor Share." *The Quarterly Journal of Economics* 129 (1): 61–103.

Kiefer, D., I. Mendieta-Muñoz, C. Rada, and R. Von Arnim. 2020. "Secular Stagnation and Income Distribution Dynamics." *Review of Radical Political Economics* 52 (2): 189–207.

Kuznets, S. 1951. "Government Product and National Income." *Review of Income and Wealth* 1 (1): 178–244.

Kuznets, S. 1953. *Economic Change: Selected Essays in Business Cycles, National Income, and Economic Growth*. Greenwood Pub Group. New York: W.W. Norton & Company.

Mazzucato, M. 2018. *The Value of Everything: Making and Taking in the Global Economy*. New York: PublicAffairs.

Nordhaus, W. D., and J. Tobin. 1972. *Economic Research: Retrospect and Prospect, Volume 5, Economic Growth*. Cambridge: NBER.

International Monetary Fund. 1993. *System of National Accounts, 1993*. Brussels/Luxembourg, New York, Paris, Washington, DC: INTERNATIONAL MONETARY FUND.

Piketty, T. 2014. *Capital in the Twenty-first Century, Trans. Arthur Goldhammer*. Cambridge: Belknap.

Pressman, S. 2015. Understanding Piketty's Capital in the Twenty-first Century. London and New York: Routledge.

Stiglitz, J. E. 2012. *The Price of Inequality: How Today's Divided Society Endangers Our Future*. New York, London: WW Norton & Company.

Stiglitz, J. E. 2019a. *Measuring What Counts: The Global Movement for Well-Being*. New York: New Press.

Stiglitz, J. E. 2019b. *People, Power, and Profits: Progressive Capitalism for an Age of Discontent*. New York, London: WW Norton & Company.

Stiglitz, J. E., A. Sen, and J.-P. Fitoussi. 2010. *Mismeasuring Our Lives: Why GDP Doesn't Add Up*. New York: New Press.

Studenski, P. 1961. *The Income of Nations*. New York: New York University Press.

Summers, L. H. 2015. "Demand Side Secular Stagnation." *American Economic Review* 105 (5): 60–65.

Taylor, L., and Ö. Ömer (2018). "Where Do Profits and Jobs Come From? Employment and Distribution in the Us Economy." *Working Paper No.72, Institute for New Economic Thinking*.

Temin, P. 2020. "Finance and Intangibles in American Economic Growth: Eating the Family Cow." *International Journal of Political Economy* 49 (1): 23–42.

Vanoli, A. 2005. *A History of National Accounting*. Washington, DC: IOS Press..

Appendix: Adjustments to GDP by the value added

First, final expenditures recorded under the finance, healthcare, and PBS rows of the Input-Output Use Tables of the BEA are subtracted from the consumption, investment, net exports, and government spending columns. Then, the deduction is distributed to intermediate consumption of industries. As a first approximation, I keep the value added shares constant because this approach does not require further information about the deduction's value added decomposition.

In Table A1, I apply my methodology to a hypothetical economy that produces only two commodities: A and B. Final expenditures on commodity A (B) are $10 billion ($20 billion), and intermediate consumption is $15 billion ($13 billion). On the value added side, each industry has a 50% share, and I use this ratio to treat expenditures on commodity B as intermediate consumption of the economy in Table A2.

In Table A2, I shift final expenditures on commodity B to intermediate consumption. GDP by the expenditures declines by $20 billion. I distribute this amount equally to intermediate consumption of industries; thus, GDP by the value added also falls to $10 billion. Values in the total output row and column stay constant because I change only the decomposition of the output.

Table A3 shows how this methodology applies to 2016 Input-Output Use Table A3 divides the economy into healthcare, finance, PBS, and the rest. Intermediate consumption on healthcare is negligible, whereas most of the PBS output (73%) is used as inputs by other industries. On the expenditures side, most of the final uses of healthcare and finance are consumption expenditures, and investment is the most significant expenditure on PBS. The US economy has a trade surplus in finance and PBS, but a small deficit in healthcare services. Owing to investments in research and development, US government spending on PBS is significant. On the value added side, PBS has a share of 12%, followed by finance (10%), and the healthcare industry (7%). These ratios are used in adjusting healthcare, finance, and PBS in Table A4.

In Table A4, all final expenditures on healthcare and finance are shifted to intermediate consumption of industries according to their value added shares. I shift only consumption expenditures and net exports of PBS to intermediate consumption because the rest are private and public investments in intellectual property products. For both the value added and the expenditures, GDP declines from $18,624 billion to $14,667 billion.

Table A1. Input-Output Use Tables before adjustments (in billions of dollars).

	Ind.A	Ind.B	C	I	NX	G	GDP	Total output
Commodity A	5	10	5	5	0	0	10	25
Commodity B	5	8	15	5	0	0	20	33
Total Value added	15	15					30	
Total output	25	33						58

Table A2. Input-Output Use Tables after adjustments (in billions of dollars).

	Ind.A	Ind.B	C	I	NX	G	GDP	Total output
Commodity A	5	10	5	5	0	0	10	25
Commodity B	15	18	0	0	0	0	0	33
Total Value added	5	5					10	
Total output	25	33						58

Table A3. Healthcare, finance, and professional-business services (PBS) in 2016 before adjustments (billions of current dollars).

	Health	Finance	PBS	Rest	C	I	NX	G	GDP	Output
Health	26	0	0	23	2,438	0	−2	0	2,436	2,485
Finance	279	847	215	908	988	143	83	0	1,214	3,463
PBS	269	404	622	1,776	254	662	51	187	1,154	4,225
Rest	381	409	533	6,769	9,141	2,252	−653	3,081	13,821	21,912
Value added	1,349	1,835	2,252	13,188					18,624	
Output	2,304	3,496	3,622	22,664						32,085

Table A4. Healthcare, finance, and professional-business services (PBS) in 2016 before adjustments (billions of current dollars).

	Health	Finance	PBS	Rest	C	I	NX	G	GDP	Output
Health	197	244	292	1753	0	0	0	0	0	2,485
Finance	364	968	361	1,770	0	0	0	0	0	3,463
PBS	290	435	659	1,993	0	662	0	187	849	4,225
Rest	381	409	533	6,769	9,141	2,252	−653	3,081	13,821	21,912
Value added	1,072	1,439	1,777	10,379					14,667	
Output	2,304	3,496	3,622	22,664						32,085

Sovereign currency and long-term interest rates

Hongkil Kim

ABSTRACT
This paper investigates the effects of government debt and deficits on long-term interest rates in 17 advanced economies over the period 1973–2016 from the perspective of currency sovereignty. The empirical findings of this paper suggest the market penalizes non-sovereign nations for the same amount of fiscal deficit with higher interest rates than sovereigns. In addition, non-sovereign countries face accelerating interest rates for an increase in the debt-to-GDP ratio beyond a certain threshold (49% to GDP) while such a pattern is not obvious among sovereign nations. Overall, the results support Modern Monetary Theory (MMT) view that a monetarily sovereign government, as a monopoly issuer of currency, can influence the prices of their liabilities to a significant extent, somewhat independent of existing public debt and market sentiment.

1. Introduction

What explains why some countries have not experienced higher interest rates even as their sovereign debt expands to ever greater proportions of GDP and fiscal deficits continue apace while others are faced with higher interest rates? Given that the marked rise in borrowing needs and the decrease in GDP during the Covid-19 driven crisis have led to the significant increase in the deficit/debt-to-GDP ratio especially for developed economics, it raises a number of important issues regarding the impact of fiscal deficits and public debt on long-term interest rates.[1]

In this paper, we highlight monetary institutions or the presence/absence of a sovereign currency in public debt and government bond rates as acknowledged most notably in Modern Monetary Theory (hereafter, MMT).[2] MMT begins with differentiating a monetarily sovereign and non-sovereign state, further emphasizing the capacity of the former with its domestic, floating currency to issue its currency at will and thus administer interest rates exogenously to their advantages (Wray 2006).[3] It was easily revealed in monetary policies pursued in US, UK, Australia, and Japan vs. eurozone economies during the Great Recession; the former kept interest rates low despite the growing debt-to-GDP ratios, thus minimizing crowding out and making it easier for the government to conduct expansionary policy at cheap borrowing costs while some of the latter, faced with spikes on government bond yield, had to rely on other entities such as the International Monetary Fund (IMF), the European Financial Stability Facility (EFSF), and the European Stability Mechanism (ESM) for external funds. Their lack of sovereign currency limited their ability to respond adequately to a crisis.

Our investigation reveals the importance of currency sovereignty in the relationship between fiscal variables and long-term interest rates in 17 advanced economies over the period 1973–2016. Specifically, we empirically find that the impact of fiscal variables on long-term interest rates hinges on the monetary and exchange rate regime of a country. The link is economically and statistically strong, and robust to alternative model specifications, the sample period, outlier countries, included control variables, and an alternative left-hand side variable. We find that the same amount of the fiscal deficit of non-sovereign countries leads to higher interest rates than that of sovereign countries. In addition, non-sovereign countries face accelerating interest rates for an increase in the debt-to-GDP ratio while such a pattern is not obvious among sovereign nations. One potential explanation for the larger financial crowding out effect and the positive, non-linear relationship between debt and long-term interest rates of the former derives from their lack of capacity to issue their own currencies at will. For that reason, non-sovereign countries are subject to additional risks and stronger market discipline, not capable of implementing accommodative monetary policy tools on its own to bring the rates down. Alternatively, the conventional theories of fiscal variables and interest rates receive little support from our data in monetarily sovereign countries.

The rest of the paper is structured as follows: Section 2 discusses currency sovereignty and the relevant empirical literature. Section 3 describes the data and the empirical technique employed in the analysis. Section 4 presents and discusses the empirical findings. The last section concludes with policy implication.

2. Sovereign currency and interest rates

In the standard neoclassical (loanable funds) model, fiscal deficits deplete national savings and increase aggregate demand, leading to an excess supply of government debt and thus raising interest rates, with other things being equal (Elmendorf and Mankiw 1998). In the conventional IS-LM model, the deterioration of fiscal balance contributes to aggregate demand and leads to an increase in interest rates regardless of the primary policy instrument (i.e. money supply vs. interest rates) adopted by a central bank (Taylor 1993). Although the immediate effect of fiscal expansion is on short-term interest rates, long-term interest rates could rise more in response to the anticipated persistence (or even worsening) of fiscal deficits (Blanchard 1984). According to the expectations hypothesis of the term-structure of interest rates, market participants' expectation of the future stance of monetary policy as well as the current stance determine long-term interest rates (Moore 1991). In addition, government debt plays an important role in determining interest rates through default risk (Manasse, Roubini, and Schimmelpfennig 2003; Eaton and Fernandez 1995; Drudi and Prati 2000). When there are concerns about the government's ability to service its debts, the market would raise default (credit) risk premia and government bond yields. Nominal interest rates on government bonds could also increase in recognition of higher risk that derive from monetization-induced inflation and depreciation particularly when public debt is considered too large (Sargent and Wallace 1981; Dai and Philippon 2005). In short, as fiscal deficits and public debt rise, bond yields should increase due to financial crowding out and the heightened risks associated with future interest rates, default, and monetization-driven depreciation and inflation.

MMT claims that much of the standard theories above regarding fiscal deficit/public debt-interest rate relation may not hold among sovereign countries because these theories were formulated in the previous monetary systems, not the current fiat monetary system (Mitchell 2009).

A sovereign currency is issued by a monetarily sovereign government. Three essential elements for a sovereign government are (1) the government issues the national currency and imposes tax liabilities in that currency, (2) the currency is fully floating and non-convertible, and (3) the nation has no debt denominated in foreign currency (Mitchell, Wray, and Watts 2019). Examples of non-sovereign countries in point are currently 19 eurozone countries that abandoned their national currencies and effectively adopted a foreign currency (the euro) that the only ECB is authorized to issue. Central banks of the eurozone, as a result, do not have the power to issue their new currency, violating the first qualification. Other important cases are European countries that, after the breakdown of the Bretton Woods system, peg their currencies to a European Currency Unit (ECU), made up of a basket of European currencies, violating the second requirement. Their central banks technically did issue their own currencies, but their commitment to maintain exchange rates within a certain range rids them of the autonomy to set the price of and quantity of their currencies.[4]

The significance in currency sovereignty has an implication on the relationship between fiscal variables and government bond yields. A sovereign government, as a monopoly issuer of currency, can influence the prices of their liabilities to a significant extent (Wray 2006). First, sovereign central banks exogenously set the overnight rate target, affecting market short-term interest rates to a large degree. Although they leave medium- and long-term interest rates on government bonds determined by 'market forces,' the market does not penalize an increase in public debt with higher default risk because monetarily sovereign governments are always solvent and can always fulfill financial obligations to their bond holders in their currencies (Wray 2012).[5] In other words, sovereign governments do not face risk of involuntary default. Second, sovereign central banks can directly intervene in government bonds markets by purchasing assets, underpinning prices and thus lowering yields, somewhat independent of existing public debt and market sentiment as clearly shown in QE policy. Such actions gave signaling effects to the market about the probability of the future policy intervention, optimizing the impact of the purchase (Valiante 2015). Sovereign central banks are also endowed with maximum room for domestically oriented monetary policy as floating exchange rates provide a shock absorber against external macroeconomic shocks, if not completely (Obstfeld 2015). With those benefits, sovereign central banks can resist (or at least minimize) the crowding out effect, and rising public debt of sovereign countries does not have to raise interest rates.

Non-sovereign governments are, however, in a quite different interest rate environment, generating constraints on both fiscal and monetary policy (Wray 2006). The eurozone countries must obtain euros first for any government spending (or asset purchase) program. Given that they, as currency users, do not have the currency-issuing power, unlike sovereign countries, their spending is limited by tax revenues and bond issuance. When non-sovereign countries run large fiscal deficits and accumulate an excessive amount of debt, the market would raise default risk premia and government bond yields because default is a legitimate possibility for non-sovereign governments. While higher interest rates may not follow if an increase in the fiscal variables is offset by an inflow of foreign capital, an increase in the marginal propensity to save, investors' preference toward liquid government bonds, or emergency funds/QE-type policy conducted by other external entities, non-sovereign countries are not immune to the market perception of default risk and it works as market discipline that calls for fiscal prudence and fiscal consolidation (Kim 2020).

Likewise, fiscal loosening in countries that peg their currencies to a foreign currency is quite dangerous especially when they do not have sufficient reserves of foreign currency. When an increment of the public debt-to-GDP ratio (or large fiscal deficits), for instance, stimulates an economy and ends up running trade deficits, non-sovereign central banks may have to slow down the outflows of foreign reserves by raising interest rates in order to defend certain exchange rates. In other words, expansionary fiscal policy of non-sovereign countries is almost necessarily followed by a tighter monetary policy (higher interest rates), and, as a consequence, monetary policy is held hostage to the exchange rate such that the interest rate is considered endogenous (Wray 2006).[6] Market anticipation of such a rising future policy rate also raises long-term interest rates (the expectations hypothesis). Moreover, 'sudden stop' or speculative capital outflows (self-fulfilling or anticipatory) that disrupt fixed exchange rates can lead to higher interest rates. Their dedication to a fixed exchange rate, therefore, overrides their intrinsic power to administer interest rates exogenously. In a nutshell, government bond yields of non-sovereign countries are positively associated with fiscal deficits/public debt, the marginal effect of which on the bond yield depends on the market perception of the default risk, future interest rates, and potential capital outflow.

Although the effect of fiscal deteriorations on interest rates has been the subject of extensive empirical literature, only a few papers have focused on the link from the perspective of currency sovereignty. Sharpe (2013) and De Grauwe and Ji (2013) found that the market did not penalize sovereign countries for high public debt-to-GDP ratios that seemed to be equally unsustainable as non-sovereign countries since the market knew the former could always issue their currency and pay back the bondholders at maturity, while the latter did not have such a guarantee and thus more vulnerable to negative market sentiments. Akram and Das (2014, 2015) and Akram and Li (2017) found that the actions of sovereign central banks primarily determined long-term interest rates through short-term interest rates and various monetary policy measures, but government debt did not have any adverse effect on yields in the long-run. These papers are distinguished from a large body of literature that focus on fiscal factors (Laubach 2009; Baldacci and Kumar 2010; Gruber and Kamin 2012; Bernoth, Hagen, and Schuknecht 2012) and non-domestic variables such as over-reaction of financial markets (Aizenman, Hutchison, and Jinjarak 2013; Ang and Longstaff 2013; De Grauwe and Ji 2013; Beirne and Fratzscher 2013; D'Agostino and Ehrmann 2014; Poghosyan 2014), global fiscal policy (Ardagna, Caselli, and Lane 2007; Dell'Erba and Sola 2016), and external competitiveness/imbalances (Costantini, Fragetta, and Melina 2014; Salem and Castelletti-Font 2016), all of which have not taken into account underlying monetary and exchange rate arrangements of countries. Once recognizing the importance of currency sovereignty, it is perhaps not surprising to see heterogeneous findings regarding the impact of fiscal deterioration on interest rates.

Given the above a priori considerations, we seek to extend the analysis of Sharpe (2013) by testing the hypothesis with an expanded panel of 17 advanced economies over a longer time period (1973–2016). Since the early 1970s breakdown of the Bretton Woods system, the share of countries with pegged exchange rates fell dramatically from about 90% in 1970 to about 40% by the 1980s (Obstfeld and Taylor 2017). Despite such a trend, many European countries established their own fixed exchange rate system in the 1970s, a precursor of the euro with Germany as the center county. Their efforts to reduce the volatility in exchange rates limit their power to issue currency and thus monetary policy space, qualifying them as non-sovereign countries. Therefore, about half of countries in our sample are non-sovereign in the post-Bretton Woods era including not only the period when the European Economic Community (EEC) members fixed their exchange rates, followed by the inception of the euro, but also the recent episode of the global financial crisis and the post-recession economies.

3. Data and methodology

3.1. Data and descriptive analysis

Our database contains annual macroeconomic and fiscal variables taken from *Jordà-Schularick-Taylor Macrohistory Database*, and covers the post-Bretton Woods period from 1973 to 2016.[7] Jordà, Schularick, and Taylor (2017) compiled a long-run macro-financial dataset that covers 17 advanced economies since 1870 on an annual basis. This paper, however, studies the period of the fiat monetary system, followed by the breakdown of Bretton Woods system so that government bond yields of monetarily sovereign governments can be contemporaneously compared and contrasted to those of non-sovereign governments. The countries included are 17: Australia, Belgium, Canada, Denmark, Finland, France,

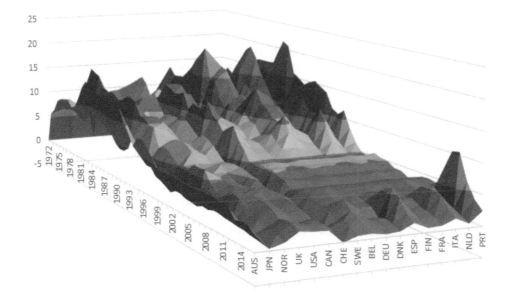

■ -5-0　■ 0-5　■ 5-10　■ 10-15　■ 15-20　■ 20-25

Figure 1. Shows the location of map of (a) Sindh province, Pakistan (b) Tharparkar district (c) taluka Dahili study area.

Germany, Italy, Japan, Netherlands, Norway, Portugal, Spain, Sweden, Switzerland, UK and US.

The importance of sovereign currency can be observed when looking at the behavior of long-term interest rates in our sample (Figure 1). We have grouped the sovereign countries from the left, followed by non-sovereign countries (Belgium, Denmark, Finland, France, Germany, Italy, Netherlands, Spain, Portugal). Long-term interest rates are higher at the beginning of the sample in all countries. Starting from the early 1990s, there appears to be a downward trend of 10-year government bond rates. In particular, the bond rates in the eurozone countries were markedly converging to the German bond rate after the inception of the euro. The bond rates, however, began to diverge from the German bond rate in 2009, most notably for Portugal, Italy, and Spain while such a phenomenon did not occur among sovereign countries. Nevertheless, throughout the sample period, the government bond rates of sovereigns were lower than those of non-sovereign countries.

This evidence is in line with the hypothesis that sovereign countries might not suffer from the adverse impact of fiscal deterioration on interest rates as much as non-sovereign countries. To verify the importance of currency sovereignty in a link between fiscal variables and long-term nominal bond yields, a variety of other factors that are likely to affect bond yields must be considered in a formal econometric model.

3.2. Model specification

The econometric specification proposed for the study follows Baldacci and Kumar (2010) by (1) accounting for the nonlinear effect of public debt, (2) allowing for an additional interactive term between fiscal variables and a dummy variable reflecting a country's characteristics, and (3) adding the same control variables to achieve identification in the baseline model below. Based on the discussion in the previous section, the impact of fiscal variables on long-term rates is shaped by an economy's currency sovereignty. In particular, the overall impact of higher deficits and higher debt on long-term interest rates is expected to be more adverse in countries without sovereign currencies. To account for this factor, we add interactive terms between both the fiscal balance (*fb*)/public debt (*D*) and a dummy variable (*peg*), indicating the absence of sovereign currency as follows[8]:

$$
\begin{aligned}
Lr_{it} = {} & \alpha_i + \beta_1 Sr_{it} + \beta_2 \pi_{it} + \beta_3 z_{it-1} + \delta_1 fb_{it} + \delta_2 fb_{it}{}^* peg_{it} \\
& + \delta_3 D_{it-1} + \delta_4 D^2{}_{it-1} + \delta_5 D_{it-1}{}^* peg_{it-1} \\
& + \delta_6 D^2{}_{it-1}{}^* peg_{it-1} + \varepsilon_{it}
\end{aligned}
$$

where Lr_{it} denotes nominal yields on 10-year government bonds for country i, period t[9]; the country intercepts α_i capture country-specific fixed effects such as a country's established record of timely repayment, growth performance, or political stability, all of which affect creditworthiness of the country. Sr is the 3-month Treasury bills interest rate (to control for monetary policy); π is CPI inflation; z is output growth (to control for a country's business cycle); fb is the fiscal balance in percent of GDP; D is the level of gross general government debt in percent of GDP; peg represents currency non-sovereignty as a proxy for a country's monetary and exchange rate regime (see Appendix A); ε is the error term. Results are based on fixed effects' least squares estimates.

3.3. Exchange-rate regime coding

One of the major challenges is to classify exchange regimes because a country's actual exchange rate regime choice often departs from its self-reported status as published by the International Monetary Fund (IMF). Every country, to some degree, manages their exchange rates to their advantage and thus there is no unambiguous line between pegged vs. floating exchange rate regime (Baxter and Stockman 1989). In this paper, we follow the de facto approach developed in Shambaugh (2004) that classify countries based on 'deeds not words'. (Obstfeld and Rogoff 1995; Calvo and Reinhart 2000; Reinhart and Rogoff 2004; Levy-Yeyati and Sturzenegger 2005) since their attempt to stabilize exchange rates constrains monetary policy independence. As a result, the dummy variable (*peg*) takes the value of 1 for economies whose exchange rate stays within a ±2% band, and is 0 otherwise. According to the approach, Germany is referred to as the base country in the pre- and post-euro times because other European countries were de facto pegging to the Deutsche Mark and following the Bundesbank.[10] Moreover, the German bonds are virtually considered free of default risk in the European regions that adopt the euro as their currency. For these reasons, Germany is categorized as a sovereign country throughout the sample period. Nevertheless, strictly following its technical status as a non-sovereign country or even completely excluding it from the sample does not change the results.

3.4. Unit root properties of the data

The stationarity properties of nominal interest rates (on 10-year government bonds and 3-month Treasury bills), the inflation rate, and the fiscal balance and public debt as a share of GDP are examined using the unit root test for panel data proposed by Im, Pesaran, and Shin (2003) and Maddala and Wu (1999). As a result, all variables are found to be stationary (see Appendix B). Hence, we estimate our models in levels. To estimate the baseline model, we mainly draw on instrumental variable LSDV. The difference and system GMM estimators that work best for small T and large N (Bond 2002) are not appropriate for our sample size of T = 44 and N = 17. Therefore, LSDV estimation is used following Judson and Owen (1999)'s recommendation and the vast majority of existing studies with the second and third period lag for the fiscal balance so as to deal with the potential endogeneity of the fiscal balance to changes in debt service.

4. Estimation results and discussion

4.1. Model estimates

The econometric analysis adopts a three-step approach: in step one, we estimate baseline models in Baldacci and Kumar (2010) and other studies that examine linear and non-linear relation between fiscal variables and interest rates. In step two, we estimate the specification of our baseline model that includes the interaction terms between the fiscal variables and the non-sovereign currency dummy variable. Third, for robustness checks, we (1) augment our baseline model with potentially relevant variables, (2) modify our sample coverage to exclude an outlier or a structurally inconsistent period, and (3) use an alternative left-hand side variable.

Table 1. Impact of deficits and debt on long-term interest rates.

	Dependent variable: 10-year government bond yields							
	[1]	[2]	[3]	[4]	[5]	[6]	[7]	[8]
Short-term Interest rates	0.70***	0.69***	0.70***	0.68***	0.69***	0.68***	0.69***	0.67***
	(0.03)	(0.03)	(0.03)	(0.03)	(0.03)	(0.03)	(0.03)	(0.03)
Inflation	10.11***	9.13***	10.56***	12.25***	10.99***	12.32***	11.06***	12.27***
	(3.16)	(3.28)	(3.29)	(2.76)	(3.17)	(2.88)	(3.14)	(2.91)
Initial GDP Growth	-7.53***	-7.51**	-7.95***	-6.22**	-7.19***	-6.58**	-7.36***	-5.84**
	(2.87)	(2.93)	(2.90)	(2.82)	(2.75)	(2.88)	(2.73)	(2.77)
Fiscal Balance	-15.69***	-17.05***	-16.12***	-11.35***	-17.18***	-12.07***	-17.24***	-12.62***
	(2.64)	(2.77)	(2.66)	(2.85)	(2.78)	(2.84)	(2.76)	(2.81)
Initial Public Debt		-0.48**	0.90*	-0.48**	-0.76***	0.59	0.23	0.27
		(0.22)	(0.50)	(0.22)	(0.21)	(0.50)	(0.51)	(0.52)
(Initial Public Debt)^2			-0.65***			-0.49***	-0.53***	-0.52***
			(0.17)			(0.17)	(0.18)	(0.18)
Fiscal Balance*Peg				-14.35***		-14.05***		-16.71***
				(5.49)		(5.41)		(6.02)
Initial Public Debt*Peg					1.04***		-0.84	-2.35**
					(0.40)		(0.90)	(0.74)
(Initial Public Debt)^2*Peg							1.47**	2.40***
							(0.73)	(0.77)
Constant	2.19***	2.53***	1.97***	2.06***	2.32***	1.96***	2.17***	2.27***
	(0.16)	(0.22)	(0.31)	(0.15)	(0.26)	(0.30)	(0.29)	(0.30)
Number of observations	714	711	711	714	711	711	711	711
R-squared	0.90	0.90	0.90	0.90	0.90	0.90	0.90	0.90
Hansen J statistic (p-value)	0.53	0.83	0.66	0.81	0.78	0.89	0.68	0.97

All specifications include country fixed effects.

The table reports the regression coefficients and, in brackets, the associated standard errors (Hubert–White sandwich correction), which are robust to heteroskedasticity and adjusted for clusters by countries.

***p < 0.01, **p < 0.05, *p < 0.1.

The main results are presented in Table 1. Columns 1–3 show that the coefficients of the fiscal balance are negative and statistically significant at conventional levels, implying that a 1-percentage deterioration in the fiscal balance as a share of GDP is associated with an increase in the 10-year government bonds rate from a minimum of 15.69 basis points to a maximum of 17.05 basis points. On the other hand, we do find a negative and statistically significant linear relationship between long-term interest rates and the stock of public debt as a share of GDP in one specification (column 2), suggesting that a 2-percentage point increase in the stock of public debt as a share of GDP is approximately associated with a decrease of a 1 basis point of the 10-year government bonds interest rate. The estimation of the specification in column 3 yields results similar to those of the baseline model in Baldacci and Kumar (2010) and Gruber and Kamin (2012) in terms of the signs of the coefficients; the positive coefficient of initial public debt and the negative coefficient of initial public debt squared indicate that the effect of public debt is non-linear, and the response of long-term interest rates is negative and statistically significant only when the stock of public debt is above 69% relative to GDP.

These negative linear and non-linear impacts of public debt on 10-year government bonds rates can be explained with a portfolio effect (Caporale and Williams 2002). Portfolio theory states that when a government issues high-quality, low-risk debt, investors switch into them from bad-quality, high-risk debt, putting downward pressure on the bond yield of the former. It implies that an increase in public debt is negatively associated with interest rates when the stock of debt is low. However, further increases beyond a given threshold are associated with higher interest rates as investors demand a higher default risk premium. If portfolio theory works, the coefficient of initial public debt and that of initial public debt squared should be negative and positive, respectively, in column 3. Our estimation results, however, provide little support for portfolio theory. Rather than the marginal impact of debt on the interest rate intensifying as might be expected, it becomes mitigated as the debt ratio increases.

This lack of evidence for portfolio theory can be addressed by introducing the concept of currency sovereignty as in our baseline model. Column 8 now has expected signs, providing strong evidence that portfolio theory works but only in non-sovereign countries. The coefficients of the interaction variables between initial public debt squared and non-sovereignty currency in columns 7 and 8 are positive and statistically significant, indicating that the impact of an increase in public debt of non-sovereign nations beyond a certain tipping point (49% to GDP) raises long-term bond yields exponentially. While in a non-sovereign country with a debt-to-GDP ratio of 131% (Italy in 2016), a 10-percentage increase in government debt leads to an increase in the nominal interest rate on 10-year government bonds of about 28 basis points, an increase by the same amount where the public debt-to-GDP ratio is 63% (Finland in 2016) leads to a 3 basis points increase in the interest rate. By contrast, this positive non-linear relationship between public debt and long-term interest rates is not found in sovereign countries as shown in negative and statistically significant coefficients of initial public debt squared in columns 6, 7, and 8. In a sovereign country with a debt-to-GDP ratio of 106% (US in 2016), a 10-percentage increase in government debt is associated with a decrease in the nominal interest rate on 10-year government bonds of about 8 basis points. Moreover, the results presented in columns 4–8 of Table 1 clearly show that the impacts of fiscal deterioration on long-term interest rates are more sizable in non-sovereign than in sovereign countries. Non-sovereign countries have to bear an additional increase in long-term interest rates from 14.05 to 16.71 basis points for a 1-percentage increase in fiscal deficit to GDP with statistical significance in columns 4, 6, and 8.

These results seem perplexing at first sight, but are actually consistent with the MMT proposition that the relationship between public debt and long-term interest rates depends on monetary and exchange rate arrangements of a country. We find evidence supporting the hypothesis that rising public debt of sovereign governments does not have to positively contribute to their bond yields because sovereign governments are not subject to default risk and their accommodative policy can exogenously dampen upward pressures in long-term market interest rates in a significant way, somewhat independent of existing public debt and market sentiment. By contrast, non-sovereign economies especially with a high level of debt are faced with rising sovereign bond yields through, for example, the default risk premium, as implied by conventional models of sovereign debt crises which link the default risk to the debt ratio (Manasse, Roubini, and Schimmelpfennig 2003). Also, the conventional theories of financial crowding out now seem more pertinent (stronger) to non-sovereign countries. These estimation results add new insight to a large but inconclusive empirical literature that fails to differentiate sovereign from non-sovereign economies and hence impose the restriction that these fiscal effects are the same across countries under different monetary and exchange rate arrangements. This finding is alternatively in congruence with De Grauwe and Ji (2013) and Sharpe (2013).

The coefficients on all other explanatory variables are of the expected sign and significant at conventional levels. A 1-percentage increase in short-term rates raises long-term bond yields by almost 70 basis points. Inflation is also positively associated with yields; yields increase from 9.13 to 12.32 basis points for each 1-percentage increase in inflation. Economic growth has an additional effect on yields; a 1-percentage increase in initial GDP growth is associated with a decrease in the 10-year government bonds yield from 5.84 to 7.95 basis points.

4.2. Robustness tests

Several robustness checks are performed to ensure our results are not unduly sensitive to a change in model specification or sample coverage. First, the baseline model is augmented separately with additional variables such as current account, exchange rate, IMF financial development index, broad money (M3) as a share of GDP, net foreign asset position as a share of GDP, capital control, total return on equity, global financial risk, and systemic financial crises.[11] Following existing studies, we use the IMF financial development index, the External Wealth of Nations dataset (Lane and Milesi-Ferretti 2007), the IMF's classifications of capital mobility restrictions, and VIX index (a measure of the volatility implied in the pricing of options on US stocks compiled by the Chicago Board Options Exchange) as proxies of financial development, net foreign asset position, capital control, and global financial risk, respectively.

Table 2 demonstrates that the results are robust to alternative model specifications. Our results interestingly do not support Salem and Castelletti-Font (2016)'s findings that decreasing (increasing) sovereign yields of core (periphery) economies are explained by their external positions after controlling for fiscal factors. In particular, financial development and broad money relative to GDP turns out to be negatively associated with the interest rate on 10-year government bonds with statistical significance. Our estimation results also suggest that the presence of systemic financial crises leads to a decrease in the long-term interest rate. Negative

Table 2. Additional variables and impact of deficits and debt on long-term interest rates.

	Dependent variable: 10-year government bond yields								
	[1]	[2]	[3]	[4]	[5]	[6]	[7]	[8]	[9]
Short-term Interest rates	0.67***	0.67***	0.53***	0.65***	0.66***	0.67***	0.66***	0.68***	0.68***
	(0.03)	(0.03)	(0.05)	(0.03)	(0.03)	(0.03)	(0.03)	(0.04)	(0.03)
Inflation	12.25***	12.41***	17.21***	11.37***	13.06***	11.65***	13.11***	9.23**	12.23***
	(2.93)	(3.07)	(3.18)	(3.01)	(2.88)	(3.40)	(2.96)	(4.10)	(2.80)
Initial GDP Growth	-5.77**	-5.77**	-3.02	-6.77**	-5.90**	-5.96**	-5.10*	-0.98	-5.85**
	(2.80)	(2.75)	(2.04)	(2.63)	(2.88)	(2.79)	(2.74)	(2.95)	(2.76)
Fiscal Balance	-12.96***	-12.60***	-10.16***	-14.24***	-10.75***	-12.96***	-11.73***	-16.33***	-12.12***
	(3.85)	(2.84)	(2.13)	(2.79)	(3.45)	(2.52)	(2.87)	(3.01)	(2.84)
Initial Public Debt	0.22	0.31	0.91*	0.33	0.40	0.09	0.26	1.28*	0.32
	(0.52)	(0.58)	(0.54)	(0.45)	(0.53)	(0.58)	(0.49)	(0.75)	(0.54)
(Initial Public Debt)^2	-0.50***	-0.53***	-0.54***	-0.24	-0.51**	-0.46**	-0.54***	-0.79***	-0.54***
	(0.18)	(0.20)	(0.18)	(0.17)	(0.22)	(0.20)	(0.18)	(0.23)	(0.19)
Fiscal Balance*Peg	-16.41***	-17.16***	-8.54*	-12.46**	-17.76***	-14.80***	-16.38***	-15.52**	-15.92***
	(6.16)	(5.65)	(4.39)	(5.86)	(5.89)	(5.38)	(6.24)	(7.32)	(5.87)
Initial Public Debt*Peg	-2.32**	-2.44**	-2.37*	-2.40**	-2.50**	-2.11*	-2.52***	-0.74	-2.21*
	(0.98)	(1.01)	(1.14)	(1.00)	(1.00)	(1.10)	(0.97)	(1.28)	(0.99)
(Initial Public Debt)^2*Peg	2.38***	2.41***	2.09***	2.19***	2.33***	2.38***	2.59***	1.45**	2.28***
	(0.77)	(0.77)	(0.74)	(0.80)	(0.85)	(0.81)	(0.81)	(0.73)	(0.78)
Current Account	0.64								
	(1.96)								
Exchange Rate		0.00							
		(0.00)							
Financial Development			-5.20***						
			(0.84)						
Money Supply (M3)				-2.38***					
				(0.38)					
Net Foreign Asset					-0.46				
					(0.32)				
Capital Control						0.53			
						(0.56)			
Return on Equity							0.48		
							(0.313)		
VIX								0.02	
								(0.02)	
Financial Crises									-0.86**
									(0.37)
Constant	2.27***	2.23***	6.03***	4.09***	2.33***	2.29***	2.28***	0.81	2.27***
	(0.30)	(0.34)	(0.76)	(0.32)	(0.30)	(0.31)	(0.31)	(0.60)	(0.29)
Number of observations	711	711	626	711	694	711	653	458	711
R-squared	0.90	0.90	0.94	0.91	0.90	0.90	0.90	0.88	0.90
Hansen J statistic (p-value)	0.98	0.95	0.63	0.95	0.83	0.95	0.96	0.33	0.99

Table 3. Impact of deficits and debt on long-term interest rates with restriction on sample.

	[1] 1973–2007	[2] 1986–2016	[3] Excluding USA	[4] Excluding Japan	[5] Excluding Germany	[6] Excluding Italy	[7] Excluding Spain	[8] Excluding Portugal
				Dependent variable: 10-year government bond yields				
Short-term Interest rates	0.60***	0.68***	0.67***	0.67***	0.67***	0.66***	0.66***	0.68***
	(0.04)	(0.04)	(0.03)	(0.03)	(0.03)	(0.03)	(0.03)	(0.03)
Inflation	14.24***	13.98***	13.06***	12.56***	12.91***	12.76***	15.99***	11.71***
	(3.12)	(3.94)	(3.02)	(2.98)	(2.91)	(3.25)	(2.19)	(2.29)
Initial GDP Growth	−6.99**	−2.11	−5.87**	−6.58**	−6.14*	−4.31	−5.50**	−6.32**
	(2.73)	(2.61)	(2.77)	(2.98)	(3.49)	(3.17)	(2.55)	(2.82)
Fiscal Balance	−15.25***	−15.86***	−12.23***	−12.43***	−11.65***	−11.60***	−11.75***	−11.48***
	(3.59)	(2.59)	(2.76)	(2.73)	(2.84)	(2.96)	(2.98)	(2.61)
Initial Public Debt	1.54***	1.31**	0.67	1.53*	1.06*	−0.19	−0.04	0.22
	(0.55)	(0.62)	(0.53)	(0.89)	(0.57)	(0.60)	(0.52)	(0.50)
(Initial Public Debt)^2	−1.25***	−0.81***	−0.64***	−1.69**	−0.79***	−0.33	−0.38**	−0.48***
	(0.31)	(0.20)	(0.19)	(0.82)	(0.20)	(0.21)	(0.18)	(0.18)
Fiscal Balance*Peg	−15.36***	−5.48	−16.98***	−15.87***	−16.97***	−24.99***	−15.04***	−13.98***
	(5.46)	(4.94)	(5.97)	(5.87)	(5.98)	(6.53)	(5.13)	(5.20)
Initial Public Debt*Peg	−3.31***	−0.96	−2.41**	−3.05***	−2.54**	−1.93	−2.82***	−1.47
	(1.21)	(0.93)	(0.95)	(1.00)	(0.98)	(1.17)	(0.94)	(0.94)
(Initial Public Debt)^2*Peg	2.73***	1.61**	2.35***	3.26***	2.35***	2.18***	2.87***	1.51*
	(1.03)	(0.62)	(0.77)	(0.94)	(0.77)	(0.84)	(0.75)	(0.78)
Constant	2.62***	1.33***	2.09***	2.07***	1.97***	2.35***	2.37***	2.24***
	(0.34)	(0.33)	(0.31)	(0.31)	(0.32)	(0.30)	(0.30)	(0.31)
Number of observations	558	526	669	669	669	669	669	669
R-squared	0.89	0.91	0.90	0.90	0.90	0.89	0.91	0.90
Hansen J statistic (p-value)	0.64	0.28	0.99	0.96	0.92	0.80	0.98	0.71

Table 4. Impact of deficits and debt on long-term real interest rates.

	Dependent variable: real 10-year government bond yields				
	[1]	[2]	[3]	[4]	[5]
Real Short-term Interest rates	0.80***	0.81***	0.81***	0.80***	0.80***
	(0.03)	(0.02)	(0.02)	(0.02)	(0.02)
Initial GDP Growth	−12.99***	−11.59***	−11.53***	−10.98***	−11.50***
	(3.10)	(2.65)	(2.26)	(2.43)	(2.23)
Fiscal Balance	−8.41***	−6.46**	−7.20**	−10.89***	−9.71***
	(3.04)	(2.94)	(2.42)	(2.43)	(2.52)
Initial Public Debt		1.30***	4.67***	0.41***	3.32***
		(0.25)	(0.54)	(0.26)	(0.64)
(Initial Public Debt)^2			−1.80***		−1.41***
			(0.26)		(0.28)
Fiscal Balance*Peg	−15.27**	−12.87**	−8.05*	−2.89	−3.61
	(6.40)	(6.08)	(4.49)	(4.92)	(5.03)
Initial Public Debt*Peg				2.12***	0.33
				(0.39)	(0.70)
(Initial Public Debt)^2*Peg					0.85*
					(0.48)
Constant	1.19***	0.47**	0.62***	0.43**	−0.27
	(0.15)	(0.20)	(0.23)	(0.18)	(0.23)
Number of observations	714	711	711	711	711
R-squared	0.81	0.83	0.83	0.83	0.84
Hansen J statistic (p-value)	0.67	0.24	0.14	0.23	0.12

All specifications include country fixed effects.
The table reports the regression coefficients and, in brackets, the associated standard errors (Hubert–White sandwich correction), which are robust to heteroskedasticity and adjusted for clusters by countries.
***p < 0.01, **p < 0.05, *p < 0.1.

estimates of systematic financial crises, however, should not be interpreted as a financial crisis would lead to lower yields, but as monetary authorities have more than offset its adverse impact on sovereign government markets with expansionary monetary policy.

Second, we re-estimate the regression by excluding influential (or outlier) countries such as the United States, Japan, Germany, Italy, Spain, and Portugal one at a time to explore whether our results are dominated by a specific country.[12] The conclusions are broadly unchanged in all cases as shown in columns from 3 to 8 of Table 3. Furthermore, we restrict our sample to the pre-crisis period (1973–2007) to avoid abnormal behavior of long-term interest rates in the post-crisis. Also, we begin the time sample at 1986 instead of 1973 in order to exclude the 1970s and the early 1980s, characterized by long-term interest rate peaks. Columns 1 and 2 of Table 3 provide similar estimates to those in Tables 1 and 2.

Finally, an alternative indicator – real long-term bond yields was used as a dependent variable.[13] Our results (shown in Table 4) are similar relatively to the ones in the specifications using nominal interest rates. Overall, the robustness checks confirm that the main results on the determinants of government bond yields remain intact to additional controls, different sub-samples, and alternative dependent variable.

5. Conclusions and policy implications

The impact of COVID-19 on fiscal deficits and public debt in many countries has raised concerns about their adverse effect on economies. Despite the call for more policy intervention to stop the crisis from spreading contagiously across countries and regions, **conventional** theories predict that a deficit-financed government stimulus inevitably crowds out private consumption and investment by raising interest rates. In this paper, such a link between fiscal variables and long-term interest rates is examined from the perspective of currency sovereignty.

When applied to a post-Bretton wood, 17-advanced country panel, we find evidence for a more adverse effect of fiscal deterioration among monetarily non-sovereign countries. First, financial crowding out seems more pronounced in non-sovereign countries. The market penalizes non-sovereign nations for the same amount of fiscal deficit with higher interest rates than sovereigns. In addition, an increase in public debt of non-sovereign nations beyond a certain threshold (49% to GDP) raises long-term bond yields exponentially in accordance with portfolio theory. By contrast, a positive link between public debt and default risk is not obvious among sovereign countries. The estimation results of our specification of the long-term bond yields are confirmed by the various robustness tests.

The results in this paper have an important policy implication for sovereign and non-sovereign countries, respectively. For monetarily sovereign countries, there is no theoretical or empirical ground for austerity measures, featured by attempts to run a balanced (or even surplus) budget to decrease public debt relative to GDP and thus bond yields. Because the crowding out argument and the principle of 'sound finance'[14] are less relevant to sovereign countries, premature fiscal consolidation has to be averted especially when the government's fiscal stimulus is still effective. However, many non-sovereign countries (the eurozone countries in our sample) due to the lack of their currency, may have to bear deficit/debt-reducing efforts and structural reforms as part of Germany's insistence on fiscal discipline to avoid exploding interest rates. One possibility to overcome crowding out and the positive, non-linear debt-interest rate relation is to introduce more aggressive policy instruments and effective communication by the ECB, who can issue their currency (the euro) at will for public purposes (Kim 2020). It is worth noting that the ECB's quick and accommodative monetary measures in response to the COVID-19 crisis has been successful in preventing surges in bond yields of the eurozone countries in comparison to what occurred during the European sovereign debt crisis (Hutchinson and Mee 2020). There, however, remains a concern that its unwinding (or even termination), once economic output returns to pre-crisis levels, initiates damaging impacts on interest rates, as empirically found in this paper, especially in view of the large increase in sovereign debts under way.

The analysis of this paper has focused on advanced economies, but it would be of interest to extend the analysis to emerging economies. Also, the time series can be expanded to include pre-fiat monetary system periods (e.g. the Bretton Woods system and the classical Gold Standard periods) to investigate any change in the relationship between the fiscal variables and interest rates in different international monetary systems.

Notes

1. OECD (2020) predicted that the central government marketable debt-to-GDP ratio for the OECD area would rise by 13.4 percentage to around 86% in 2020.
2. Currency sovereignty will be discussed more in detail in the next section.
3. Wray (2006) meant exogeneity in the 'control' sense in that the government can 'control' the variables such as the money supply, the interest rate, the price level, and so on. It implies that that a central bank can independently set (or administer) the target interest rate where it seems best for a domestic economy. Furthermore, accommodative policy of a sovereign central bank can exogenously affect medium and long-term market interest rates in a significant way, somewhat independent of existing public debt and market sentiment.

4. Most advanced economies, unlike emerging economies, have minimal or no foreign currency denominated public debt liabilities, which automatically satisfies the third qualification for a sovereign government. Thus, we do not discuss a case of the breach of the last qualification in this paper.

5. Nersisyan and Wray (2010) critiqued Reinhart and Rogoff's monumental studies (2009, 2010) by arguing that they misplaced emphases on the size of the public debt and its (domestic or foreign) ownership to calculate default risk. Nersisyan and Wray advocated that the correct way to analyze government finances must start from currency sovereignty because no sovereign country has defaulted on their own currency debt in Reinhart and Rogoff's sample. Japan, for example, has not defaulted despite its high debt to GDP ratio (230% in 2018) not because a majority of bonds are held by its citizen but because its debt is denominated in its own floating, nonconvertible currency and thus the central bank of Japan can always honor its liabilities by creating its currency. Second, Nersisyan and Wray (ibid.) found it problematic that Reinhart and Rogoff made an analogy between public debt and private debt. The main difference between the two is that the sovereign government always services debt liabilities insofar as they are willing, while the households or private firms have to earn money to make payments as they come due. Non-sovereign countries can be rightly compared to the private sector.

6. Exchange rates are important to sovereign countries as well, but they do not need to raise the policy interest rate in response to government deficits. The floating rate regime provides an additional degree of autonomy to monetary policy that is not available in a fixed exchange rate regime.

7. Jordà, Schularick, and Taylor consulted a broad range of sources, such as economic and financial history volumes and journal articles, and various publications of statistical offices and central banks. For more information about a source of each variable for a country, refer to their website: http://www.macrohistory.net/JST/JSTdocumentationR4.pdf.

8. It is noted that both the deficit and the debt are, in contrast to other studies, included in our baseline specification because the interaction between them is expected. Given the current stock of debt, including the deficit, for instance, may help control for the expected future path of the debt itself (Ardagna, Caselli, and Lane 2007).

9. We focus on nominal, not real interest rates because the former is generally the main policy instrument of the central bank and thus a proxy for monetary policy space. However, we also use real interest rates as an alternative left-hand side variable for a robustness test.

10. The author appreciates an anonymous reviewer for his or her comment on this.

11. Current account, exchange rate, broad money supply (M3), total return on equity, and systemic financial crises are retrieved from *Jordà-Schularick-Taylor Macrohistory Database*.

12. Non-outlier countries are additionally excluded from the sample one by one for re-estimation without a significant difference in estimation results. Estimation results that exclude non-outlier countries for robustness test are available from the author upon request.

13. Real long-term and short-term bond yields are calculated using Fisher's formula.

14. The principle of 'sound finance' is the necessity of balanced budgets over some time period or over the course of a business cycle that must be applied to households, firms, and non-sovereign governments. It is irrelevant to sovereign nations, but wrongly, widely used by politicians to justify fiscal consolidation.

Disclosure statement

No potential conflict of interest was reported by the author(s).

References

Aizenman, J., M. Hutchison, and Y. Jinjarak. 2013. "What is the Risk of European Sovereign Debt Defaults? Fiscal Space, CDS Spreads and Market Pricing of Risk." *Journal of International Money and Finance* 34 (1): 37–59. doi:10.1016/j.jimonfin.2012.11.011.

Akram, T., and A. Das. 2014. "Understanding the Low Yields of the Long-Term Japanese Sovereign Debt." *Journal of Economic Issues* 48 (2): 331–340. doi:10.2753/JEI0021-3624480206.

Akram, T., and A. Das. 2015. "A Keynesian Explanation of Indian Government Bond Yields." *Journal of Post Keynesian Economics* 38 (4): 565–587. doi:10.1080/01603477.2015.1090294.

Akram, T., and H. Li. 2017. "What Keeps Long-Term U.S. Interest Rates so Low?" *Economic Modelling* 60: 380–390. doi:10.1016/j.econmod.2016.09.017.

Ang, A., and F. A. Longstaff. 2013. "Systemic Sovereign Credit Risk: Lessons from the U.S. and Europe." *Journal of Monetary Economics* 60 (5): 493–510. doi:10.1016/j.jmoneco.2013.04.009.

Ardagna, S., F. Caselli, and T. Lane. 2007. "Fiscal Discipline and the Cost of Public Debt Service: Some Estimates for OECD Countries." *The B.E. Journal of Macroeconomics* 7 (1): 1–35. doi:10.2202/1935-1690.1417.

Baldacci, E., and M. S. Kumar. 2010. "Fiscal Deficits, Public Debt, and Sovereign Bond Yields." IMF Working Papers 10/184. International Monetary Fund.

Baxter, M., and A. C. Stockman. 1989. "Business Cycles and the Exchange-Rate Regime: Some International Evidence." *Journal of Monetary Economics* 23 (3): 377–400. doi:10.1016/0304-3932(89)90039-1.

Beirne, J., and M. Fratzscher. 2013. "The Pricing of Sovereign Risk and Contagion during the European Sovereign Debt Crisis." *Journal of International Money and Finance* 34 (1): 60–82. doi:10.1016/j.jimonfin.2012.11.004.

Bernoth, K., J. V. Hagen, and L. Schuknecht. 2012. "Sovereign Risk Premiums in the European Government Bond Market." *Journal of International Money and Finance* 31 (5): 975–995. doi:10.1016/j.jimonfin.2011.12.006.

Blanchard, O. 1984. "Current and Anticipated Deficits, Interest Rates, and Economic Activity." NBER Working Paper No. 1265. Cambridge, MA: National Bureau of Economic Research.

Bond, S. R. 2002. "Dynamic Panel Data Models: A Guide to Micro Data Methods and Practice." *Portuguese Economic Journal* 1: 141–162. doi:10.1007/s10258-002-0009-9.

Calvo, G. A., and C. M. Reinhart. 2000. ""Fixing for Your Life," NBER Working Paper No. 8006, Cambridge, MA.

Caporale, G., and G. Williams. 2002. "Long-Term Nominal Interest Rates and Domestic Fundamentals." *Review of Financial Economics* 11 (2): 119–130. doi:10.1016/S1058-3300(02)00038-1.

Costantini, M., M. Fragetta, and G. Melina. 2014. "Determinants of Sovereign Bond Yield Spreads in the EMU: An Optimal Currency Area Perspective." *European Economic Review* 70: 337–349. doi:10.1016/j.euroecorev.2014.06.004.

D'Agostino, A., and M. Ehrmann. 2014. "The Pricing of G7 Sovereign Bond Spreads – The Times, They are A-Changin." *Journal of Banking & Finance* 47 (1): 155–176. doi:10.1016/j.jbankfin.2014.06.001.

Dai, Q., and T. Philippon. 2005. "Fiscal Policy and the Term Structure of Interest Rates." NBER Working Paper No. 11574. Cambridge, MA: National Bureau of Economic Research.

De Grauwe, P., and Y. Ji. 2013. "Self-Fulfilling Crises in the Eurozone: An Empirical Test." *Journal of International Money and Finance* 34 (1): 15–36. doi:10.1016/j.jimonfin.2012.11.003.

Dell'Erba, S., and S. Sola. 2016. "Fiscal Discipline and the Cost of Public Debt Service: Some Estimates for OECD Countries." *The B.E. Journal of Macroeconomics* 16 (2): 395–437.

Drudi, F., and A. Prati. 2000. "Signaling Fiscal Regime Sustainability." *European Economic Review* 44 (10): 1897–1930. doi:10.1016/S0014-2921(99)00035-5.

Eaton, J., and R. Fernandez. 1995. "Sovereign Debt." NBER Working Paper No. 5131. Cambridge, MA: National Bureau of Economic Research.

Elmendorf, D. W., and N. G. Mankiw. 1998. "Government Debt." NBER Working Paper No. 6470. Cambridge, MA: National Bureau of Economic Research.

Gruber, J. W., and S. B. Kamin. 2012. "Fiscal Positions and Government Bond Yields in OECD Countries." *Journal of Money, Credit and Banking* 44 (8): 1563–1587. doi:10.1111/j.1538-4616.2012.00544.x.

Hutchinson, J., and S. Mee. 2020. "The Impact of the ECB's Monetary Policy Measures Taken in Response to the COVID-19 Crisis." In *Economic Bulletin Boxes*. Vol. 5. Frankfurt, Germany: European Central Bank. https://www.ecb.europa.eu/pub/economic-bulletin/html/eb202005.en.html

Im, K. S., M. Pesaran, and Y. Shin. 2003. "Testing for Unit Roots in Heterogeneous Panels." *Journal of Econometrics* 115 (1): 53–74. doi:10.1016/S0304-4076(03)00092-7.

Jordà, Ò., M. Schularick, and A. M. Taylor. 2017. "Macrofinancial History and the New Business Cycle Facts." *NBER Macroeconomics Annual* 31 (1): 213–63. doi:10.1086/690241.

Judson, R. A., and A. Owen. 1999. "Estimating Dynamic Panel Data Models: A Guide for Macroeconomists." *Economics Letters* 65 (1): 9–15. doi:10.1016/S0165-1765(99)00130-5.

Kim, H. 2020. "The Relationship between Public Debt Accumulation and Default Risk under the ECB's Conventional vs. Non-Standard Monetary Policy: A Panel Data Analysis of 9 Eurozone Countries (2000–2015)." *Journal of Post Keynesian Economics* 43 (1): 112–130. doi:10.1080/01603477.2019.1673176.

Lane, P. R., and M. Milesi-Ferretti. 2007. "The External Wealth of Nations Mark II: Revised and Extended Estimates of Foreign Assets and Liabilities, 1970–2004." *Journal of International Economics* 73 (2): 223–250. doi:10.1016/j.jinteco.2007.02.003.

Laubach, T. 2009. "New Evidence on the Interest Rate Effects of Budget Deficits and Debt." *Journal of the European Economic Association* 7 (4): 858–885. doi:10.1162/JEEA.2009.7.4.858.

Levin, A., C.-F. Lin, and C.-S. J. Chu. 2002. "Unit Root Tests in Panel Data: Asymptotic and Finite-Sample Properties." *Journal of Econometrics* 108 (1): 1–24.

Levy-Yeyati, E., and F. Sturzenegger. 2005. "Classifying Exchange Rate Regimes: Deeds vs. Words." *European Economic Review* 49 (6): 1603–1635. doi:10.1016/j.euroecorev.2004.01.001.

Maddala, G. S., and S. Wu. 1999. "A Comparative Study of Unit Root Tests with Panel Data and a New Simple Test." *Oxford Bulletin of Economics and Statistics* 61 (S1): 631–652. doi:10.1111/1468-0084.0610s1631.

Manasse, P., N. Roubini, and A. Schimmelpfennig. 2003. "Predicting Sovereign Debt Crises." IMF Working Paper 03/221.

Mitchell, W. F. 2009. "A Modern Money Perspective on the Crisis and a Reform Agenda." CofFEE Working Paper 2. Newcastle, NSW: Centre of Full Employment and Equity.

Mitchell, W. F., L. R. Wray, and M. Watts. 2019. *Macroeconomics*. London: MacMillan.

Moore, B. J. 1991. "Money Supply Endogeneity: "Reserve Price Setting" or "Reserve Quantity Setting"?" *Journal of Post Keynesian Economics* 13 (3): 404–413. doi:10.1080/01603477.1991.11489857.

Nersisyan, Y., and L. R. Wray. 2010. "Does Excessive Sovereign Debt Really Hurt Growth? A Critique of This Time Is Different". Levy Economics Institute Working Paper 603. Annandale-On-Hudson, NY: Levy Economics Institute.

Obstfeld, M. 2015. "Trilemmas and Tradeoffs: Living with Financial Globalization." In *Global Liquidity, Spillovers to Emerging Markets and Policy*, edited by C. Raddatz, D. Saravia, and J. Ventura, 13–79. Santiago, Chile: Central Bank of Chile.

Obstfeld, M., and A. M. Taylor. 2017. "International Monetary Relations: Taking Finance Seriously." *The Journal of Economic Perspectives* 31 (3): 3–28. doi:10.1257/jep.31.3.3.

Obstfeld, M., and K. Rogoff. 1995. "The Mirage of Fixed Exchange Rates." *Journal of Economic Perspectives* 9 (4): 73–96. doi:10.1257/jep.9.4.73.

OECD. 2020. "Sovereign Borrowing Outlook for OECD Countries." In *OECD Sovereign Borrowing Outlook 2020*. Paris: OECD Publishing. https://doi.org/10.1787/68622280-en.

Poghosyan, T. 2014. "Long-Run and Short-Run Determinants of Sovereign Bond Yields in Advanced Economies." *Economic Systems* 38 (1): 100–114. doi:10.1016/j.ecosys.2013.07.008.

Reinhart, C. M., and K. S. Rogoff. 2004. "The Modern History of Exchange Rate Arrangements: A Reinterpretation." *Quarterly Journal of Economics* 119 (1): 1–48. doi:10.1162/003355304772839515.

Salem, M. B., and B. Castelletti-Font. 2016. "Which Combination of Fiscal and External Imbalances to Determine the Long-run Dynamics of Sovereign Bond Yields?" Working Papers 606. Banque de France.

Sargent, T. J., and N. Wallace. 1981. "Some Unpleasant Monetarist Arithmetic." *Quarterly Review* 5 (Fall). Federal Reserve Bank of Minneapolis. doi:10.21034/qr.531

Shambaugh, J. C. 2004. "The Effect of Fixed Exchange Rates on Monetary Policy." *Quarterly Journal of Economics* 119 (1): 301–352. doi:10.1162/003355304772839605.

Sharpe, T. P. 2013. "A Modern Money Perspective on Financial Crowding-Out." *Review of Political Economy* 25 (4): 586–606. doi:10.1080/09538259.2013.837325.

Taylor, J. B. 1993. "Discretion versus Policy Rules in Practice." *Carnegie-Rochester Conference Series on Public Policy* 39 (1): 195–214. doi:10.1016/0167-2231(93)90009-L.

Valiante, D. 2015. "The 'Visible Hand' of the ECB's Quantitative Easing." CEPS Working Document No. 407.

Wray, L. R. 2006. "When are Interest Rates Exogenous?" In *Complexity, Endogenous Money, and Macroeconomic Theory: Essays in Honor of Basil J. Moore*, edited by M. Setterfield, 271–289. Cheltenham, UK: Edward Elgar Publishing.

Wray, L. R. 2012. *Modern Money Theory; A Primer on Macroeconomics for Sovereign Monetary Systems*. New York: Palgrave MacMillan.

Appendices

Appendix A. Sovereign country categorization

	AUS	BEL	CAN	DNK	FIN	FRA	DEU	ITL	JPN	NLD	NOR	PRT	ESP	SWE	CHE	UK	USA
1973	1	1	0	1	1	1	0	1	1	1	0	0	1	0	0	0	0
1974	1	1	0	1	1	1	0	1	1	1	0	0	1	0	0	0	0
1975	1	1	0	1	1	1	0	1	1	1	0	0	1	0	0	0	0
1976	1	1	0	1	1	1	0	0	1	1	0	0	1	0	0	0	0
1977	1	1	0	1	1	1	0	0	1	1	0	0	1	0	0	0	0
1978	1	1	0	1	1	1	0	0	0	1	0	0	1	0	0	0	0
1979	1	1	0	1	1	1	0	0	0	1	0	0	1	0	0	0	0
1980	1	1	0	1	1	1	0	0	0	1	0	0	1	0	0	0	0
1981	1	1	0	1	1	1	0	0	0	1	0	1	1	0	0	0	0
1982	1	1	0	1	1	1	0	0	0	1	0	1	1	0	0	0	0
1983	0	1	0	1	1	1	0	1	0	1	0	1	1	0	0	0	0
1984	0	1	0	1	1	1	0	1	0	1	0	1	1	0	0	0	0
1985	0	1	0	1	1	1	0	1	0	1	0	1	1	0	0	0	0
1986	0	1	0	1	1	1	0	1	0	1	0	1	1	0	0	0	0
1987	0	1	0	1	1	1	0	1	0	1	0	1	1	0	0	0	0
1988	0	1	0	1	1	1	0	1	0	1	0	1	1	0	0	0	0
1989	0	1	0	1	1	1	0	1	0	1	0	1	1	0	0	0	0
1990	0	1	0	1	1	1	0	1	0	1	0	1	1	0	0	0	0
1991	0	1	0	1	1	1	0	1	0	1	0	1	1	0	0	1	0
1992	0	1	0	1	1	1	0	1	0	1	0	1	1	0	0	1	0
1993	0	1	0	1	1	1	0	1	0	1	0	1	1	0	0	0	0
1994	0	1	0	1	1	1	0	1	0	1	0	1	1	0	0	0	0
1995	0	1	0	1	1	1	0	1	0	1	0	1	1	0	0	0	0
1996	0	1	0	1	1	1	0	1	0	1	0	1	1	0	0	0	0
1997	0	1	0	1	1	1	0	1	0	1	0	1	1	0	0	0	0
1998	0	1	0	1	1	1	0	1	0	1	0	1	1	0	0	0	0
1999	0	1	0	1	1	1	0	1	0	1	0	1	1	0	0	0	0
2000	0	1	0	1	1	1	0	1	0	1	0	1	1	0	0	0	0
2001	0	1	0	1	1	1	0	1	0	1	0	1	1	0	0	0	0
2002	0	1	0	1	1	1	0	1	0	1	0	1	1	0	0	0	0
2003	0	1	0	1	1	1	0	1	0	1	0	1	1	0	0	0	0
2004	0	1	0	1	1	1	0	1	0	1	0	1	1	0	0	0	0
2005	0	1	0	1	1	1	0	1	0	1	0	1	1	0	0	0	0
2006	0	1	0	1	1	1	0	1	0	1	0	1	1	0	0	0	0
2007	0	1	0	1	1	1	0	1	0	1	0	1	1	0	0	0	0
2008	0	1	0	1	1	1	0	1	0	1	0	1	1	0	0	0	0
2009	0	1	0	1	1	1	0	1	0	1	0	1	1	0	0	0	0
2010	0	1	0	1	1	1	0	1	0	1	0	1	1	0	0	0	0
2011	0	1	0	1	1	1	0	1	0	1	0	1	1	0	0	0	0
2012	0	1	0	1	1	1	0	1	0	1	0	1	1	0	1	0	0
2013	0	1	0	1	1	1	0	1	0	1	0	1	1	0	1	0	0
2014	0	1	0	1	1	1	0	1	0	1	0	1	1	0	1	0	0
2015	0	1	0	1	1	1	0	1	0	1	0	1	1	0	0	0	0
2016	0	1	0	1	1	1	0	1	0	1	0	1	1	0	0	0	0

Sovereign country categorizations is based on Shambaugh (2004). Sovereign country takes the value of 0 and non-sovereign country 1.

Appendix B. Panel unit root tests

	Im, Pesaran and Shin W-stat	ADF-Fisher Chi-square
10-yearm Government Bond Yields	−6.00*	98.87*
Short-term Interest Rate	−5.22*	89.01*
Inflation	−3.59*	71.39*
GDP Growth	−15.20*	253.91*
Fiscal Balance	−6.08*	98.33*
Public Debt	−2.89*	62.30*

The panel integration tests are based are on Levin, Lin and Chu (2002) and Maddala and Wu (1999). *Indicates that the null-hypothesis that the time series is I(1) is rejected at the 5% level.

Government expenditure and economic growth: a post-Keynesian analysis

Pintu Parui

ABSTRACT
In a post-Keynesian growth model with positive saving propensity out of wages, this paper analyses the implication of different kinds of government expenditures on aggregate demand and economic growth. We distinguish between government expenditure on consumption and investment. The basic idea is that certain kinds of government investment expenditure influences labour productivity. In a formal model, we incorporate this idea by assuming labour productivity as an increasing function of government investment expenditure. When the economy is in a profit-led demand regime, under the balanced budget assumption, we show that a shift in government expenditure from consumption to investment leads to an unambiguous rise in both aggregate demand and economic growth. However, the result is ambiguous in the wage-led demand regime. Once the balanced budget assumption is dropped, while in a wage-led demand regime a rise in government investment expenditure may decrease aggregate demand and growth, it unambiguously raises both aggregate demand and growth in a profit-led demand regime. On the other hand, in the absence of a balanced budget assumption, a rise in government consumption expenditure has a positive effect in both regimes. We also show that allowing the government to run a deficit and incur debt does not necessarily lead to the public debt rising without bounds.

1. Introduction

The recent financial crisis has forced economists and policymakers to rethink conventional wisdom regarding economic theories and policies. The old debate as to whether government expenditure can stimulate economic growth has once again emerged, in a new way. The view generally held by Keynesians is that government involvement in economic activity is vital for growth while others say that government operations are inherently bureaucratic and inefficient and therefore rather than promoting growth, stifles it.[1]

Starting from Keynes, several institutions in the literature discuss the relationship between government expenditure and growth rate, but we can find more formal analysis in the literature beginning with the work of Barro (1990). His work explains the role of public expenditure in economic growth from the supply side of the economy. The demand side analysis incorporating the effect of effective demand on economic growth is absent there. In the Keynesian analysis, although sufficient attention was paid regarding the implication of various kinds of government expenditures by its pioneers, the subject was largely overlooked later on. According to Commendatore and Pinto (2011), though Kaldor presented interesting insights regarding the relationship between the composition of government expenditure and long-run growth, there was no formal analysis. The formal analysis of the impact of government expenditure on growth more or less starts with You and Dutt (1996). While aiming to address the question of whether government debt worsens income distribution, You and Dutt show that fiscal expansion has a significant effect on the government debt-capital ratio, economic growth, and income distribution. Their analysis implies a positive relationship between fiscal expansion and the rate of economic growth in the short-run. As fiscal expansion increases, aggregate demand and the degree of capacity utilisation rises which in turn enhances the growth rate. But in the long run, the effect of fiscal expansion on the growth rate is ambiguous. This is because while fiscal expansion raises the growth rate through an increase in aggregate demand and the degree of capacity utilisation, it can either increase or decrease the government debt-capital ratio. An increase in the government debt-capital ratio has a positive impact on the growth rate. When a rise in fiscal expansion raises the government debt-capital ratio, fiscal expansion unambiguously enhances the growth rate. However, when due to a rise in the fiscal expansion, the government debt-capital ratio falls then its effect on the growth rate is ambiguous and depends on the strength of change in debt-capital ratio due to change in the ratio of government expenditure to capital.

In a later contribution in the neo-Kaleckian tradition,[2] Commendatore and Pinto (2011) analyse the impact of different kinds of government expenditure on capacity utilisation and growth. In a single-good closed economy framework with numeraire good price, they introduce two different types of public expenditure (namely government consumption expenditure and public provision of capital) to analyse the impact of those different kinds of government expenditure on capacity utilisation and growth. According to them, government consumption expenditure through an increase in effective demand increases the equilibrium degree of capacity utilisation. An increase in the equilibrium degree of capacity utilisation on the other hand increases the equilibrium growth rate. Public investment expenditure influences the level of capacity utilisation in three ways. First, it increases the effective demand. Second, it crowds-in private investment. Third, they assume the public provision of capital positively affects capital productivity by enhancing the potential output-capital ratio. On the other hand, capital productivity itself has a negative impact on the equilibrium degree of capacity utilisation. Public provision of capital influences capital productivity which in turn has a negative effect on the level of capacity utilisation. So, the final effect of a rise in public investment expenditure on equilibrium degree of capacity utilisation and capital accumulation is ambiguous and depends on the strength of negative effect on aggregate demand which comes through the enhancement of capital productivity, and the strength of positive effect on aggregate demand which comes from the increase in investment demand due to crowding-in effect.

But the ratio of physical capital to output is nearly constant. It is one of the stylized facts given by Kaldor (1963). Long-term data also show the same result (Barro and Sala-i-Martin 2004). Moreover, government investment expenditure such as expenditure on streets and highways, electricity, gas and water supply, hospitals, and education, can enhance labour productivity as well, but the analysis of the impact of government expenditure on labour productivity is absent here (in Commendatore and Pinto (2011)).

Considering a Kaleckian and a Classical-Harrodian framework, Commendatore, Panico, and Pinto (2011) explore the effect of different kinds of government expenditure on capacity utilisation and growth – they consider a non-linear investment function which depends positively on the capacity utilisation rate. When the capacity utilisation rate is below the 'normal' degree of capacity utilisation,[3] the investment function rises at an increasing rate with respect to capacity utilisation rate. On the other hand, if the capacity utilisation rate is above the 'normal' degree of capacity utilisation, it rises at a decreasing rate. This non-linearity assumption leads the investment to be an 'S-shaped' function of the degree of capacity utilisation in the sense that for a low capacity utilisation the investment propensity is weak and investment improves when capacity utilisation rises.[4] Finally, the investment slows down again when capital utilisation is high. This non-linearity of invest-ment function allows the occurrence of multiple equilibria. Two types of government expenditure are there: 'unproductive' expenditure that does not affect labour productivity and the 'productive' expenditure that has a positive effect on it. In their model, government expenditure[5] may affect labour productivity which in turn, depending on the bargaining power of the unions, affects the profit share. In the Kaleckian analysis, when the govern-ment expenditure enhances labour productivity, and when wages rise more than that of labour productivity (because of high bargaining power), government expenditure has an expansionary effect on the capacity utilisation and growth. Moreover, for both 'productive' as well as 'unproductive' government expenditure, the rate of capacity utilisation and growth moves in the opposite direction to after-tax profits. However, in the Classical-Harrodian analysis, the opposite tendency occurs. Further, in the Kaleckian analysis, the 'unproductive' expenditure always has a positive effect on growth, whereas in the Classical-Harrodian analysis the opposite occurs.

However, a profit-led demand regime is not possible in Commendatore, Panico, and Pinto (2011). This is due to the investment function they consider. Second, they assume that for a high bargaining power, a rise in labour productivity translates into a rise in the wage share. Although it is mathematically possible, given the world experience for the last several decades, it is highly an unrealistic assumption. Stockhammer (2013, 40) points out that from 73.4 in 1980, the (adjusted) wage share in the advanced economies on average has fallen to 64.0% in 2007. The decline is more pronounced in Japan (77.2 to 62.2) and relatively weaker in the United States (70.0 to 64.9). Developing and emerging economies also witness a marked decline in the wage share, at least since 1990 (Stockhammer 2013, 43). For the last several decades the US economy has observed institutional changes which have been less supportive to workers bargaining power. This is due to a reduction in the incidence of unionism and the credibility of the 'threat effect' of unionism (Stansbury and Summers 2020). There has been a decline in private-sector union membership rate from over one-third at its peak in the 1950s to 6% in 2019 (Stansbury and Summers 2020, 9). Therefore, if we drop the assumption that a rise in labour productivity translates more than proportionately into a rise in wage share, in the Kaleckian analysis, a 'productive' government expenditure would always lead to a deterioration in capacity utilisation rate and growth rate in their model.

Dutt (2013) analyses the impact of different kinds of government expenditure on aggregate demand and growth in the short run as well as in the long run in a single-good closed economy framework. Unlike Commendatore and Pinto (2011), in Dutt's (2013) analysis, the potential output to capital ratio is fixed and is not influenced by the government investment expenditure. Thus, actual output to capital ratio can be used as a proxy for the degree of capacity utilisation. In his analysis, he assumes that the government budget is balanced and the government does not carry any debt. In the short run, both kinds of government expenditure – government consumption and investment expenditure – enhance aggregate demand and hence the degree of capacity utilisation. An increase in capacity utilisation increases the growth rate, so both kinds of government expenditure have positive effects on aggregate demand, degree of capacity utilisation and accumulation rate. But government investment expenditure due to its 'crowding-in' effect on private investment increases investment and hence aggregate demand further. Thus, the degree of capacity utilisation and the growth rate both are higher in this case compared to the case of an increase in government consumption expenditure. In a balanced budget situation, when total revenue is given in the short run, a switch from government consumption to investment expenditure does not increase the level of aggregate demand directly, but its indirect effect through 'crowding-in' of private investment increases aggregate demand, the degree of capacity utilisation, and the growth rate. Dutt (2013) introduces the endogenous technological change where the long-run rate of growth of the economy is determined by both demand and supply forces. In the long run, both kinds of government expenditure have a positive effect on growth rate. But a switch from government consumption to government investment expenditure increases the growth rate. In other words, government investment expenditure is more effective in the long run too. The reason is two-fold. First, it 'crowds-in' private investment. Second, it influences the speed of adjustment of technological change positively. Dutt (2013) relaxes the balanced budget assumption by allowing the government to run a deficit and incur debt. Here again, both kinds of government expenditure have positive effects on aggregate demand, degree of capacity utilisation, and accumulation rate. But again, the degree of capacity utilisation and the growth rate both are higher in government investment expenditure compared to that of an increase in government consumption expenditure. However, this analysis does not take into account the issue of income distribution; taking account of the issues related to income distribution may affect the results in several ways.

In the next section following a 'post-Keynesian' framework of growth theory, we want to verify whether government expenditure influences the growth rate at all, and if so, how it differs from the previous literature. The general structure of the model is taken from Dutt (2013). However, our model deviates from Dutt (2013) in the following ways. First, unlike Commendatore and Pinto (2011), Commendatore, Panico, and Pinto (2009; 2011), and Dutt (2013), we assume that along with the degree of capacity utilisation (u), investment function also depends on the profit rate. Second, in our model, we incorporate the fact that certain kinds of investment expenditure can influence labour productivity. Labour productivity on the other hand through its impact on the share of profit influences the current profitability of the private capital formation. The novelty of the model of this paper lies in taking account of this fact. Third, unlike Dutt (2013), in our model, we assume that workers save a positive fraction s_W of their wage income. We introduce s_W for two purposes: first, it is a more general case than that of where only capitalists save, and second, by introducing savings out of wages we are able to open up the possibility of profit-led demand regime along with the wage-led demand regime in the economy. Section 3 is about the impact of changes in fiscal policy on the equilibrium employment rate from the short run perspective. We will show that an increase in either of autonomous investment and tax rate leads to an increase in the equilibrium employment rate, but the effect of a rise in government investment expenditure at the cost of government consumption expenditure on the equilibrium employment rate is ambiguous. In Section 4, we consider the effect of deficit and government debt on the economy in the short run whereas Section 5 considers the long-run where we investigate whether allowing the government to run in deficit and incur debt necessarily leads to public debt rising without bounds. The last section (Section 6) discusses some concluding remarks.

2. The model

We assume a simple one-sector post-Keynesian growth model in which the economy consists of two classes: capitalists and workers. The economy is closed, the labour supply is constant and there is no technological change in the economy. Workers save a fraction s_W of their wage income while capitalist's saving propensity is s_P. We also assume capitalists saving propensity (s_P) is higher than that of workers. We introduce s_W for two purposes. Firstly, it is a more general case than that of where only capitalists save. Secondly, introducing savings out of wages, along with the wage-led demand regime, we can provide the possibility of profit-led demand regime in the economy.[6]

Income is distributed between wages and profits in the following way

$$pY = WL + rpK \tag{2.1}$$

where p is price level, Y is real income, W is nominal wage rate, L is total amount of labour employment, K is the existing capital stock, r is the real rate of profit. There is excess supply of labour and no depreciation of capital in the economy. The production function is of Leontief type, i.e.

$$Y = \min\{aL, bK\} = aL, \quad b = \frac{Y^P}{K} > \frac{Y}{K} \tag{2.2}$$

where Y^P is the potential output level. So, the actual output is below the potential output level.[7] The market is oligopolistic in nature where price is determined by mark-up on prime cost. For simplicity, we assume away cost of raw materials and overhead cost.[8] We assume here that the only cost is the labour cost. So, price is given by

$$p = (1 + \lambda)\frac{WL}{Y} = (1 + \lambda)\frac{W}{a} \tag{2.3}$$

where λ is the rate of mark-up and $a = \frac{Y}{L}$ is labour productivity. Total wage share $= \frac{WL}{pY} = \frac{w}{a}$, where w is real wage rate. So, the share of profit is $\pi = \left(1 - \frac{w}{a}\right)$.[9] From this equation, we can conclude that share of profit depends on labour productivity and real wage rate. Real wage rate itself depends on labour productivity, i.e. $w = w(a)$. But the rate of change in the real wage rate with respect to labour productivity depends on the bargaining power of the workers which in turn depends on the prevailing employment rate and the extent of unionization. We assume $\varepsilon_{w,a} < 1$, i.e. elasticity of real wage rate with respect to labour productivity is less than one.[10] As a consequence, if labour productivity increases, wage share (w/a) decreases. Consequently, the share of profit rises. Thus, $\pi'(a) > 0$, i.e. change in share of profit due to change in labour productivity is positive.[11]

We assume that there are two types of government expenditure: government consumption expenditure, denoted by C_G, and government investment expenditure, denoted by I_G. We also assume that government investment expenditure is proportional to the aggregate real income, i.e. $I_G = \theta Y$, where θ represents government investment-output ratio. Government raises revenue through an income tax. Total tax revenue is $T = tY$, where t is the tax rate. For simplicity, we assume that there is no transfer payments, the government budget is balanced, and there is homogeneous tax rate in the economy. So,

$$tY = C_G + I_G \tag{2.4}$$

For a given level of t, Y, and θ, this equation can be satisfied through adjustment in C_G. Given the tax rate, if θ increases then for a given income level, government consumption expenditure must fall. Thus, for a given aggregate government expenditure, a change in the parameter θ represents a change in fiscal policy, i.e. here change in θ represents change in fiscal policy related to the government's decision as how much to spend on consumption and how much to spend on investment purposes.

Total savings in the economy are expressed as

$$S = s_P(1-t)P + s_W(1-t)W = [(s_P - s_W)(1-t)r + s_W(1-t)u]K \tag{2.5}$$

where $r =$ the rate of profit $= \frac{P}{K} = \frac{P}{Y}\frac{Y}{K} = \pi(a).u$, P represents total profit, $\frac{P}{Y} =$ share of profit $= \pi(a)$, and u is the output-capital ratio which is used as a proxy for degree of capacity utilization (Dutt 1984, 1987, 1990).[12] We assume that there is excess capacity in the economy (i.e. $u < 1$). The investment function (normalized by capital stock) is given by

$$\frac{I}{K} = \gamma_0 + \gamma_1 u + \gamma_2(1-t)r + \gamma_3\left(\frac{I_G}{K}\right) = \gamma_0 + \gamma_1 u + \gamma_2(1-t)\pi(a)u + \gamma_3\theta u \tag{2.6}$$

where $\gamma_0, \gamma_1, \gamma_2, \gamma_3$ all are positive parameters. γ_0 represents the autonomous part of the investment function. We assume that investment depends positively on the degree of capacity utilisation (u), the rate of profit (r) and the ratio of government investment to capital stock $\left(\frac{I_G}{K}\right)$. γ_1 indicates the responsiveness of investment to a change in u. Similarly, γ_2 and γ_3 indicate the responsiveness of investment due to a change in the rate of profit and the ratio of government investment to capital, respectively. The positive effect of u is the static equivalent of the accelerator effect.

The argument for the rate of capacity utilisation entering in the investment function comes from Steindl (1952). According to Steindl, because of indivisibilities in capital equipment, it is profitable for profit-maximizing firms to have a certain desired level of excess capacity due to fluctuations in demand or expected growth in demand. Thus, when capacity utilisation rises above the desired level, firms would like to invest more; while the capacity utilisation falls below the desired level, the firm would like to increase utilisation by dis-investing and hence by reducing the stock of capital. The rate of profit enters the investment function as a proxy for the expected rate of return. It also provides internal funding for accumulation plans. For firms depending on external finance, it is also easier to raise external finance while the rate of profit is higher. For simplicity, we assume that the actual rate of profit is equal to the expected profit rate. Now let us focus on the last variable in the investment function – the ratio of government investment expenditure to capital stock. Following Dutt (2013) and Taylor (1991), we can say that government investment expenditure has a positive impact on private investment because of a 'crowding in' effect.[13] Certain kinds of government investment expenditure (like expenditure on part of infrastructure, education and health facilities, water and electricity supply) have a positive impact on labour productivity as well. Therefore, we can say $a = a(\theta)$ and $a'(\theta) > 0$ i.e. labour productivity depends positively on the ratio of government investment to output.[14]

In the short run, the goods market is cleared through changes in the level of output and capacity utilisation. In equilibrium, saving must be equal to investment and so after some rearrangement we get,

$$u^* = \frac{\gamma_0}{\left\{ \left(s_P - s_W - \gamma_2\right)(1-t)\pi(a) + s_W(1-t) - \gamma_1 - \gamma_3\theta \right\}} \tag{2.7}$$

where u^* is the equilibrium degree of capacity utilisation. The equilibrium is stable if and only if the induced increase in saving as u rises is greater than the induced increase in investment, i.e.

$$(s_P - s_W)(1-t)\pi(a) + s_W(1-t) > \gamma_1 + \gamma_2(1-t)\pi(a) + \gamma_3\theta$$

$$\Rightarrow \left(s_P - s_W - \gamma_2\right)(1-t)\pi(a) + s_W(1-t) - \gamma_1 - \gamma_3\theta > 0 \tag{2.8}$$

In other words, for the equilibrium to be stable the denominator of u^* must be positive. So, the stability condition can be satisfied if $s_W > \frac{\gamma_1 + \gamma_3\theta - \left(s_P - \gamma_2\right)(1-t)\pi}{(1-t)(1-\pi)}$. Thus, the stability condition imposes a lower bound to the savings propensity of the workers.

We get the equilibrium growth rate[15] in terms of equilibrium degree of capacity utilisation as,

$$g^* = \gamma_0 + \gamma_1 u^* + \gamma_2(1-t)\pi(a)u^* + \gamma_3\theta u^* \tag{2.9}$$

Now let's discuss when the economy is in a wage-led demand regime and when it is in a profit-led demand regime. The following proposition due to Blecker (2002) provides the sufficient conditions for the economy to be in either of those regimes.

Proposition 1. *Whether the economy is in a wage-led demand regime or in a profit-led demand regime depends on the value of s_W as follows: (i) if $s_W < \left(s_P - \gamma_2\right)$ then the economy is in a wage-led demand regime and (ii) if $s_W > \left(s_P - \gamma_2\right)$ then the economy is in a profit-led demand regime.*

Proof. See Appendix A.1. □

For every unit redistribution of income from workers to capitalist, consumption demand falls by $(s_P - s_W)$ unit. For the same, however, investment demand rises by γ_2 unit. If the latter effect dominates the former, for a rise in π, the equilibrium degree of capacity utilisation increases. In this case, the economy is said to be in a profit-led demand regime. Following a similar argument, we can say that when $\left(s_P - s_W - \gamma_2\right) > 0$ then the economy is in a wage-led demand regime (i.e. $\frac{du^*}{d\pi} < 0$). So, depending on the sign of $\left(s_P - s_W - \gamma_2\right)$ the economy can be either in a wage-led or in a profit-led demand regime.

In the next proposition, we discuss the impact of a change in θ (a switch in government expenditure from consumption to investment purposes) on different demand regimes.

Proposition 2. *An increase in θ converts the profit-led demand regime into a stronger profit-led demand regime, while in the wage-led demand regime, the effect of a rise in θ on $\frac{du^*}{d\pi}$ depends on the product of $\varepsilon_{\pi,a}$ and $\varepsilon_{a,\theta}$ as follows: $\frac{d}{d\theta}\left(\frac{du^*}{d\pi}\right) \gtreqless 0$ according to whether $\varepsilon_{\pi,a}.\varepsilon_{a,\theta} \gtreqless \psi$, where $\psi = \frac{\gamma_3\theta}{\left(s_P - s_W - \gamma_2\right)(1-t)\pi(a)}$, $\varepsilon_{\pi,a} = \left(\frac{d\pi}{da}\right)\left(\frac{a}{\theta}\right)$ and $\varepsilon_{a,\theta} = \left(\frac{da}{d\theta}\right)\left(\frac{\theta}{a}\right)$.*

Proof. See Appendix A.2. $\square$

Proposition 2 implies that when the economy is in a profit-led demand regime, an increase in θ always has a positive impact on $\frac{du^*}{d\pi}$, and hence, a rise in θ converts the profit-led demand regime into a stronger profit-led demand regime. On the other hand, when the economy is in a wage-led demand regime if the product of $\varepsilon_{\pi,a}$ and $\varepsilon_{a,\theta}$ exceeds (less than) the critical value (let's say ψ) then the effect of a rise in θ on $\frac{du^*}{d\pi}$ is positive (negative). In other words, if the product of $\varepsilon_{\pi,a}$ and $\varepsilon_{a,\theta}$ exceeds the critical value ψ, a rise in θ converts the wage-led demand regime into a stronger wage-led demand regime. In contrast, if the product of $\varepsilon_{\pi,a}$ and $\varepsilon_{a,\theta}$ is less than the critical value ψ, a rise in θ converts the wage-led demand regime into a weaker wage-led demand regime.

Note that an increase in either of $\gamma_0, \gamma_1, \gamma_2, \gamma_3$ leads to an increase in the equilibrium degree of capacity utilisation. But the effect of a rise in t on u^* is ambiguous in the profit-led demand regime. Although in the wage-led demand regime, a rise in t raises u^*. As γ_1 increases, the accelerator effect of u on investment demand rises, which in turn raises aggregate demand and hence u^*.[16] As γ_2 increases, u^* also increases. This is because, an increase in γ_2, for a given profit rate, leads to an increase in investment demand which in turn increases the aggregate demand and hence the degree of capacity utilisation. Similarly, as γ_3 increases, for a given θ, investment demand increases which in turn raises the aggregate demand and hence the degree of capacity utilisation.[17]

The tax rate has a positive impact on the equilibrium degree of capacity utilisation in the wage-led demand regime.[18] This is mainly because of the balanced budget assumption. Each unit rise in tax rate decreases capitalists' consumption demand (normalized by the capital stock) by $(1 - s_P)\pi u$ unit, while the consumption of workers decreases by $(1 - s_W)(1 - \pi)u$ unit. By reducing the after-tax profit rate, it also reduces investment demand by $\gamma_2\pi u$ unit, but the entire tax revenue is spent by the government and so the aggregate demand increases by u unit.[19] As the increase in government spending is higher than the reduction in consumption and investment demand,[20] an increase in the tax rate increases the equilibrium degree of capacity utilisation in a wage-led demand regime.

The next proposition discusses the effect of an increase in the government investment expenditure (at the cost of government consumption expenditure) on the aggregate demand and hence on the degree of capacity utilisation.

Proposition 3. *An increase in θ leads to an unambiguous increase in the equilibrium degree of capacity utilisation in the profit-led demand regime, while in the wage-led demand regime, the effect of a rise in θ on the equilibrium degree of capacity utilisation depends on the product of $\varepsilon_{\pi,a}$ and $\varepsilon_{a,\theta}$ as follows: $\frac{du^*}{d\theta} \lessgtr 0$ according to whether $\varepsilon_{\pi,a} \cdot \varepsilon_{a,\theta} \gtrless \psi$, where $\psi = \frac{\gamma_3 \theta}{(s_P - s_W - \gamma_2)(1-t)\pi(a)}$, $\varepsilon_{\pi,a} = \left(\frac{d\pi}{da}\right)\left(\frac{a}{\theta}\right)$ and $\varepsilon_{a,\theta} = \left(\frac{da}{d\theta}\right)\left(\frac{\theta}{a}\right)$.*

Proof. See Appendix A.3. $\square$

Proposition 3 implies that while in the profit-led demand regime an increase in θ always has a positive impact on the equilibrium degree of capacity utilisation, in the wage-led demand regime the effect of a rise in θ on u^* depends on some critical value of the product of $\varepsilon_{\pi,a}$ and $\varepsilon_{a,\theta}$. If the product of the elasticity of profit share with respect to labour productivity $(\varepsilon_{\pi,a})$ and the elasticity of labour productivity with respect to the government investment to output ratio $(\varepsilon_{a,\theta})$ exceeds (less than) the critical value (let's say ψ) then the effect of a rise in θ on u^* is negative (positive). But when the product of $\varepsilon_{\pi,a}$ and $\varepsilon_{a,\theta}$ is equal to the critical value ψ, then government consumption and investment expenditures both have the same degree of impact on u^*. Consequently, a switch in government expenditure from consumption to investment does not raise the equilibrium degree of capacity utilisation.

The economic intuition behind the result is that a rise in θ raises labour productivity which in turn raises profit share. Due to a rise in profit share, the investment demand rises. But on the other hand, due to the redistribution of income from wages to profits, consumption demand decreases. In the wage-led demand regime, the latter effect dominates the previous one and so the rise in the share of profit reduces the degree of capacity utilisation. On the other hand, a rise in θ directly raises the investment demand through the crowding-in effect leading to a rise in aggregate demand and the degree of capacity utilisation. The final impact of a change in θ on u^* depends on the relative strength of the direct effect of θ on u and its indirect effect on u through the change in profit share. When the elasticities have lower values then a change in θ has a lower impact on labour productivity and/or a change in labour productivity has a lower impact on profit share. So, the indirect effect of θ on u through the change in the profit share is comparatively lower and consequently, the direct effect of θ on u dominates the indirect effect. On the other hand, when the above elasticities have sufficiently high values then the indirect effect of θ on u dominates the direct effect and thus the impact of θ on u is negative.

Proposition 4. *An increase in θ leads to an unambiguous increase in the equilibrium rate of capital accumulation in the profit-led demand regime, while in the wage-led demand regime, the effect of a rise in θ on the equilibrium rate of capital accumulation* *depends on the product of* $\varepsilon_{\pi,a}$ *and* $\varepsilon_{a,\theta}$ *as follows:* $\frac{dg^*}{d\theta} \gtrless 0$ *according to whether*

$\varepsilon_{\pi,a} \cdot \varepsilon_{a,\theta} \lessgtr \rho$, *where* $\rho = \frac{\{(s_P - s_W)\pi + s_W\}\gamma_3\theta}{\pi\{(s_P - s_W)(\gamma_1 + \gamma_3\theta) - s_W\gamma_2(1-t)\}}$, $\varepsilon_{\pi,a} = \left(\frac{d\pi}{da}\right)\left(\frac{a}{\pi}\right)$ *and* $\varepsilon_{a,\theta} = \left(\frac{da}{d\theta}\right)\left(\frac{\theta}{a}\right)$.

Proof. See Appendix A.4. □

So, when the economy is in a profit-led demand regime, due to a rise in θ, the equilibrium rate of capital accumulation unambiguously rises. But when the economy is in a wage-led demand regime, the effect of a rise in θ on g^* depends on the value of the product of $\varepsilon_{\pi,a}$ and $\varepsilon_{a,\theta}$. If the product of $\varepsilon_{\pi,a}$ and $\varepsilon_{a,\theta}$ exceeds (less than) the critical value (let's say ρ) then the effect of a rise in θ on g^* is negative (positive). θ affects g^* in three ways: first, its direct effect on g^* which we call crowding in effect; second, its effect through the profit share, and finally its effect through u^*. So even if θ has a negative effect on u^*, if the positive effect of θ on the profit share and its direct effect on g^* (crowding in effect) are very high, these two can more than compensate for the negative effect.

Note that for $s_W = 0$, we get only the wage-led demand regime. If we focus on the wage-led demand regime, our results in Propositions 3 and 4 are not qualitatively different from the situation where workers do not save. Only the critical values change. In the next section, we focus on the employment rate in the economy.

3. Issues regarding the employment rate

Here we discuss about the impact of different kinds of government expenditures on labour employment rate from the short-run perspective. Equilibrium employment rate e^* can be written as:

$$e^* = \frac{L}{N} = \frac{Y}{K}\frac{L}{Y}\frac{K}{N} = u^*\frac{1}{a}k_0 = u^*k \qquad (3.1)$$

where N is the total supply of labour which is fixed in the short-run, k_0 is the ratio of capital stock to total supply of labour, and $k\left(= \frac{K}{aN}\right)$ is the ratio of capital stock to the product of labour supply and labour productivity. An increase in either of $\gamma_0, \gamma_1, \gamma_2, \gamma_3$, and t leads to an increase in the e^*. As in the short-run K and N both are fixed, k_0 is fixed too. Therefore, as long as the labour productivity is not influenced by any change in parameters, k is also fixed. Thus, a change in any parameter which does not have an impact on a, can change the equilibrium rate of employment only through change in u^*.

Proposition 5: *An increase in* θ *leads to an ambiguous effect on the equilibrium employment rate both in the profit-led demand regime and in the wage-led demand regime. The effect of a rise in* θ *on the equilibrium employment rate depends on the product of* $\varepsilon_{\pi,a}$ *and* $\varepsilon_{a,\theta}$ *as follows:* $\frac{de^*}{d\theta} \gtrless 0$ *according to whether* $\varepsilon_{\pi,a} \lessgtr \Omega$, *where*

$\Omega = \frac{\gamma_3\theta + (\gamma_1 + \gamma_3\theta - s_W(1-t))\varepsilon_{a,\theta}}{(s_P - s_W - \gamma_2)(1-t)\pi(a)\varepsilon_{a,\theta}} - 1, \varepsilon_{\pi,a} = \left(\frac{d\pi}{da}\right)\left(\frac{a}{\pi}\right)$ *and* $\varepsilon_{a,\theta} = \left(\frac{da}{d\theta}\right)\left(\frac{\theta}{a}\right)$.

Proof. See Appendix A.5. □

A change in θ affects the equilibrium employment rate in two ways. First, a rise in θ raises the labour productivity which in turn leads to a fall in the employment rate. Second, its impact on employment rate through change in u^*. The effect of θ on u^* has already been discussed in Proposition 3. Thus, the final effect of a change in θ on e^* depends on its effect on u^* and labour productivity.

In the profit-led demand regime, $\frac{du^*}{d\theta} > 0$. So, the effect of a rise in θ on e^* is ambiguous. In the wage-led demand regime, if $\frac{du^*}{d\theta} > 0$ the effect of a rise in θ on e^* is ambiguous. But if $\frac{du^*}{d\theta} < 0$, a rise in θ always decreases e^*.

4. Effects of government deficits and debt

So far, we have assumed that the government budget is balanced and there is no government debt. Now we will relax the assumption. Let's assume there is a budget deficit and the government incurs debt. The government borrows from capitalists at the interest rate i. So, capitalists have two sources of income (1) profit income (P) and (2) the interest income (iD) whereas workers earn only wage income (W). Let us assume that the aggregate government tax revenue is given as,

$$T = t \left(\overbrace{P + iD}^{\text{capitalists' income}} + \overbrace{W}^{\text{workers' income}} \right) = t(Y + iD) \tag{4.1}$$

where t is the tax rate, Y is the real aggregate productive income, i is the interested rate that is paid by the government, and D is the real stock of government debt. In this section, we assume that government consumption and investment expenditure both depend on the income level of the economy as $C_G = \eta Y$ and $I_G = \theta Y$ respectively. Following Dutt (2013) for the sake of simplicity, we ignore monetary and other assets. The entire government deficit is financed by issuing government debt. So, the change in debt with respect to time is given by,

$$\frac{dD}{d\tau} = (C_G + I_G) - T + iD \tag{4.2}$$

Aggregate private saving in the economy is given by,

$$S = (1-t)[s_P P + s_W W + s_P iD] = [(s_P - s_W)(1-t)r + s_W(1-t)u + (1-t)s_P i\delta]K \tag{4.3}$$

where $\delta = \frac{D}{K}$ is the debt-capital ratio. The investment function in the economy is given by,

$$I = [\gamma_0 + \gamma_1 u + \gamma_2(1-t)\pi(a)u + \gamma_3\theta u - \gamma_4\delta]K \tag{4.4}$$

where $\gamma_0, \gamma_1, \gamma_2, \gamma_3$ and γ_4 all are positive parameters. γ_4 is the coefficient measuring responsiveness of investment due to a change in δ. Here, the fifth term entering in the investment function, represents the financial crowding out effect.[21] In the short run equilibrium, the following equation must be satisfied,

$$\frac{S}{K} + \frac{T}{K} = \frac{I}{K} + \frac{G}{K}$$

Putting the values we get the equilibrium degree of capacity utilisation as

$$u^* = \frac{\gamma_0 - \left[\{s_P(1-t)+t\}i + \gamma_4\right]\delta}{(s_P - s_W - \gamma_2)(1-t)\pi(a) + s_W(1-t) + t - \gamma_1 - \eta - \theta(1+\gamma_3)} \qquad (4.5)$$

The equilibrium is stable when the induced increase in private savings and revenue income as u rises must be greater than the induced increase in private investment and government expenditure. That is when the following equation is satisfied:

$$(s_P - s_W)(1-t)\pi(a) + s_W(1-t) + t > \gamma_1 + \gamma_2(1-t)\pi(a) + \gamma_3\theta + (\eta + \theta)$$

$$\Rightarrow (s_P - s_W - \gamma_2)(1-t)\pi(a) + s_W(1-t) + t - \gamma_1 - \eta - (1+\gamma_3)\theta > 0$$

In other words, for the equilibrium to be stable the denominator of u^* must be positive. But for a meaningful positive equilibrium degree of capacity utilisation, the numerator also should be positive. So, we need,

$$\gamma_0 - \left[\{s_P(1-t)+t\}i + \gamma_4\right]\delta > 0$$

We get the equilibrium growth rate in terms of u^* as,

$$g^* = \gamma_0 + \gamma_1 u^* + \gamma_2(1-t)\pi(a)u^* + \gamma_3\theta u^* - \gamma_4\delta \qquad (4.6)$$

Here again, whether the economy is in a wage-led demand regime or in a profit-led demand regime depends on the value of s_W. If $s_W < (s_P - \gamma_2)$ then the economy is in a wage-led demand regime and if $s_W > (s_P - \gamma_2)$ then the economy is in a profit-led demand regime.

An increase in either of δ, i, γ_4 causes a fall in u^* while an increase in either of $\gamma_0, \gamma_1, \gamma_2, \gamma_3$ leads to an increase in u^* independent of the regime the economy is in. An increase in the government consumption expenditure (η) leads to an increase in the aggregate demand and hence the equilibrium degree of capacity utilisation. However, the effect of an increase in government investment expenditure (θ) on u^* depends on which regime the economy is in (discussed in Proposition 6). Note that here a rise in θ means solely an increase in government investment expenditure, not the increase in government investment expenditure at the cost of government consumption expenditure.[22]

Proposition 6. *An increase in θ leads to an unambiguous increase in the equilibrium degree of capacity utilisation in the profit-led demand regime, while in the wage-led demand regime, the effect of a rise in θ on the equilibrium degree of capacity utilisation depends on the product of $\varepsilon_{\pi,a}$ and $\varepsilon_{a,\theta}$ as follows: $\frac{du^*}{d\theta} \gtrless 0$ according to whether $\varepsilon_{\pi,a}.\varepsilon_{a,\theta} \lessgtr \phi$, where $\phi = \frac{(1+\gamma_3)\theta}{(s_P - s_W - \gamma_2)(1-t)\pi(a)}$, $\varepsilon_{\pi,a} = \left(\frac{d\pi}{da}\right)\left(\frac{a}{\theta}\right)$ and $\varepsilon_{a,\theta} = \left(\frac{da}{d\theta}\right)\left(\frac{\theta}{a}\right)$.*

Proof. See Appendix A.6. □

It follows that in the wage-led demand regime if the product of $\varepsilon_{\pi,a}$ and $\varepsilon_{a,\theta}$ is greater than (less than) a critical value (let's say ϕ) then the effect of θ on u^* is negative (positive). A rise in θ raises labour productivity which in turn raises profit share. In the wage-led demand regime, a change in profit share has a negative impact on u^*. On the other hand, a rise in θ directly raises the investment demand through the crowding-in effect leading to a rise in aggregate demand and the degree of capacity utilisation. So, the final impact of a change in θ on u^* depends on the relative strength of the direct effect of θ on u and its indirect effect on u through the change in the share of profit. When the elasticities have lower values, the indirect effect of θ on u through the change in the profit share is comparatively lower and as a result, the direct effect of θ on u dominates the indirect effect and thus the impact of θ on u^* is positive. Finally, in the profit-led demand regime, a change in profit share has a positive impact on u^*. Hence the overall effect is also positive. Let's focus on the effect of θ on g^* now. The following proposition captures the relationship.

Proposition 7. *An increase in θ leads to an unambiguous increase in the equilibrium rate of accumulation in the profit-led demand regime, while in the wage-led demand regime, the effect of a rise in θ on the equilibrium rate of capital accumulation depends on the product of $\varepsilon_{\pi,a}$ and $\varepsilon_{a,\theta}$ as follows: $\frac{dg^*}{d\theta} \gtrless 0$ according to whether $\varepsilon_{\pi,a}\cdot\varepsilon_{a,\theta} \lessgtr \rho'$, where $\varepsilon_{\pi,a} = \left(\frac{d\pi}{da}\right)\left(\frac{a}{\theta}\right)$, $\varepsilon_{a,\theta} = \left(\frac{da}{d\theta}\right)\left(\frac{\theta}{a}\right)$, and $\rho' =*

$$\frac{\theta\left[\gamma_1+\gamma_3(t-\eta)+(1-t)\pi\left\{\gamma_2+\left(s_p-s_w\right)\gamma_3+s_w(1-t)\gamma_3\right\}\right]}{(1-t)\pi\left[\left(s_p-s_w-\gamma_2\right)\left(\gamma_1+\theta\gamma_3\right)-\gamma_2\left\{t-\gamma_1\left(1+\gamma_3\right)\theta-\eta+s_w(t-t)\right\}\right]}$$

Proof. See Appendix A.7. $\square$

When the economy is in a profit-led demand regime, a rise in θ unambiguously raises the equilibrium rate of capital accumulation. But when the economy is in a wage-led demand regime, the effect of a rise in θ on g^* depends on the value of the product of $\varepsilon_{\pi,a}$ and $\varepsilon_{a,\theta}$. If the product of $\varepsilon_{\pi,a}$ and $\varepsilon_{a,\theta}$ exceeds (less than) the critical value (let's say ρ') then the effect of a rise in θ on g^* is negative (positive). θ affects g^* in three ways: first, its direct effect on g^* which we call crowding in effect; second, its effect through the profit share and finally its effect through u^*. So even if θ has a negative effect on u^*, if the positive effect of θ on the profit share and its direct effect on g^* (crowding in effect) are very high, these two can more than compensate for the negative effect. The next proposition focuses on the effect of a rise in δ on g^*.

Proposition 8. *An increase in δ decreases the equilibrium rate of capital accumulation.*

Proof. See Appendix A.8. $\square$

An increase in δ decreases g^* in two ways: (1) directly through financial crowding-out effect and (2) indirectly through decrease in the equilibrium degree of capacity utilisation. Let's discuss its (δ) effect on u^* first. In the short run an increase in δ decreases the equilibrium degree of capacity utilisation. Due to one unit increase in δ, the ratio of private saving to capital stock increases by $(1-t)s_p i$ unit while the ratio of government revenue income to capital stock increases by it unit. Thus, due to one unit increase in δ, the consumption demand decreases by $\{(1-t)s_p + t\}iK$ unit. On the other hand, due to a unit rise in δ, investment rate decreases by γ_4 unit.[23] Thus aggregate demand decreases by $\left[\{s_p(1-t) + t\}i + \gamma_4\right]$ unit. As a result, the equilibrium degree of capacity utilisation falls. On the other hand, a rise in debt-capital ratio because of the financial crowding out effect, leads directly to a fall in the equilibrium growth rate, therefore the effect of a change in δ on the equilibrium rate of capital accumulation is unambiguously negative.

Hungerford (2016) points out that in the US economy, Federal public investment spending was roughly 2.2% of GDP between 1965 and 1981 whereas, after 1986, it fell to 1.5% of GDP. From 0.8% of GDP in 1980, public investments in physical capital (primarily infrastructure – bridges, roads, sewage systems, drinking water systems, waterways, etc.) have plummeted to less than 0.5% in 2015. There is also a dramatic rise in the concentration of income and wealth at the top of income distribution. Hungerford (2016) advocates for a rise in the progressive tax-and-transfer system which according to him will reduce the after-tax income of the richest 1%. He also advocates for a rise in public investment spending that, as he says, will assist in reducing income inequality by creating more higher-paying jobs and boosting productivity growth. As in our model, there is a homogeneous income tax rate – we investigate what happens if the rise in government investment expenditure is financed completely by a rise in the tax rate.[24] Proposition 9 discusses this issue.

Proposition 9. *Suppose an increase in θ is financed completely by a rise in the tax rate. When the economy is in a profit-led demand regime, the effect of a rise in θ on the equilibrium degree of capacity utilisation depends on the product of $\varepsilon_{\pi,a}$ and $\varepsilon_{a,\theta}$ as follows: $\frac{du^*}{d\theta} \gtreqless 0$ according to whether $\varepsilon_{\pi,a} \cdot \varepsilon_{a,\theta} \gtreqless \sigma$. In the wage-led demand regime, on the other hand, the effect of a rise in θ on the equilibrium degree of capacity utilisation depends on the product of $\varepsilon_{\pi,a}$ and $\varepsilon_{a,\theta}$ as follows: $\frac{du^*}{d\theta} \gtreqless 0$ according to whether $\varepsilon_{\pi,a} \cdot \varepsilon_{a,\theta} \lesseqgtr$ σ, where $\sigma = \frac{\left[\left\{\left((s_P - s_W - \gamma_2)\pi + s_W\right)\left(\frac{u^*}{u^*+i\delta}\right) + \gamma_3\right\} + s_P\left(\frac{i\delta}{u^*+i\delta}\right)\right]\theta}{\left(s_P - s_W - \gamma_2\right)(1-t)\pi}$, $\varepsilon_{\pi,a} = \left(\frac{d\pi}{da}\right)\left(\frac{a}{\theta}\right)$ and $\varepsilon_{a,\theta} = \left(\frac{da}{d\theta}\right)\left(\frac{\theta}{a}\right)$.*

Proof. See Appendix A.9.□

So, a rise in θ, financed completely by a rise in tax rate, need not necessarily increase the aggregate demand. If the economy is in a profit-led demand regime, as long as the product of $\varepsilon_{\pi,a}$ and $\varepsilon_{a,\theta}$ is sufficiently high, a rise in θ without increasing the debt level of the economy further, can boost aggregate demand in the economy. In the wage-led demand regime, however, as long as the product of $\varepsilon_{\pi,a}$ and $\varepsilon_{a,\theta}$ is sufficiently low, a rise in public investment spending can increase aggregate demand.

Proposition 10 discusses the economic impact of a rise in government consumption expenditure when it is financed completely by a rise in the tax rate.

Proposition 10. *Suppose an increase in η is financed completely by a rise in the tax rate. An increase in η leads to an unambiguous increase in the equilibrium degree of capacity utilisation in the wage-led demand regime, while in the profit-led demand regime, the effect of a rise in η on the equilibrium degree of capacity utilisation depends as follows: $\frac{du^*}{d\eta} \gtreqless 0$ according to whether $\left(\frac{s_P i\delta + s_W u^*}{\pi u^*}\right) \gtreqless -\left(s_P - s_W - \gamma_2\right).$*

Proof. See Appendix A.10.□

When the economy is in a wage-led demand regime $(s_P - s_W - \gamma_2) > 0$ and therefore $\frac{du^*}{d\eta}\big|_{udn = (u+i\delta)dt}$ is unambiguously positive. On the other hand, if the economy is in a profit-led demand regime, as long the profit-led demand regime is relatively weak, a rise in government consumption expenditure financed with a rise in the tax revenue increases the equilibrium degree of capacity utilisation.

5. The long-run analysis

Now we analyse the long-run dynamics of the government debt and the capital stock. The long-run equilibrium is attained when the government debt-capital ratio (δ) remains constant over time. We know, $\delta = \frac{D}{K}$. So, $\hat{\delta} = \hat{D} - \hat{K}$. Further,

$$\frac{dD}{d\tau} = (C_G + I_G) - T + iD \tag{5.1}$$

Inserting the value of T from Equation (4.1), and inserting $C_G = \eta Y$, and $I_G = \theta Y$ in Equation (5.1) we get,

$$\hat{D} = \frac{\frac{dD}{d\tau}}{D} = (\eta + \theta - t)\frac{u^*}{\delta} + i(1 - t)$$

$$\Rightarrow \hat{\delta} = \hat{D} - \hat{K} = (\eta + \theta - t)\frac{u^*}{\delta} + i(1 - t) - g^*$$

$$\Rightarrow \frac{d\delta}{d\tau} = \delta[\hat{D} - \hat{K}] = (\eta + \theta - t)u^* + i(1 - t)\delta - g^*\delta$$

$$\Rightarrow \frac{d\delta}{d\tau} = D_0 + D_1\delta + D_2\delta^2$$

where $D_0 = \frac{(\eta+\theta-t)\gamma_0}{\Lambda}$, $D_1 = \left[-\frac{(\eta+\theta-t)\Gamma}{\Lambda} + i(1 - t) - \gamma_0 - \frac{\{\gamma_1+\gamma_2(1-t)\pi+\gamma_3\theta\}\gamma_0}{\Lambda}\right]$ and $D_2 = \left[\frac{\{\gamma_1+\gamma_2(1-t)\pi+\gamma_3\theta\}\Gamma}{\Lambda} + \gamma_4\right]$. Here $D_2 > 0$. Let's assume $D_0 > 0$. $D_0 > 0$ implies that the government expenditure is more than the revenue collection and hence the government incurs debt. This assumption also ensures that even when $\delta = 0$, the government runs a deficit and so δ increases over time. In Dutt (2013), both D_1 and D_2 do not have any definite sign. So, various possibilities regarding $\frac{d\delta}{d\tau}$ may occur depending on the sign of D_1 and D_2. But in our analysis D_2 is unambiguously positive. Then depending on the sign of D_1, δ can either have a stable equilibrium value or it can rise without bound (see Figure 1 for δ rising without bound).

Now let us discuss the conditions for existence and stability of equilibrium. If the interest rate (i) is not too high (i.e. if $i < \left(\frac{\frac{(\eta+\theta-t)\Gamma}{\Lambda}+\gamma_0+\frac{\{\gamma_1+\gamma_2(1-t)\pi+\gamma_3\theta\}\gamma_0}{\Lambda}}{(1-t)}\right)$ then D_1 can have a negative value). Then, the change in debt-capital ratio with respect to δ would be "U"shaped and there is a possibility of existence of equilibrium.[25] On the other hand, if

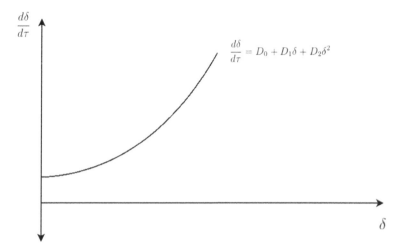

Figure 1. When $D_1 > 0$.

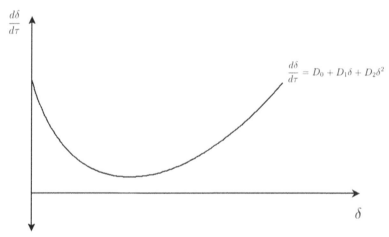

Figure 2. When $D_1 < 0$ but $\left(D_0 - \frac{D_1^2}{4D_2}\right) > 0$.

the interest rate is too high, the debt-capital ratio will rise without bound. Therefore, a relatively low rate of interest is desirable for ensuring a steady state in the economy.

In Figure 2 there is no equilibrium. Figure 4 represents a unique but unstable equilibrium. Figure 3 represents existence of multiple equilibria where one of them[26] is stable. Even allowing the neo-classical argument of financial crowding-out of private investment (because of a rise in public debt),[27] the model does not necessarily become unstable and δ does not rise boundlessly. Rather, for a relatively low interest rate and for a relatively low debt-capital ratio, a stable equilibrium debt-capital ratio can be achieved.

Suppose $\left(D_0 - \frac{D_1^2}{4D_2}\right) < 0$ holds. In that case, a rise in the interest rate leads to a rise in the stable debt-capital ratio (δ^*) and a fall in the unstable debt-capital ratio (δ^{**}).[28] The upwards corridor of stability for the stable equilibrium debt-capital ratio (δ^*) therefore

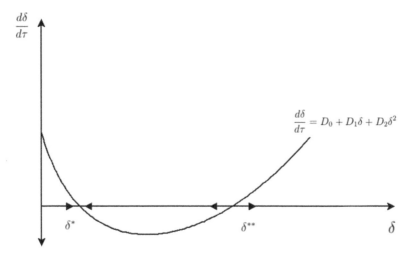

Figure 3. When $D_1 < 0$ and $\left(D_0 - \frac{D_1^2}{4D_2}\right) < 0$.

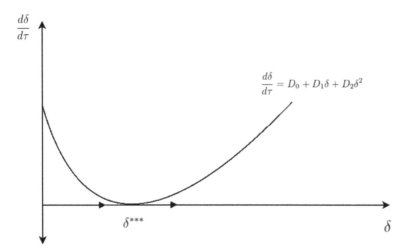

Figure 4. When $D_1 < 0$ and $\left(D_0 - \frac{D_1^2}{4D_2}\right) = 0$.

shrinks, as is shown in Figure 5. On the other hand, a fall in the interest rate increases the upwards stability corridor of the stable steady-state debt-capital ratio (δ^*).

6. Conclusion

Following a post-Keynesian framework, we have tried to analyze the impact of expansionary fiscal policy on aggregate demand and economic growth. In the short-run, we have found that an increase in the government consumption expenditure increases the aggregate demand, equilibrium degree of capacity utilisation and the equilibrium growth rate.

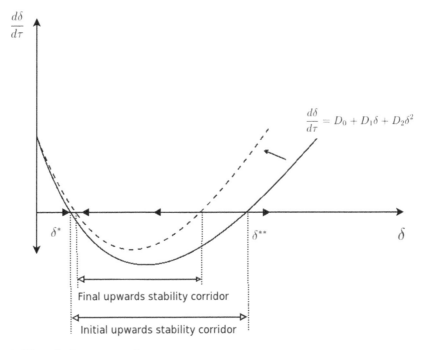

$$\frac{d\delta}{d\tau} = D_0 + D_1\delta + D_2\delta^2$$

Figure 5. When the interest rate rises.

Following Blecker (2002) we have shown that when workers also save, the possibility of a profit-led demand regime arises. When the economy is in a profit-led demand regime, an increase in θ (government investment expenditure at the cost of government consumption expenditure) unambiguously raises both the equilibrium degree of capacity utilisation and the equilibrium growth rate. But when the economy is in the wage-led demand regime, the results are not the same. Unlike Dutt (2013), a switch in government expenditure from consumption to investment purposes does not always lead to a rise in the equilibrium degree of capacity utilisation and the equilibrium growth rate in the wage-led demand regime. As a rise in θ represents simply a switch in government expenditure from consumption to investment purposes, it does not increase aggregate demand and capacity utilisation directly. Rather, it may raise the aggregate demand through its indirect 'crowding in' effect. On the other hand, public investment expenditure through its effect on labour productivity can lead to a rise in the share of profit in the economy which in turn decreases[29] aggregate demand and the degree of capacity utilisation. Thus, the final outcome of a rise in θ depends on the relative magnitudes of these opposing effects. In this case, although our findings are similar to Commendatore and Pinto (2011), the reason is different. Unlike a change in capital productivity, in our analysis, it is a change in labour productivity that influences the equilibrium degree of capacity utilisation which in turn has an impact on the equilibrium growth rate.

When the balanced budget assumption is dropped, an increase in government debt-capital ratio leads to a decrease in the equilibrium degree of capacity utilisation and the equilibrium growth rate. This is in contrast to the analysis by Dutt (2013), where a rise in the government debt-capital ratio has an ambiguous effect on the equilibrium levels of capacity utilisation and the accumulation rate. We also find that a rise in the current government consumption expenditure to output ratio raises the aggregate demand, capacity utilisation, and the accumulation rate. But a rise in θ^{30} has an ambiguous effect on both the equilibrium level of capacity utilisation and the accumulation rate in the wage-led demand regime. This result differs from Dutt (2013) as there is a positive relation between θ and u^* and θ and g^* in his analysis. This is because a rise in public investment expenditure through its effect on labour productivity leads to a rise in the share of profit which in turn mitigates the positive effect of a rise in θ on u^* and g^*. Regarding the employment issue, we also have seen that a change in any model parameter, which increases the equilibrium degree of capacity utilisation without affecting labour productivity, necessarily increases the employment rate. But the effect of a rise in θ on the employment rate is ambiguous and it depends on the elasticity of the share of profit with respect to the labour productivity and the elasticity of labour productivity with respect to θ. Even allowing the neo-classical argument of financial crowding-out of private investment, we have shown that the model does not necessarily become unstable and the debt-capital ratio does not rise without bound. Rather, for a relatively low interest rate and for a relatively low debt-capital ratio, a stable steady-state debt-capital ratio can be attained. Further, a fall in the interest rate may increase the upwards stability corridor of the stable steady-state debt-capital ratio (δ^*).

Landau (1983), Aschauer (1989), Barro (1990, 1991), etc., find that output growth is negatively related with government consumption expenditure whereas Aschauer (1989) and Barro (1990, 1991) also find that government investment expenditure has a positive effect on output growth. On the other hand, using data from 43 developing countries over 20 years, Devarajan, Swaroop, and Zou (1996) show that a rise in the share of current expenditure has a positive effect on economic growth. On the contrary, the capital component of government expenditure has a negative effect on growth. Our finding, however, suggests that both types of expenditure enhance aggregate demand and economic growth. However, whether government investment expenditure (as compared to government consumption expenditure) is more effective in increasing aggregate demand and economic growth depends on (1) which regime the economy is in and (2) if the economy is in a wage-led demand regime it depends on some critical values of the product of $\varepsilon_{\pi,a}$ and $\varepsilon_{a,\theta}$. Barbosa-Filho and Taylor (2006) find that the US is in a profit-led demand regime. Naastepad and Storm (2007) also find a profit-led regime in the US economy. However, Hein and Vogel (2008) find wage-led growth regimes in France, Germany, the UK, and the USA and profit-led growth regimes in Austria and the Netherlands. Onaran, Stockhammer, and Grafl (2011) find that the USA economy is in a moderately wage-led growth regime. Onaran and Galanis (2013) find that all 16 major developed and developing countries of the G20 sample (Argentina, Australia, Canada, China, France, Germany, India, Italy, Japan, Mexico, South Africa, Turkey, the European Union, the UK, the US, and the Republic of Korea) are in a domestic wage-led demand regime. Of these, under open economy consideration, only Argentina, Australia, Canada, China, India, Mexico, and South Africa exhibit a profit-led demand regime. Therefore, without a prior investigation on whether the economy is in a wage-led demand regime, an emphasis on government expenditures in favor of capital expenditures (at the expense of current expenditures) can be highly misleading.

It should be noted that the results of our analysis are based on a very simple model. We have taken a homogeneous tax rate for different classes in the economy. Introduction of different tax rates may change the results. Further, our model is based on the closed economy assumption. Introduction of an open economy framework may significantly change our findings. In the long-run, we only have considered the dynamics of the government debt-capital ratio and the capital stock. If instead of assuming constant level of labour supply, profit share and the technological growth, we allow these to vary in the long-run, then the analysis will be more interesting and the results may vary.

Notes

1. According to Olson (1982), as pointed out by Hansson and Henrekson (1994), organized interest groups evolve and seek to acquire advantages for their own group in the form of legislation or transfers which have the growth-retarding side effect. The scope for interest group action of this kind is positively related to larger public sectors. Keefer and Knack (2007) show that a rise in various public investment expenditure in countries with low quality of governance will have little impact on economic growth. Cameron (1982) and Landau (1983) observe that government expenditure and taxation crowd out private investment in human and physical capital and therefore negatively affect economic growth. Considering 19 nations, Cameron (1982, 51) finds that a rise in government spending to GDP ratio by 1 percentage point between the early 1960s and late 1970s lowered the growth rate in the late 1970s by 0.05 percentage point. Koskela and Viren (2000) say that a rise in government demand for labour raises real wages and therefore crowds out private employment and output. However, the effect of a rise in public employment on total output and aggregate demand is positive for a small public sector and negative when the public sector is large (Koskela and Viren 2000).
2. The post-Keynesian growth model was originally developed independently by Robert Rowthorn (1981) and Amitava Krishna Dutt (1984). Taylor (1983, 1985), Amadeo (1986), Blecker (1989), Bhaduri and Marglin (1990), Marglin and Bhaduri (1990) are others among the initial contributors. The post-Keynesian growth model is also recognized as neo-Kaleckian or Kalecki-Steindl or structuralist growth model. Results similar to the neo-Kaleckian growth model, as Lavoie (2014, 359) points out, were arrived by Del Monte (1975) in a paper published in Italian.
3. Given the existing technology, the 'normal' degree of capacity utilisation, as they interpret, is the optimal degree of capacity utilisation.
4. A similar kind of analysis can be observed in Commendatore, Panico, and Pinto (2009).
5. Depending on whether it is productive expenditure or not.
6. Blecker (2002) shows that in the presence of positive saving out of wages, the exhilarationist regime (or profit-led demand regime) can be achieved even using an investment function that would otherwise imply stagnationism (or wage-led demand regime).
7. This is a standard assumption in Kalecki-Steindl type growth theory. See Hein (2014, pp. 181 & 242) for further.
8. Overhead labour is an important feature of the post-Keynesian growth model. For example, Nah and Lavoie (2018) introduce the overhead labour in the neo-Kaleckian model of growth and distribution, and find the profit share as an endogenous variable which positively depends on the rate of capacity utilisation. Rowthorn (1981), Nichols and Norton (1991), Lavoie (1995, 2009), Palley (2005), Nikiforos (2015) are among others who incorporated overhead labour in the post-Keynesian model of growth and distribution. Lavoie (2014, pp. 322–339 and 416–423) provides a nice discussion of this issue. However, for simplicity, we exclude the overhead labour and overhead cost in our model.
9. For simplicity, we do not endogenize the income distribution. However, it is worth remembering that distribution can be affected by the aggregate demand. For example,

Dutt (1992, 583) says that firms attempt to increase markups to take advantage of buoyant markets and decrease markups when sales are low. Rowthorn (1977, 119) argues that when there is a sufficient amount of unutilised capacity, firms cannot aspire for higher profit share in the fear that other firms may invade their markets. However, higher the capacity utilisation rate, firms become stronger and can use their market power asking for higher profit share. For further on this issue see Dutt (2012), Assous and Dutt (2013), and Blecker (2016).

10. In developing countries, a large number of workers are employed in unorganized sectors (eg. India, Pakistan, Bangladesh) where either they don't have any organized labour union or the union is too weak to have strong bargaining power. On the other hand, in developed countries as well, the workers may not be able to fully internalize the increase in productivity through proportionate increases in the real wage rate. (Carter 2007; Setterfield 2013, pp.163; Sharpe, Arsenault, and Harrison 2008a, 2008b).

11. Dutt (2013, 108) assumes the rate of growth of labor productivity growth as a function of labor market conditions. However, as our prime objective is to analyse the implication of different kinds of government expenditures on aggregate demand and economic growth, for simplicity we are not considering modeling productivity as a function of labour market conditions.

12. As long as the potential output-capital ratio is fixed, actual output-capital ratio can be used as a proxy for the degree of capacity utilisation.

13. Government investment in infrastructure, education, water supply, health facilities, etc. boosts private investment through its complementary and other external effects. Again, it raises the future profitability of private capital formation. Note that the government consumption expenditure through its accelerator effect may also generate a type of 'crowding in' effect. However, government investment expenditure is more effective at generating 'crowding in'. Therefore, for simplicity, in line with Dutt (2013) and Taylor (1991), we assume that only the government investment expenditure causes the 'crowding in' effect. I am grateful to the anonymous referee for pointing this out.

14. Government investment expenditures like expenditure on roads, electricity, and water supply have an impact on labour productivity within a fairly short period. Aschauer (1989) points out that a 'core' infrastructure of streets, highways, airports, mass transit, sewers, water systems, etc. has the most explanatory power for productivity in the United States for the period 1949–1985. According to Aschauer (1989), the elasticity for the core infrastructure equals 0.24 and is highly significant whereas public capital expenditure on public-sector hospitals carries only elasticity of 0.06. Finally, public capital expenditure on educational buildings carries the elasticity of −0.01 which is statistically insignificant. According to the U.S. Bureau of Economic Analysis (https://www.bea.gov/resources/learning-center/what-to-know-government), 'Government consumption expenditures include spending by governments to produce and provide services to the public, such as national defense and education. Government gross investment consists of spending on fixed assets that directly benefit the public, such as highway construction, or that assist government agencies in doing their jobs, such as military hardware. Consumption expenditures and gross investment are the measures of government spending included in calculations of gross domestic product or GDP.' However, Devarajan, Swaroop, and Zou (1996, 321) point out that while Kormendi and Meguire (1985), Grief and Tullock (1989), and Summers and Heston (1988) classify defense and education as government consumption and hence 'unproductive', Barro (1991) models them as productive. Barro (1991) treats spending on public education as investing in human capital. According to Barro (1989, 1990, 1991), government consumption expenditure has no direct effect on private productivity. On the other hand, expenditures on education and defense, as Barro (1991, 430) points out, are more like a public investment than public consumption. These expenditures are likely to affect private-sector productivity or property rights, which matter for private investment. A sizeable segment of government 'consumption expenditure', as Landau (1983, 784) points out, is in fact investment in the broader sense, especially education and healthcare.

Therefore, we consider government expenditure on education and healthcare as public investment expenditure. However, it is worthy to remember that, irrespective of whether it is a government consumption expenditure or an investment expenditure, government expenditures on education affects labour productivity with a long time-lag. Therefore, for simplicity, we assume that government investment expenditure is mainly of a 'core' infra-structure type. This type of government expenditure has the potential to influence the current profitability of private investment and thus this is different from the crowding in effect where the future profitability of private capital formation is influenced.

15. Note that g^* represents the equilibrium value of investment to capital ratio $(\frac{I}{K})$.

16. For a give capital stock, a rise in capacity utilization rate $(u = \frac{Y}{K})$ is associated with a rise in aggregate demand. Therefore, *ceteris paribus*, a discussion on the effect of various parameters on u is sufficient for taking care of the formal discussion of the effect of various changes on aggregate demand.

17. Note that for a given capital stock K, a rise in aggregate demand is always associated with a rise in the degree of capacity utilisation $u = \frac{Y}{K}$.

18. Note that $\frac{du^*}{dt} = \frac{\gamma_0\left[(s_P - s_W - \gamma_2)\pi + s_W\right]}{\left[(s_P - s_W - \gamma_2)(1-t)\pi + s_W(1-t) - \gamma_1 - \gamma_3\theta\right]^2}$. In the wage-led demand regime $(s_P - s_W - \gamma_2) > 0$ and therefore $\frac{du^*}{dt}$ is unambiguously positive.

19. As total tax revenue is tY, for every unit rise in the tax rate, tax revenue (normalized by the capital stock) rises by $\frac{Y}{K} = u$ unit. Therefore, government expenditure and hence the aggregate demand rises by u unit.

20. A simple algebra shows $\left[(1 - s_W)(1 - \pi) + (1 - s_P)\pi + \gamma_2\pi\right]u < u$.

21. Following Dutt (2013), we introduce it to show that, even allowing the neo-classical argument of financial crowding-out of private investment due to rise in public debt, when we introduce government deficits and the dynamics of the government debt into our analysis, the model does not necessarily become unstable and δ does not rises without bound. We also will show that our long-run result differs from Dutt (2013) as here are two equilibrium values of δ and the smaller one represents the stable equilibrium value while in Dutt (2013) this is not necessarily the case.

22. Note that the fiscal policy is pro-cyclical in our model. As one of the main objectives of our paper is to show that even considering Dutt's (2013) model we get a different result (that a rise in government investment expenditure need not necessarily increases aggregate demand and economic growth), we stick to the assumption that $I_G = \theta Y$ and $C_G = \eta Y$. However, instead of these if we modify our assumptions like $\frac{I_G}{K} = \frac{\bar{I}_G + \theta Y}{K} = \bar{I}_g + \theta u$ and $\frac{C_G}{K} = \frac{\bar{C}_g + \eta Y}{K} = \bar{C}_g + \eta u$, we get some autonomous components of fiscal expenditure in forms of government autonomous investment expenditure $(\bar{I}_g K)$ and government autonomous consumption expenditure $(\bar{C}_g K)$. The equilibrium degree of capacity utilization will change to $u^* = \frac{\gamma_0 + \bar{I}_g + \bar{C}_g - \left[\{s_P(1-t) + t\}i + \gamma_4\right]\delta}{(s_P - s_W - \gamma_2)(1-t)\pi(a) + s_w(1-t) + t - \gamma_1 - \eta - \theta(1+\gamma_3)}$. Consequently, we get $\frac{du^*}{d\bar{I}_g} = \frac{1}{(s_P - s_W - \gamma_2)(1-t)\pi(a) + s_w(1-t) + t - \gamma_1 - \eta - \theta(1+\gamma_3)} > 0$, and $\frac{du^*}{d\bar{C}_g} = \frac{1}{(s_P - s_W - \gamma_2)(1-t)\pi(a) + s_w(1-t) + t - \gamma_1 - \eta - \theta(1+\gamma_3)} > 0$. Apart from this, however, other results do not change qualitatively. Further, if we assume that when the actual capacity utilization is lower than its normal level (u_n), $\bar{I}_g$ rises and the vice versa i.e. if

$$\bar{I}_g = \zeta(u_n - u), \zeta > 0$$

then we may be able to introduce the government autonomous expenditure as an automatic stabilizer of aggregate demand and growth. However, for simplicity, I am not considering this issue. This is left for future work.

23. This is happening because of the financial crowding out effect.

24. I am indebted to an anonymous reviewer from this journal for pointing out this possibility.

25. Note that the necessary and sufficient condition for existence of equilibrium is that the minimum value of $\frac{d\delta}{d\tau}$ must be negative. Minimum value of $\frac{d\delta}{d\tau}$ can be attained at $\delta = -\frac{D_1}{2D_2}$. Then the minimum value of $\frac{d\delta}{d\tau} = D_0 - \frac{D_1^2}{4D_2}$. Thus, the necessary and sufficient condition for existence of equilibrium is $\left(D_0 - \frac{D_1^2}{4D_2}\right) \leq 0$. On the other hand, the necessary and sufficient condition for existence of a stable equilibrium is that the minimum value of $\frac{d\delta}{d\tau}$ must be <0. Thus $\left(D_0 - \frac{D_1^2}{4D_2}\right) < 0$ ensures the necessary and sufficient condition for existence of a stable equilibrium.

26. Note that between those two equilibrium values of δ, the low equilibrium value of δ (i.e. δ^*) gives the stable equilibrium.

27. See Equation (4.4). Also see Footnote 21.

28. A rise in the interest rate decreases the absolute value of D_1 Consequently, the absolute value of $\left(D_0 - \frac{D_1^2}{4D_2}\right)$ falls. On the other hand, as $D_1 > \sqrt{D_1^2 - 4D_2D_0}$, $\frac{d\delta^*}{dD_1} = \frac{1}{2D_2}\left\{-1 - \frac{D_1}{\sqrt{D_1^2-4D_2D_0}}\right\} > 0$ Further, $\frac{d\delta^{**}}{dD_1} = \frac{1}{2D_2}\left\{-1 + \frac{D_1}{\sqrt{D_1^2-4D_2D_0}}\right\} < 0$.

29. When the economy is in a wage-led demand regime.

30. When balanced budget assumption is dropped, a rise in θ does not imply a rise in government investment expenditure at the cost of consumption expenditure. It simply implies a rise in the ratio of government investment expenditure to income.

31. From Equation (2.7) we get $\frac{\partial u^*}{\partial \theta} = \frac{\gamma_0\gamma_3}{\left[\left(s_P - s_W - \gamma_2\right)(1-t)\pi + s_W(1-t) - \gamma_1 - \gamma_3\theta\right]^2} > 0$.

32. From Equation (2.9) we get $\frac{\partial g^*}{\partial u^*} = \left\{\gamma_1 + \gamma_2(1-t)\pi + \gamma_3\theta\right\} > 0$, $\frac{\partial g^*}{\partial \pi} = \gamma_2(1-t)u^* > 0$ and $\frac{\partial g^*}{\partial \theta} = \gamma_3 u^* > 0$..

33. $\frac{du^*}{d\theta} = \frac{[\gamma_0 - \Gamma\delta](1+\gamma_3)}{\Lambda^2} > 0$.

Acknowledgements

This paper formed my M.Phil. dissertation. I am grateful to Subrata Guha, Gogol Mitra Thakur, Debarshi Das, C. Saratchand and an anonymous referee of this journal for their helpful comments and suggestions. I am thankful to Gokulananda Nandan for his help. However, I am solely responsible for all the shortcomings of the paper.

Disclosure statement

No potential conflict of interest was reported by the author(s).

References

Amadeo, E. J. 1986. "Notes on Capacity Utilization, Distribution and Accumulation." *Contributions to Political Economy* 5 (1): 83–94. doi:10.1093/oxfordjournals.cpe.a035705.

Aschauer, D. 1989. "Is Public Expenditure Productive?" *Journal of Monetary Economics* 23 (2): 177–200. doi:10.1016/0304-3932(89)90047-0.

Assous, M., and A. K. Dutt. 2013. "Growth and Income Distribution with the Dynamics of Power in Labour and Goods Markets." *Cambridge Journal of Economics* 37 (6): 1407–1430. doi:10.1093/cje/bes086.

Barbosa-Filho, N., and L. Taylor. 2006. "Distributive and Demand Cycles in the US Economy–A Structuralist Goodwin Model." *Metroeconomica* 57 (3): 389–411. doi:10.1111/j.1467-999X.2006.00250.x.

Barro, R., and X. Sala-i-Martin. 2004. *Economic Growth*. New Delhi, India: Prentice-Hall of India.

Barro, R. J. 1989. "A Cross-Country Study of Growth, Savings, and Government" *NBER, working paper no. 2855*, February, 1989.

Barro, R. J. 1990. "Government Spending in a Simple Model of Endogenous Growth." *Journal of Political Economy* 98 (5): 103–125. doi:10.1086/261726.

Barro, R. J. 1991. "Economic Growth in a Cross Section of Countries." *The Quarterly Journal of Economics* 106 (2): 407–444. doi:10.2307/2937943.

Bhaduri, A., and S. A. Marglin. 1990. "Unemployment and the Real Wage: The Economic Basis for Contesting Political Ideologies." *Cambridge Journal of Economics* 14 (4): 375–393. doi:10.1093/oxfordjournals.cje.a035141.

Blecker, R. A. 1989. "International Competition, Income Distribution and Economic Growth." *Cambridge Journal of Economics* 13 (3): 395–412.

Blecker, R. A. 2002. "Distribution, Demand and Growth in neo-Kaleckian Macro Models." In *The Economics of Demand- Led Growth: Challenging the Supply Side Vision of the Long Run*, edited by M. Settereld, 129–152. Cheltenham, UK: Edward Elgar.

Blecker, R. A. 2016. "Wage-led versus Profit-led Demand Regimes: The Long and the Short of It." *Review of Keynesian Economics* 4 (4): 373–390. doi:10.4337/roke.2016.04.02.

Cameron, D. R. 1982. "On the Limits of the Public Economy." *The ANNALS of the American Academy of Political and Social Science* 459 (1): 46–62. doi:10.1177/0002716282459001004.

Carter, S. 2007. "Real Wage Productivity Elasticity across Advanced Economies, 1963–1996." *Journal of Post Keynesian Economics* 29 (4): 573–600. doi:10.2753/PKE0160-3477290403.

Commendatore, P., C. Panico, and A. Pinto. 2009. "Government Spending, Effective Demand, Distribution and Growth: A Dynamic Analysis." In *Institutional and Social Dynamics of Growth and Distribution*, edited by N. Salvadori, 118–138. Aldershot: Elgar.

Commendatore, P., C. Panico, and A. Pinto. 2011. "The Influence of Different Forms of Government Spending on Distribution and Growth." *Metroeconomica* 62 (1): 1–23. doi:10.1111/j.1467-999X.2009.04081.x.

Commendatore, P., and A. Pinto. 2011. "Public Expenditure Composition and Growth: A neo-Kaleckian Analysis." *Papers in Political Economy*, L'Harmattan 61: 187–222.

Del Monte, A. 1975. "Grado di monopolio e sviluppo economico." *Rivista Internazionale di Scienze Sociali* 83 (3): 231–263.

Devarajan, S., V. Swaroop, and H. Zou. 1996. "The Composition of Public Expenditure and Economic Growth." *Journal of Monetary Economics* 37 (2–3): 313–344. doi:10.1016/0304-3932(96)01249-4.

Dutt, A. K. 1984. "Stagnation, Income Distribution and Monopoly Power." *Cambridge Journal of Economics* 8 (1): 25–40.

Dutt, A. K. 1987. "Alternative Closures Again: A Comment on Growth, Distribution and Inflation." *Cambridge Journal of Economics* 11 (1): 75–82. doi:10.1093/oxfordjournals.cje.a035017.

Dutt, A. K. 1990. *Growth, Distribution and Uneven Development*. Cambridge: Cambridge University Press.

Dutt, A. K. 1992. "Conflict Inflation, Distribution, Cyclical Accumulation and Crisis." *European Journal of Political Economy* 8 (4): 579–597. doi:10.1016/0176-2680(92)90042-F.

Dutt, A. K. 2012. "Distributional Dynamics in post-Keynesian Growth Models." *Journal of Post Keynesian Economics* 34 (3): 431–451. doi:10.2753/PKE0160-3477340303.

Dutt, A. K. 2013. "Government Spending, Aggregate Demand, and Economic Growth." *Review of Keynesian Economics* 1 (1): 105–119. doi:10.4337/roke.2013.01.06.

Grief, K., and G. Tullock. 1989. "An Empirical Analysis of Cross-national Economic Growth, 1951–1980." *Journal of Monetary Economics* 24 (2): 259–276. doi:10.1016/0304-3932(89)90006-8.

Hansson, P., and M. Henrekson. 1994. "A New Framework for Testing the Effect of Government Spending on Growth and Productivity." *Public Choice* 81 (3–4): 381–401. doi:10.1007/BF01053239.

Hein, E. 2014. *Distribution and Growth after Keynes: A Post-Keynesian Guide*. Cheltenham: Edward Elgar.

Hein, E., and L. Vogel. 2008. "Distribution and Growth Reconsidered– Empirical Results for Six OECD Countries." *Cambridge Journal of Economics* 32 (3): 479–511. doi:10.1093/cje/bem047.

Hungerford, T. L. 2016. "We're Not Broke: America's Real Spending Problem and How to Fix It." *Challenge* 59 (4): 279–297. doi:10.1080/05775132.2016.1201391.

Kaldor, N. 1963. "Capital Accumulation and Economic Growth." In *Proceedings of a Conference Held by the International Economics Association*, edited by A. L. Friedrich and C. Douglas, 177–222. Hague London: Macmillan.

Keefer, P., and S. Knack. 2007. "Boondoggles, Rent-seeking, and Political Checks and Balances: Public Investment under Unaccountable Governments." *Review of Economics and Statistics* 89 (3): 566–572. doi:10.1162/rest.89.3.566.

Kormendi, R. C., and P. G. Meguire. 1985. "Macroeconomic Determinants of Growth: Cross-country Evidence." *Journal of Monetary Economics* 16 (2): 141–164. doi:10.1016/0304-3932(85)90027-3.

Koskela, E., and M. Viren. 2000. "Is There a Laffer Curve between Aggregate Output and Public Sector Employment?" *Empirical Economics* 25 (4): 605–621. doi:10.1007/s001810000036.

Landau, D. 1983. "Government Expenditure and Economic Growth: A Cross-country Study." *Southern Economic Journal* 49 (3): 783–792. doi:10.2307/1058716.

Lavoie, M. 1995. "The Kaleckian Model of Growth and Distribution and Its neo-Ricardian and neo-Marxian Critiques." *Cambridge Journal of Economics* 19 (6): 789–818.

Lavoie, M. 2009. "Cadrisme within a Kaleckian Model of Growth and Distribution." *Review of Political Economy* 21 (3): 371–393. doi:10.1080/09538250903073396.

Lavoie, M. 2014. *Post-Keynesian Economics: New Foundations*. Cheltenham, UK; Northampton, MA: Edward Elgar.

Marglin, S. A., and A. Bhaduri. 1990. "Profit Squeeze and Keynesian Theory." In *The Golden Age of Capitalism*, edited by S. A. Marglin and J. B. Schor, 153–186. Oxford: Oxford University Press.

Naastepad, R., and S. Storm. 2007. "OECD Demand Regimes (1960–2000)." *Journal of Post Keynesian Economics* 29 (2): 211–246. doi:10.2753/PKE0160-3477290203.

Nah, W. J., and M. Lavoie. 2018. "Overhead Labour Costs in a neo-Kaleckian Growth Model with Autonomous Expenditures," *IPE Working Papers 111/2018*. Berlin School of Economics and Law, Institute for International Political Economy (IPE). https://www.econstor.eu/bitstream/10419/188924/1/1041257600.pdf

Nichols, L. M., and B. Norton. 1991. "Overhead Workers and Political Economy Macro Models." *Review of Radical Political Economy* 23 (1–2): 47–54. doi:10.1177/048661349102300107.

Nikiforos, M. 2015. "Uncertainty and Contradiction: An Essay on the Business Cycle." *Review of Radical Political Economics*. doi:10.1177/0486613415621748.

Olson, M. 1982. *The Rise and Decline of Nations*. New Haven, CN: Yale University Press.

Onaran, Ö., and G. Galanis. 2013. "Is Aggregate Demand Wage-led or Profit-led? A Global Model." In *Wage-Led Growth: An Equitable Strategy for Economic Recovery*, edited by M. Lavoie and E. Stockhammer, 71–99. Hampshire, UK: Palgrave Macmillan.

Onaran, Ö., E. Stockhammer, and L. Grafl. 2011. "The Finance-dominated Growth Regime, Distribution, and Aggregate Demand in the US." *Cambridge Journal of Economics* 35 (4): 637–661. doi:10.1093/cje/beq045.

Palley, T. 2005. "Class Conflict and the Cambridge Theory of Distribution." In *Joan Robinson's Economics: A Centennial Celebration*, edited by B. Gibson, 203–224. Cheltenham, UK and Northampton, MA, USA: Edward Elgar.

Rowthorn, R. E. 1977. "Conflict, Inflation and Money." *Cambridge Journal of Economics* 1 (1): 215–239.

Rowthorn, R. E. 1981. "Demand, Real Wages and Economic Growth." *Thames Papers in Political Economy* 1981: 1–39.

Setterfield, M. 2013. "Wages, Demand and Us Macroeconomic Travails: Diagnosis and Prognosis." In *After the Great Recession: The Struggle for Economic Recovery and Growth*, edited by B. Z. Cynamon, S. M. Fazzari, and M. Setterfield, 158–184. Cambridge: Cambridge University Press.

Sharpe, A., J. Arsenault, and P. Harrison. 2008a. "The Relationship between Productivity and Real Wage Growth in Canada and OECD Countries: 1961–2006," *CSLS Research Report 2008-8,* December.

Sharpe, A., J. Arsenault, and P. Harrison. 2008b. "Why Have Real Wages Lagged Labour Productivity Growth in Canada?" *International Productivity Monitor* 17: 16–27.

Stansbury, A., and L. H. Summers. 2020. "The Declining Worker Power Hypothesis: An Explanation for the Recent Evolution of the American Economy." *NBER working paper series, Working Paper 27193.* http://www.nber.org/papers/w27193

Steindl, J. 1952. *Maturity and Stagnation in American Capitalism.* New York: Monthly Review Press.

Stockhammer, E. 2013. "Why Have Wage Shares Fallen? An Analysis of the Determinants of Functional Income Distribution." In *Wage-Led Growth: An Equitable Strategy for Economic Recovery,* edited by M. Lavoie and E. Stockhammer, 40–70. Hampshire, UK: Palgrave Macmillan.

Summers, R., and A. Heston. 1988. "A New Set of International Comparisons of Real Product and Price Levels: Estimates for 130 Countries." *Review of Income and Wealth* 34 (1): 1–25. doi:10.1111/j.1475-4991.1988.tb00558.x.

Taylor, L. 1983. *Structuralist Macroeconomics: Applicable Models for the Third World.* New York: Basic Books.

Taylor, L. 1985. "A Stagnationist Model of Economic Growth." *Cambridge Journal of Economics* 9 (4): 383–403. doi:10.1093/oxfordjournals.cje.a035588.

Taylor, L. 1991. *Income Distribution, Inflation, and Growth: Lectures on Structuralist Macroeconomic Theory.* Cambridge, MA: MIT Press, USA.

You, J.-I.-I., and A. K. Dutt. 1996. "Government Debt, Income Distribution and Growth." *Cambridge Journal of Economics* 20 (3): 335–351. doi:10.1093/oxfordjournals.cje.a013619.

A Appendix

A.1 Proof of Proposition 1

Proof. Differentiating u^* with respect to π we get,

$$\frac{du^*}{d\pi} = \frac{-\gamma_0(s_P - s_W - \gamma_2)(1-t)}{\left[(s_P - s_W - \gamma_2)(1-t)\pi + s_W(1-t) - \gamma_1 - \gamma_3\theta\right]^2} \tag{A.1}$$

Thus, if $s_W < (s_P - \gamma_2)$ then $\frac{du^*}{d\pi} < 0$ and if $s_W > (s_P - \gamma_2)$ then $\frac{du^*}{d\pi} > 0$.

A.2 Proof of Proposition 2

Proof. Differentiation of equation (A.1) w.r.t. θ yields,

$$\frac{d\left(\frac{du^*}{d\pi}\right)}{d\theta} = \frac{2\gamma_0(s_P - s_W - \gamma_2)(1-t)\left[(s_P - s_W - \gamma_2)(1-t)\frac{d\pi}{da}\frac{da}{d\theta} - \gamma_3\right]}{\left[(s_P - s_W - \gamma_2)(1-t)\pi + s_W(1-t) - \gamma_1 - \gamma_3\theta\right]^3} \tag{A.2}$$

In the profit-led demand regime $(s_P - s_W - \gamma_2) < 0$ and therefore the numerator of equation (A.2) is positive. Consequently, $\frac{d}{d\theta}\left(\frac{du^*}{d\pi}\right) > 0$. On the contrary, in the wage-led demand regime $(s_P - s_W - \gamma_2) > 0$ and therefore the sign of numerator of equation (A.2) is ambiguous. A further rearrangement of equation (A.2) yields,

$$\frac{d\left(\frac{du^*}{d\pi}\right)}{d\theta} = \frac{2\gamma_0\left(s_P - s_W - \gamma_2\right)(1-t)\left[\left(s_P - s_W - \gamma_2\right)(1-t)\varepsilon_{\pi,a}\varepsilon_{a,\theta}\frac{\pi}{\theta} - \gamma_3\right]}{\left[\left(s_P - s_W - \gamma_2\right)(1-t)\pi + s_W(1-t) - \gamma_1 - \gamma_3\theta\right]^3}$$

So, $\frac{d}{d\theta}\left(\frac{du^*}{d\pi}\right) \gtreqless 0$ according to whether $\varepsilon_{\pi,a}\varepsilon_{a,\theta} \gtreqless \frac{\gamma_3\theta}{\left(s_P - s_W - \gamma_2\right)(1-t)\pi} = \psi$.

A.3 Proof of Proposition 3

Proof. We know $\frac{du^*}{d\theta} = \frac{\partial u^*}{\partial \pi}\frac{d\pi}{da}\frac{da}{d\theta} + \frac{\partial u^*}{\partial \theta}$. If the economy is in a profit-led demand regime then $\frac{\partial u^*}{\partial \pi} > 0$. $\frac{d\pi}{da}, \frac{da}{d\theta}$, and $\frac{\partial u^*}{\partial \theta}$ are all positive.[31] So, $\frac{du^*}{d\theta}$ is unambiguously positive. Now suppose the economy is in a wage-led demand regime. Then,

$$\frac{du^*}{d\theta} = \frac{\partial u^*}{\partial \pi}\frac{d\pi}{da}\frac{da}{d\theta} + \frac{\partial u^*}{\partial \theta} = \frac{-\gamma_0\left[\left(s_P - s_W - \gamma_2\right)(1-t)\varepsilon_{\pi,a}\varepsilon_{a,\theta}\frac{\pi}{\theta} - \gamma_3\right]}{\left[\left(s_P - s_W - \gamma_2\right)(1-t)\pi + s_W(1-t) - \gamma_1 - \gamma_3\theta\right]^2}.$$

So, $\frac{du^*}{d\theta} \gtreqless 0$ according to whether $\varepsilon_{\pi,a}\varepsilon_{a,\theta} \lesseqgtr \frac{\gamma_3\theta}{\left(s_P - s_W - \gamma_2\right)(1-t)\pi} = \psi$.

A.4 Proof of Proposition 4

Proof. We know $\frac{dg^*}{d\theta} = \frac{\partial g^*}{\partial u^*}\frac{du^*}{d\theta} + \frac{\partial g^*}{\partial \pi}\frac{d\pi}{d\theta} + \frac{\partial g^*}{\partial \theta}$. If the economy is in a profit-led demand regime then $\frac{\partial g^*}{\partial u^*}, \frac{\partial g^*}{\partial \pi}, \frac{d\pi}{d\theta}$, and $\frac{\partial g^*}{\partial \theta}$ are all positive.[32] From Proposition 3 we know that when the economy is in a profit-led demand regime, $\frac{du^*}{d\theta}$ becomes positive. So, $\frac{dg^*}{d\theta} > 0$. In the wage-led demand regime, we get

$$\frac{dg^*}{d\theta} = \frac{\partial g^*}{\partial u^*}\frac{du^*}{d\theta} + \frac{\partial g^*}{\partial \pi}\frac{d\pi}{d\theta} + \frac{\partial g^*}{\partial \theta}$$

$$= \left[\gamma_1 + \gamma_2(1-t)\pi + \gamma_3\theta\right]\frac{du^*}{d\theta} + \gamma_2(1-t)u^*\frac{d\pi}{d\theta} + \gamma_3 u^*$$

$$= \frac{-\gamma_0(1-t)\left[\{(s_P - s_W)\pi + s_W\}\gamma_3 - \frac{\pi}{\theta}\{(\gamma_1 + \gamma_3\theta)(s_P - s_W) - s_W\gamma_2(1-t)\}\varepsilon_{\pi,a}\varepsilon_{a,\theta}\right]}{\left[\left(s_P - s_W - \gamma_2\right)(1-t)\pi + s_W(1-t) - \gamma_1 - \gamma_3\theta\right]^2}$$

So, $\frac{dg^*}{d\theta} \gtreqless 0$ depending on $\varepsilon_{\pi,a}\varepsilon_{a,\theta} \lesseqgtr \frac{\{(s_P - s_W)\pi + s_W\}\gamma_3\theta}{\pi\{(s_P - s_W)(\gamma_1 + \gamma_3\theta) - s_W\gamma_2(1-t)\}} = \rho$.

A.5 Proof of Proposition 5

Proof. Differentiating Equation (3.1) with respect to θ we get,

$$\frac{de^*}{d\theta} = \frac{\partial e^*}{\partial u^*}\frac{du^*}{d\theta} + \frac{\partial e^*}{\partial a}\frac{da}{d\theta}$$

$$= \frac{k_0}{a}\left[\frac{-\gamma_0\{(s_P - s_W - \gamma_2)(1-t)\frac{d\pi}{da}\frac{da}{d\theta} - \gamma_3\}}{\left[(s_P - s_W - \gamma_2)(1-t)\pi + s_W(1-t) - \gamma_1 - \gamma_3\theta\right]^2}\right]$$

$$\quad - \frac{k_0}{a^2}\left[\frac{\gamma_0}{(s_P - s_W - \gamma_2)(1-t)\pi + s_W(1-t) - \gamma_1 - \gamma_3\theta}\right]\frac{da}{d\theta}$$

$$= \frac{\nabla a}{\theta}\left[\pi\{(s_P - s_W - \gamma_2)(1-t)(\varepsilon_{\pi,a}\varepsilon_{a,\theta} + \varepsilon_{a,\theta})\} - \{(\gamma_1 + \gamma_3\theta)\varepsilon_{a,\theta} - s_W(1-t) + \gamma_3\theta\}\right]$$

where

$$\nabla = \frac{-\gamma_0 k_0}{\left[a\left\{\left(s_P - s_W - \gamma_2\right)(1-t)\pi + s_W(1-t) - \gamma_1 - \gamma_3\theta\right\}\right]^2} < 0$$

So, $\frac{de^*}{d\theta} \gtrless 0$ according to whether $\varepsilon_{\pi,a} \lessgtr \Omega = \frac{\gamma_3\theta + \left\{\gamma_1 + \gamma_3\theta - s_W(1-t)\right\}\varepsilon_{a,\theta}}{\left(s_P - s_W - \gamma_2\right)(1-t)\pi\varepsilon_{a,\theta}} - 1$.

A.6 Proof of Proposition 6

Proof. Differentiation of Equation (4.5) w.r.t. θ yields,

$$\frac{du^*}{d\theta} = \frac{\partial u^*}{\partial \pi}\frac{d\pi}{da}\frac{da}{d\theta} + \frac{\partial u^*}{\partial \theta}$$

If the economy is in a profit-led demand regime then $\frac{\partial u^*}{\partial \pi} > 0$. $\frac{d\pi}{da}, \frac{da}{d\theta}$, and $\frac{\partial u^*}{\partial \theta}$ are all positive.[33] So, $\frac{du^*}{d\theta}$ is unambiguously positive. Now suppose the economy is in a wage-led demand regime. Then, $\left(s_P - s_W - \gamma_2\right) > 0$. Hence,

$$\frac{du^*}{d\theta} = \frac{-\left\{\gamma_0 - \Gamma\delta\right\}\left\{\left(s_P - s_W - \gamma_2\right)(1-t)\varepsilon_{\pi,a}\varepsilon_{a,\theta}\frac{\pi}{\theta} - \left(1 + \gamma_3\right)\right\}}{\Lambda^2}$$

where $\Gamma = \left[\left\{s_P(1-t) + t\right\}i + \gamma_4\right] > 0$ and $\Lambda = \left[\left(s_P - s_W - \gamma_2\right)(1-t)\pi + s_W(1-t) + t - \gamma_1 - \eta - \left(1 + \gamma_3\right)\theta\right] > 0$. So, $\frac{du^*}{d\theta} \gtrless 0$ according to whether $\varepsilon_{\pi,a}\varepsilon_{a,\theta} \lessgtr \frac{\left(1+\gamma_3\right)\theta}{\left(s_P - s_W - \gamma_2\right)(1-t)\pi} = \phi$.

A.7 Proof of Proposition 7

Proof. Differentiating Equation (4.6) w.r.t. θ we get,

$$\frac{dg^*}{d\theta} = \frac{\partial g^*}{\partial u^*}\frac{du^*}{d\theta} + \frac{\partial g^*}{\partial \pi}\frac{d\pi}{d\theta} + \frac{\partial g^*}{\partial \theta}$$

We know $\frac{\partial g^*}{\partial u^*} = \left[\gamma_1 + \gamma_2(1-t)\pi(a) + \gamma_3\theta\right] > 0$; $\frac{\partial g^*}{\partial \pi} = \gamma_2(1-t)u^* > 0$; $\frac{\partial g^*}{\partial \theta} = \gamma_3 u^* > 0$; $\frac{d\pi}{d\theta} = \frac{d\pi}{da}\frac{da}{d\theta} > 0$ and from Proposition 6 we know that in the profit-led demand regime $\frac{du^*}{d\theta} > 0$. Hence, in the profit-led demand regime, we get $\frac{dg^*}{d\theta} > 0$. Now suppose the economy is in wage-led demand regime. Then $\left(s_P - s_W - \gamma_2\right) > 0$. Hence,

$$\frac{dg^*}{d\theta} = \frac{\partial g^*}{\partial u^*}\frac{du^*}{d\theta} + \frac{\partial g^*}{\partial \pi}\frac{d\pi}{d\theta} + \frac{\partial g^*}{\partial \theta}$$

$$= \left\{\gamma_1 + \gamma_2(1-t)\pi(a) + \gamma_3\theta\right\}\frac{du^*}{d\theta} + \gamma_2(1-t)u^*\frac{d\pi}{d\theta} + \gamma_3 u^*$$

$$= \frac{\left(\gamma_0 - \Gamma\delta\right)}{\Lambda^2}\frac{\pi}{\theta}\left[\left[\left\{\gamma_1 + \gamma_3(t-\eta) + (1-t)\pi\left\{\gamma_2 + (s_P - s_W)\gamma_3\right\} + s_W(1-t)\gamma_3\right\}\frac{\theta}{\pi}\right]\right.$$

$$\left. - \left\{(1-t)\varepsilon_{\pi,a}\varepsilon_{a,\theta}\left\{\left(s_P - s_W - \gamma_2\right)\left(\gamma_1 + \gamma_3\theta\right)\right\}\right\}\right]$$

$$+ \left\{(1-t)\varepsilon_{\pi,a}\varepsilon_{a,\theta}\left\{\left(t - \gamma_1 - \left(1 + \gamma_3\right)\theta - \eta + s_W(1-t)\gamma_2\right\}\right\}\right]$$

So, $\frac{dg^*}{d\theta} \gtrless 0$ according to whether $\varepsilon_{\pi,a}\varepsilon_{a,\theta} \lessgtr \rho'$ where

$$\rho' = \frac{\theta\left[\gamma_1 + \gamma_3(t-\eta) + (1-t)\pi\left\{\gamma_2 + (s_P - s_W)\gamma_3 + s_W(1-t)\gamma_3\right\}\right]}{(1-t)\pi\left[\left(s_P - s_W - \gamma_2\right)\left(\gamma_1 + \theta\gamma_3\right) - \gamma_2\left\{t - \gamma_1\left(1 + \gamma_3\right)\theta - \eta + s_W(1-t)\right\}\right]}.$$

□

A.8 Proof of Proposition 8

Proof. Differentiating Equation (4.6) w.r.t. δ we get, $\frac{dg^*}{d\delta} = \frac{\partial g^*}{\partial u^*}\frac{du^*}{d\delta} + \frac{\partial g^*}{\partial \delta} = \frac{-\{\gamma_1 + \gamma_2(1-t)\pi + \gamma_3\theta\}\Gamma}{\Lambda}$ $\delta - \gamma_4 < 0$. □

A.9 Proof of Proposition 9

Proof. For a given output level and a given debt level, if the rise in government investment expenditure is to be financed completely by a rise in tax rate, $ud\theta = (u + i\delta)dt$ must hold. In that case, from Equation (4.5) we get,

$$\frac{du^*}{d\theta}\Big|_{ud\eta=(u+i\delta)dt} = \frac{-(1-s_P)\left(\frac{i\delta u^*}{u^*+i\delta}\right)}{\left[(s_P - s_W - \gamma_2)(1-t)\pi + s_W(1-t) + t - \gamma_1 - \eta - (1+\gamma_3)\theta\right]}$$

$$- \frac{u^*\left[(s_P - s_W - \gamma_2)(1-t)\frac{d\pi}{d\theta} - \left\{\left((s_P - s_W - \gamma_2)\pi + s_W\right)\left(\frac{u^*}{u^*+i\delta}\right) + \left(\frac{i\delta}{u^*+i\delta} + \gamma_3\right)\right\}\right]}{\left[(s_P - s_W - \gamma_2)(1-t)\pi + s_W(1-t) + t - \gamma_1 - \eta - (1+\gamma_3)\theta\right]}$$

When the economy is in a wage-led demand regime $(s_P - s_W - \gamma_2) > 0$ and therefore as long as $\varepsilon_{\pi,a}\varepsilon_{a,\theta} < \sigma = \frac{\left[\left\{\left((s_P - s_W - \gamma_2)\pi + s_W\right)\left(\frac{u^*}{u^*+i\delta}\right) + \gamma_3\right\} + s_P\left(\frac{i\delta}{u^*+i\delta}\right)\right]\theta}{(s_P - s_W - \gamma_2)(1-t)\pi}$, $\frac{du^*}{d\theta}\Big|_{ud\eta=(u+i\delta)dt} > 0$ holds. On the other hand, if the economy is in a profit-led demand regime, $(s_P - s_W - \gamma_2) < 0$ and therefore as long as $\varepsilon_{\pi,a}\varepsilon_{a,\theta} > \sigma$, $\frac{du^*}{d\theta}\Big|_{ud\eta=(u+i\delta)dt} > 0$ holds.

A.10 Proof of Proposition 10

Proof. For a give output level and a given debt level, if the rise in government consumption expenditure is to be financed completely by a rise in tax rate, $ud\eta = (u + i\delta)dt$ must hold. In that case, from Equation (4.5) we get,

$$\frac{du^*}{d\eta}\Big|_{ud\eta=(u+i\delta)dt} = \frac{-(1-s_P)\left(\frac{i\delta u^*}{u^*+i\delta}\right)}{\left[(s_P - s_W - \gamma_2)(1-t)\pi + s_W(1-t) + t - \gamma_1 - \eta - (1+\gamma_3)\theta\right]}$$

$$+ \frac{u^*\left[\left\{\left((s_P - s_W - \gamma_2)\pi + s_W\right)\left(\frac{u^*}{u^*+i\delta}\right) + \left(\frac{i\delta}{u^*+i\delta}\right)\right\}\right]}{\left[(s_P - s_W - \gamma_2)(1-t)\pi + s_W(1-t) + t - \gamma_1 - \eta - (1+\gamma_3)\theta\right]}$$

$$= \frac{\left(\frac{u^*}{u^*+i\delta}\right)\left[\left((s_P - s_W - \gamma_2)\pi + s_W\right)u^* + s_P i\delta\right]}{\left[(s_P - s_W - \gamma_2)(1-t)\pi + s_W(1-t) + t - \gamma_1 - \eta - (1+\gamma_3)\theta\right]}$$

When the economy is in a wage-led demand regime $(s_P - s_W - \gamma_2) > 0$ and therefore

$\frac{du^*}{d\eta}\Big|_{ud\eta=(u+i\delta)dt}$ is unambiguously positive. On the other hand, if the economy is in a profit-led

demand regime, $(s_P - s_W - \gamma_2) < 0$ and therefore, as long as $\left(\frac{s_P i\delta + s_W u^*}{\pi u^*}\right) > -(s_P - s_W - \gamma_2)$, $\frac{du^*}{d\eta}\Big|_{ud\eta=(u+i\delta)dt} > 0$ holds. In other words, as long as there is a relatively weak profit-led demand regime, a rise in government consumption expenditure financed with a rise in tax revenue can increase the equilibrium degree of capacity utilisation.

Interpreting the world, in various ways – and changing it

Jonathan Michie

Malcolm Sawyer's work has played and continues to play an important role in both developing a better understanding of the economy and how it functions, and in developing policies to change it for the better – that is, to change the dominant economic system for the better, and hence thereby improve the state of the economy and society. These grand challenges – of analysing the current economic system, and considering how it might be changed for the better – are ones that in various ways are taken up by the three books here reviewed, Milanovic's *Capitalism Alone*, Blakeley's *Stolen: How to Save the World from Financialisation*, and Pettifor's *The Case for the Green New Deal*. Milanovic's focus is more on understanding the current state of the world economy, and of the capitalist system; Blakeley's more on advocating the replacement of capitalism itself; and Pettifor's, as her title signals, a combination of the two, with a particular focus on environmental sustainability – and how unsustainable the current system is, and how we might change it to achieve a sustainable future.

Thus, Milanovic (2019) focuses more on the former aspect, of seeking to interpret the 'laws of motion' of the capitalist system – which, he argues, now rules the world, and Blakeley (2019) and Pettifore (2019) more on the latter aspect, of seeking to change the world.

1. Capitalism alone?

Milanovic's starting point appears at first sight to be not unlike Fukuyama's *End of History*,[1] namely that with the collapse of the Soviet Union and the other socialist or centrally planned economies of Eastern Europe, capitalism had asserted itself – or would assert itself – as the dominant economic system globally. Milanovic sees this outcome as less stable than Fukuyama did, in that capitalism as a system contains within it a fundamental problem of creating, relying on and generating a 'single-minded pursuit of wealth' which 'might end in encouragement of amoral behaviour' (178), the quotes being his paraphrasing of the quandary that he claimed Adam Smith faced, when in his *The Theory of Moral Sentiments* he was concerned that the pursuit of wealth might dissolve the vitally important – to society – relations and bonds between people, involving trust and honesty, whilst in *The Wealth of Nations* Adam Smith perhaps recognised that market forces would indeed incentivise people to act in ways which were damaging to society, driven by their individual drive for personal enrichment, even if at the cost of others.[2] Or as the Wall Street trader put it, 'greed is good'.[3]

Milanovic argues that capitalism as a system has become less moral, more driven by the 'greed is good' mentality. When Marx wrote of 'accumulate, accumulate' being the 'Moses and the prophets', he was describing an era where capitalism as a system was expanding the productive forces, infrastructure and possibilities through high rates of reinvestment, including on research and development. Indeed – Milanovic argues – wealthy capitalists reinvesting rather than living ostentatiously has historically played an important part in maintaining a degree of social cohesion, and prevented opposition to the capitalist system from growing greater than might otherwise have been the case. But today, such restraint on ostentatious displays of wealth has been abandoned, along with reinvesting to build the business over the long term, replaced by a focus on removing the regulatory blocks on financial speculation, and on pursuing tax evasion and avoidance. From wealth creation to rent seeking. From entrepreneurialism to zombie capitalism.

Despite all this, Milanovic argues that 'The domination of capitalism as the best, or rather the only, way to organize production and distribution seems absolute. No challenger appears in sight' (196). So, the question for Milanovic is what type [or variety] of capitalism should be aimed for – which need not be limited to what currently exists. Thus, he describes five future scenarios, including 'social democratic capitalism', with 'significant redistribution through the tax and transfer system, including free or accessible public health care and education. Interpersonal inequality is moderate. Relatively equal access to education allows intergenerational income mobility' (215); and 'egalitarian capitalism' where 'Everyone has approximately equal amounts of both capital and labor income, such that a large increase in the capital share does not translate into greater inequality. Interpersonal inequality is low. The role of the state in redistribution is limited to social insurance. Relative equality of incomes ensures equality of opportunity. Libertarianism, capitalism, and socialism come close to each other' (216).

To move towards such an egalitarian capitalism would require increased taxation of the rich, including on inheritance; a significant increase in funding for an improvement in the quality of public schools; and 'Strictly limited and exclusively public funding of political campaigns. The objective is to reduce the ability of the rich to control the political process and form a durable upper class' (217). Milanovic rejects the idea that there is any inevitability towards progress; so, any more desirable form of capitalism will need to be proactively campaigned for. And there is a certain imperative within his analysis for this to be done, given his argument that the current form of globalised capitalism has strong tendencies to create the vast degree of corruption and illegal tax evasion that characterises the current era.

2. How to save the world from financialisation?

Blakeley's starting point is that capitalism has evolved from its previous productive form into 'financialised capitalism', that this creates inequality and environmental and economic crises, and that the capitalist system itself needs to be replaced by socialism. Despite this presentation, there are similarities with Milanovic, most obviously in their critique of the current form of globalised and financialised capitalism – a critique shared by Pettifor and others.[4]

The different conclusions might appear stark, or even diametrically opposed, with Blakeley advocating the replacement of capitalism with socialism, while Milanovic declares capitalism to be the only game in town. But much hinges on how the future economy which they both see as necessary – to replace the current version of globalised and financialised capitalism, which is regarded as not only associated with but actually causing corruption and other economic and social ills – is depicted: whether as socialism, or as a new form of capitalism. This also depends in part on timeframes. The specifics that Blakeley proposes are largely reforms of capitalism, rather than constituting an alternative socialist economic system, but she sees these as part of a process towards transitioning from one system to the other, while not being specific on the timeframes involved. Conversely, whilst Milanovic asserts that capitalism rules the world, he gives the impression that this is simply a description of the current era, rather than claiming any 'end of history'; indeed, he is explicit about the likelihood of the current system changing and being replaced, and he doesn't speculate what might follow the various new forms of capitalism he considers, and whether those subsequent systems would still be capitalist or not.

Blakeley is more critical of mainstream economics – for example citing Queen Elizabeth's well known question whilst visiting the London School of Economics (LSE) as to why economists had failed to see the financial crisis coming (148). Blakely could have gone further, by citing the response from the self-selected representatives of mainstream economics to the Queen, which was pitifully weak, particularly when seen alongside their usually confident assertions as to the way the world works. The explanation to the Queen from the mainstream economists was that 'the failure to foresee the timing, extent and severity of the crisis and to head it off, while it had many causes, was principally a failure of the collective imagination of many bright people [meaning presumably themselves], both in this country and internationally, to understand the risks to the system as a whole". They failed to make clear that it was neither all the 'bright people' within economics, nor a representative sample; rather, the 'bright people' they had in mind were themselves – namely mainstream economists who had deliberately created the problem they described by excluding from their economic departments and journals all the other bright people who had been warning of precisely these dangers. Economists in the broader political economy tradition would have always understood the risks to the system as a whole – from the likes of Keynes, Kalecki[5] and Kaldor, through to the likes of Minsky, Malcolm Sawyer and other non-mainstream economists. Indeed, Geoffrey Hodgson convened an alternative group of economists to write a more honest response to Queen Elizabeth, informing her that 'part of the responsibility lies at the door of leading and influential economists in the United Kingdom and elsewhere', and pointing out that the letter from the mainstream economists 'overlooks the part that many leading economists have had in turning economics into a discipline that is detached from the real world, and in promoting unrealistic assumptions that have helped to sustain an uncritical view of how markets operate'.[6]

Blakeley gives copious references to the literatures from which she draws – on economics and politics – for her account of the 2007–2009 international financial crash, and 'bubble economics' more generally, drawing on the various works of Minsky.[7]

3. The case for the Green New Deal

Unlike Milanovic and Blakeley, Ann Pettifor's focus is specifically on the climate crisis and the need for a Green New Deal. However, she identifies the same phenomenon as representing the problem which needs to be dealt with, the obstacle that needs to be removed, and the system that needs to be changed or replaced, namely the current system of globalised and financialised capitalism. For Pettifor, it is this system which is driving the sort of economic decision making that threatens to make the planet uninhabitable.

Thus, Chapter 1 is 'System Change, Not Climate Change', followed by Chapter 2, 'Winning the Struggle with Finance', and the concluding Chapter 6 is entitled 'The Green New Deal: Transforming Our World'. Given her focus on ensuring that the planet remains inhabitable, Pettifor includes more than do Milanovic or Blakeley on the need to challenge the assumption – or ideology – of driving continually for infinite growth on a finite planet. Hence her penultimate Chapter being on 'A Steady State Economy'. Ironically, the main argument against a steady state economy would until recently – and perhaps remains – that it would be politically impossible, as the electorate in the 'Western' world had become accustomed to annual growth delivering continual improvements in their standards of living, and yet the three authors variously document or refer to the fact that this 'fundamental truth' has actually ceased to be for a good many years now in many countries. And to the extent that people are interested in their relative standard of living rather than absolute, that assumed progress has been put in reverse. The 'capitalism unleashed' era of globalisation has led to increased inequality of income and wealth within most countries across the globe. The majority in most countries have suffered stagnant living standards – and have thus fallen further behind the wealthy elite who have continued to benefit from the globalised financialised form of capitalism that they created.

Pettifor's book is an excellent call to arms for a global Green New Deal, the importance of which is being increasingly recognised as necessary for the world economy. The 2021 World Economic Forum called for the 'great reset' to deliver real progress along these lines. It has to be hoped that the 'Cop26' United Nations (UN) climate summit in November 2021 in Glasgow will lead Governments and companies to set – and commit to achieve – the challenging targets for carbon neutrality that are needed for us to prevent disastrous climate change.

4. Conclusion

These three books share a common strength in setting the current era as just that – an era of capitalism, which superseded previous eras, and which will itself be superseded. Such eras might be broadly depicted as the initial form of capitalist globalisation in the 19th Century and leading up to – and causing – the World War One. In between the two world wars was a retreat from – or a collapse of – this previous era of globalisation, with instead the Great Depression, the widespread view that capitalism was either failing or in crisis or both, the rise of fascism, and the resulting World War Two. Then came the Bretton Woods era of regulated globalisation, depicted as the 'Golden Age of Capitalism', through to the 1970s. From the 1980s grew the current era, of deregulated, financialised globalised capitalism. What had been a constraint of a third of the globe under an alternative model, with socialist or centrally planned economies, was lifted. This opened the floodgates for what Andrew Glyn depicted as 'Capitalism Unleashed' – a no holds barred orgy of financial speculation in pursuit of private financial enrichment. The result has been increased inequality of income, wealth and power, with a huge increase in the shares generally going to the top one per cent of the population as measured by wealth, and even more so to the top 0.1%.[8]

The Great Depression of the 1930s led to the publication in 1936 of Keynes's *General Theory of Employment, Interest and Money*. This argued that there were many aspects of the capitalist economic system that were likely to give rise to crises such as the 1929 Wall Street Crash, which had sparked the subsequent depression that spread across much of the world. These aspects included the nature of speculative behaviour on stock exchanges and elsewhere, where to be successful the speculator had to guess not what the 'fundamentals' might be, but on what the other speculators were likely to do next. Keynes's *General Theory* is perhaps best known for its 'call to arms', in arguing that left to itself the capitalist economic system can get stuck in recession, with widespread unemployment – but that there was an alternative, as Government could create the necessary effective demand to create jobs, 'crowd in' private investment, and get the economy moving again, out of depression and back to full employment. But Keynes's greatest contribution – which could still play an important role today, if re-told – was to change the way people thought about economics and the economy. He was well aware that the economy did not function in the manner believed by the 'bright people' referred to in the letter from the representatives of mainstream economics to Queen Elizabeth. Keynes was aware that the economy was driven by expectations and 'animal spirits' whose future pattern was unknowable, and which could not therefore be modelled in the precise way much of mainstream economics assumes or asserts. Keynes was challenging the orthodoxy of his day, which he characterised as the 'Treasury view' – an orthodoxy that was never fully defeated.

We need another such revolution in economic thinking today. This needs to re-establish a 'political economy' approach to understand how the economy works, and to consider what sort of economic policies might best tackle the problems of crises, inequality, and environmental disaster. The work of Malcolm Sawyer has consistently done – and continues to do – precisely that. And the three books reviewed here contribute in various ways – even Milanovic's which makes clear that the current capitalist era does need to be overcome with some urgency, with a more socially sustainable variant developed in its place. Blakeley stresses the need to overcome the current era of global 'financialisation'. And Pettifor makes the case for the Green New Deal to set the world economy on a course that could actually sustain life on the planet into the future – in place of the present drive towards environmental disaster.

Notes

1. Fukuyama (1992).
2. Smith (1759) and Smith (1776) respectively.
3. Gordon Gekko in the film 'Wall Street'. The phrase re-appeared in the UK media in March 2021 after Prime Minister Boris Johnson ascribed the development of a vaccine for Covid-19 to 'capitalism and greed' – before hurriedly retracting and asking those present (at a private meeting of Conservative Party Members of Parliament) to delete his remarks from their memory banks.
4. Including for example Sawyer (2013).
5. On the economics of Michael Kalecki, see Sawyer (1985).
6. Those signing included Malcolm Sawyer, as the then Managing Editor of the *International Review of Applied Economics*.
7. In particular Minsky (1986) and Minsky (1993).
8. See Piketty (2014).

Disclosure statement

No potential conflict of interest was reported by the author.

References

Fukuyama, F. 1992. *The End of History and the Last Man*. New York: Free Press.
Minsky, H. P. 1986. *Stabilizing an Unstable Economy*. New York: McGraw Hill.
Minsky, H. P. 1993. "The Financial Instability Hypothesis." In *The Elgar Companion to Political Economy*, edited by P. Arestis and M. Sawyer. London: Edward Elgar Publishing.
Piketty, T. 2014. *Capital in the Twenty-First Century*. Cambridge, MA: Belknap Press.
Sawyer, M. 1985. *The Economics of Michal Kalecki*. London: Macmillan.
Sawyer, M. 2013. "What Is Financialisation?" *International Journal of Political Economy* 42 (4): 5–18. doi:10.2753/IJP0891-1916420401.
Smith, A. 1759. *Theory of Moral Sentiments, or an Essay Towards an Analysis of the Principles by Which Men Naturally Judge Concerning the Conduct and Character, First of Their Neighbours, and Afterwards of Themselves, to Which Is Added a Dissertation on the Origin of Languages*. 6th ed. Edinburgh: T. Creech and J. Bell.
Smith, A. 1776. *An Enquiry into the Nature and Causes of the Wealth of Nations*. Chicago: University of Chicago Press.

Index

Note: Figures are indicated by *italics*. Tables are indicated by **bold**. Endnotes are indicated by the page number followed by 'n' and the endnote number e.g., 20n1 refers to footnote 1 on page 20.